SAMUEL P. WALLACE

Probability and Statistics

with applications

Probability and Statistics with applications

Y. LEON MAKSOUDIAN

California State Polytechnic College
(San Luis Obispo)

INTERNATIONAL TEXTBOOK COMPANY
Scranton, Pennsylvania

Standard Book Number 7002 2235 9

Library of Congress Catalog Card Number: 69-15356

To My Teachers

Preface

This book is primarily directed, but not necessarily limited to, students in engineering, sciences, and mathematics. The word student is used in its broad sense to imply "any individual who studies." Hence any person interested in probability and statistics can benefit from it. The many illustrative examples are particularly helpful to students who are on a self-study program.

The encouragement and the cooperations received in writing this book have come basically from two sources—industrial and academic. During the summers of 1963 and 1964 while the author was employed at Honeywell Inc. of Minneapolis, among his duties were the projects of preparing notes on statistics and experimental designs intended specifically for Honeywell in-house use. Mr. Carl A. Edstrom, the author's then supervisor, suggested the possibility of writing a textbook of this kind, but the idea could not be pursued further at the time. Subsequent to joining the mathematics faculty of California State Polytechnic College, San Luis Obispo, and teaching many courses in statistics, the idea of compiling the lecture notes into a book form was again suggested by several colleagues. During the Winter quarter of 1967 the author was on a special research leave to investigate pedagogical improvements in the area of probability and statistics. The general outline of this volume was found to provide the student with an excellent foundation in concepts of probability and statistics.

This book is an introductory text for which differential and integral calculus are preferable prerequisites, although students may be permitted to pursue calculus concurrently. For this reason, notions of calculus are purposely minimized in the first part of the book, thus allowing the reader both the time and the opportunity to learn the tools of calculus before the need arises in the present course. Furthermore, if the reader comprehends the concepts of the noncontinuous probability models discussed in the first three chapters, the carryover to the continuous case will be accomplished with further ease.

The content of this volume has proven satisfactory for a three quarter or a two semester course meeting three hours per week. However, omitting certain sections and/or chapters could provide a textbook for a shorter course. The examples, problems, and the equations are numbered according to chapter-section-sequence scheme. For instance, Example 5.3.1 refers to the first illustrative example in Section 3 of Chapter 5. Again, problem 4.5.3 refers to the third problem in Section 5 of Chapter 4.

A word is in order regarding the solutions of the problems for exercise. For pedagogical reasons, problems for exercise are included at the conclusion of each section instead of being placed as a long list of problems at the end of each chapter. Consequently, it is often necessary to solve segments of a given problem and refer to it in the light of added information in order to demonstrate the interrelationships of the topics being discussed. Therefore, *it is suggested and recommended* that the student make duplicate copies of each assigned problem and keep one copy in a master notebook to avoid unnecessarily working problems a second time.

The author is grateful to the Chemical Rubber Company of Cleveland, Ohio for permission granted to include Tables II and III of the Appendix. He is also indebted to the Cambridge University Press and the *Biometrika* Trustees for their permission to include Tables V, VI, VII, and VIII of the Appendix. The author wishes to express his gratitude to Professors K. G. Fuller and A. D. Miller for their many valuable suggestions. Special thanks are due those who had a part in typing the manuscript, particularly to Miss Gail Norfolk.

Y. LEON MAKSOUDIAN

San Luis Obispo, California
December, 1968

Contents

Probability and Statistics

with applications

1

Probability Theory

1.0 INTRODUCTION

Probability theory should definitely be considered both the cornerstone and the foundation of statistical studies. However, a thorough and rigorous study of it is neither our interest nor intent in this volume, yet without the introduction of some basic concepts of probability, topics in statistics could not be developed. Statistics without probability theory might perhaps be compared to a factory where men's suits are manufactured and where there are no available models (patterns) to which the suits can be fitted. Probability theory is the study of mathematical models, to which we fit results of random phenomena. Statisticians look for models that describe the random phenomena best, just as salesmen of men's suits try to fit a size 38 regular person with a suit that fits him best. When salesmen cannot find a 38-regular suit in stock, they may suggest the possible use of a 40-short, which may be the next best size. Similarly, statisticians may consider the possible use of another mathematical model if a suggested model does not fit the random phenomena.

Also, as a tailor often alters the suit to fit a particular person, the statistician considers the possibility and the consequences of altering the mathematical model slightly. If the reader keeps this analogy in mind during the first few chapters of this book, he may well appreciate the reason why so much space is given to the formulation of mathematical models.

The reader should also bear in mind that probability and statistics although closely related are not synonymous terms, just as a model or a pattern in the factory does not represent the actual person who wishes to purchase a suit. In probability as in the suit patterns we assume certain numeric values to represent certain constants, whereas in statistics we infer from an actual measurement the appropriateness of the assumed constants. For instance, if we have the 44-regular suit the following may represent plausible numeric values associated with it. (1) chest measurement of 44 inches, (2) waist measurement of 34 inches, (3) sleeve measurement of 33 inches. These are fixed constants for the particular pattern in question. However, the chest measurement, the waist measurement, and sleeve length of the buyer represent actual measurements from which we can infer the proper choice of the suit model. Furthermore, it should be clear from this analogy that choosing suit patterns with specified constants and making actual measurements do not constitute the same process. Similarly, building or constructing probability models and measuring specific characteristics do not imply the same process. Therefore in differentiating between these two processes we shall devote our attention to the construction of probability models first, and then to the measurements of specific characteristics.

A final remark should also be made at this juncture. The various probability models we are to discuss in the next few chapters do not have the same number of numeric constants. A particular model may have one, another two, and still another three or more. These notions will be clarified as we present specific cases.

1.1 SAMPLE SPACE, EVENTS, AND ALGEBRA OF SETS

A repetitive process or operation that results in any one of a number of possible outcomes such that the particular outcome is determined by chance and is impossible to predict *a priori* is called a *random experiment.* Tossing a coin constitutes an example of a random experiment because we can repeat tossing the coin, obtain either of two possible outcomes heads or tails and not be able to predict prior to tossing whether heads or tails will show for certain. A transformer tested for breakdown voltage constitutes another example of a random experiment. Firing of a missile would be still another example if the outcomes are success or failure, depending upon whether or not the launch is accomplished. The set of points representing all possible outcomes of an experiment is called the *sample space* or the *event space*, of an experiment. In the examples above, the sample spaces have two outcomes each, namely heads, tails; breakdown, no breakdown; success or failure, respectively. If in the example of tossing a coin we let 0 represent heads and 1 tails, the result of the experiment might be presented graphically as in Fig. 1.1.

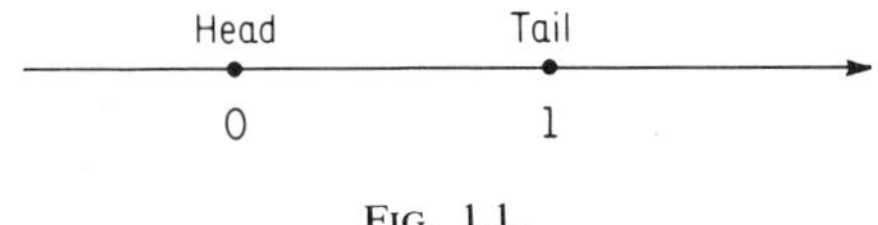

FIG. 1.1.

Similarly, if we consider a throw of a regular tetrahedron with sides colored blue, red, white, and yellow, the random experiment will have four outcomes: falling on side blue, side red, side white, or side yellow. We can represent these sample points graphically as in Fig. 1.2.

Let us observe in the above graphs that the important issue is not how the outcomes are labeled but rather that a proper and well-defined repre-

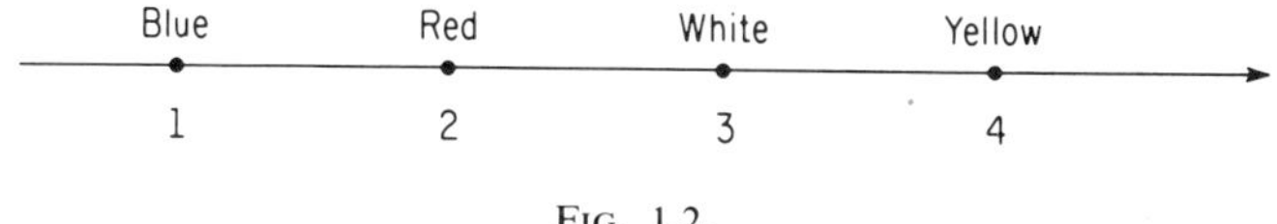

FIG. 1.2.

sentation exists. It is the one-to-one correspondence between the outcomes and their representation with which we are concerned.

Suppose that two coins are tossed or that one coin is tossed two times. Assigning points of the coordinate plane to the possible outcomes, we have four ordered pairs designating (head, head), (head, tail), (tail, head) and (tail, tail) represented as in Fig. 1.3. In a throw of two dice,

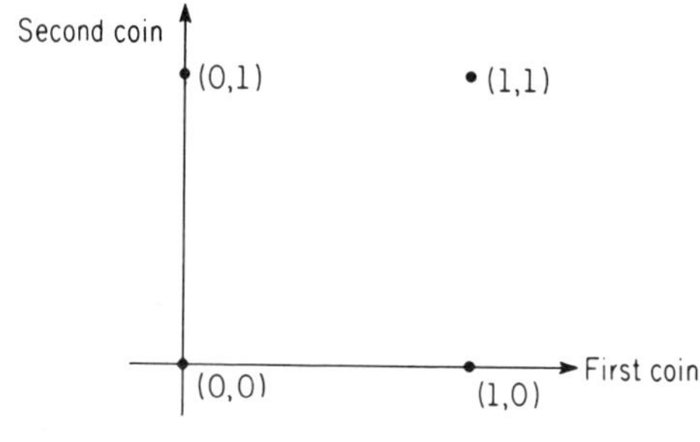

FIG. 1.3.

or equivalently, the throw of one dice twice, the sample space will have 36 distinct points, namely (1,1), (2,1), . . . , (6,5) and (6,6), which represent all possible outcomes of the random experiment. The totality of all possible pairs will be called the *sample space* of 2-*tuples*.

Another example of 2-tuples is all possible combinations of ages of college students and their IQ scores within certain limits. If $10 \leq x \leq 79$ is an interval representing the ages of college students and $100 \leq y \leq 175$ an interval representing the IQ scores, then (x,y) represents all possible age and IQ combinations of college students. If only integers are used for measurements (19,128) would be a point in the sample space of 5320 (70×76) 2-tuples.

An example of 3-tuples would be the outcomes of three coins tossed once. Again, if 0 represents heads and 1 represents tails, and the sample space is graphed in the three-dimensional cartesian coordinate system, a possible representation might be the eight vertices of the unit cube in the first octant.

Next let us consider the situation in which two or more random phenomena take place simultaneously. For example, consider the toss of a coin together with a throw of a die. The sample space has twelve points corresponding to the 12 distinct outcomes, (head, 1), (head, 2),

. . . , (tail, 6). This particular sample space can be represented as in Fig. 1.4. Similarly, if a coin, a die, and the above-mentioned tetrahedron are tossed together, the sample space will consist of 48 distinct points which when graphed in the three-dimensional cartesian coordinate system might correspond to the vertices of 48 unit cubes contained in a rectangular box of dimensions 2, 6, and 4. In general then, if $E_1, E_2, \ldots, E_n$ are n distinct experiments with $k_1, k_2, \ldots, k_n$ distinct outcomes

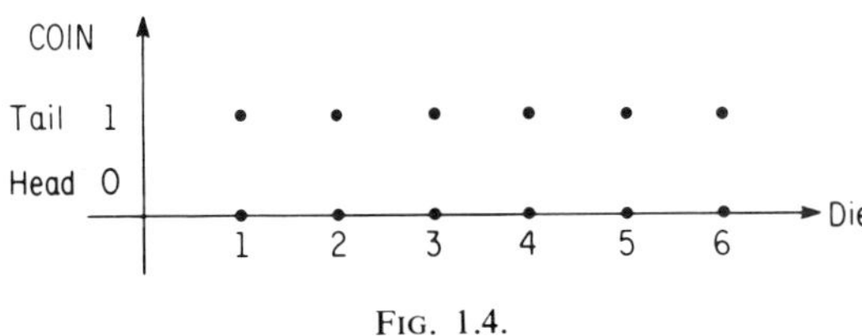

FIG. 1.4.

respectively, the sample space of a composite of these experiments will have $k_1 \cdot k_2 \cdots k_n$ sample points. The totality of all possible outcomes in a random composite of n experiments may be represented by n-tuples. If the outcomes $k_1, k_2, \ldots, k_n$ are equal to a constant k, then the sample space has k^n n-tuples. For example, if a coin is repeatedly tossed ten times, the 10-tuples representing the sample space will consist of $2^{10} = 1{,}024$ sample points, and if a die is rolled three times, the 3-tuples representing the sample space will consist of $6^3 = 216$ sample points.

A *tree diagram* is often used in representing all possible n-tuples of a composite experiment. The tree would have k_1 branches at the trunk, k_2 branches at the end of each k_1 branch, k_3 branches at the end of each k_2 branch, etc. The total number of branches, therefore, would be equal to $k_1 \cdot k_2 \cdots k_n$ which represents the totality of sample points, that is, the sample space. If we were to use a tree diagram for the composite experiment of tossing a die, a coin, and a tetrahedron, the total number of branches would be $6 \cdot 2 \cdot 4 = 48$ and the tree representation would be as in Fig. 1.5.

An *event* is a collection of outcomes in a random experiment, when the outcome has been clearly defined and precisely stated. In the toss of a coin an event could be the showing of a head, or in the throw of two dice an event could be the showing of a pair, that is, one of (1,1) (2,2) $\cdots$ (6,6) takes place. To say that an event E has occurred is to imply that a specific outcome of a random experiment has taken place. It may be apparent that any event may be well described in the notion of subsets. For this reason a brief presentation of algebra of sets is given at this juncture.

A *set* is any well-defined collection of objects. The objects in the collection can be anything—books, transistors, people, numbers, etc. These

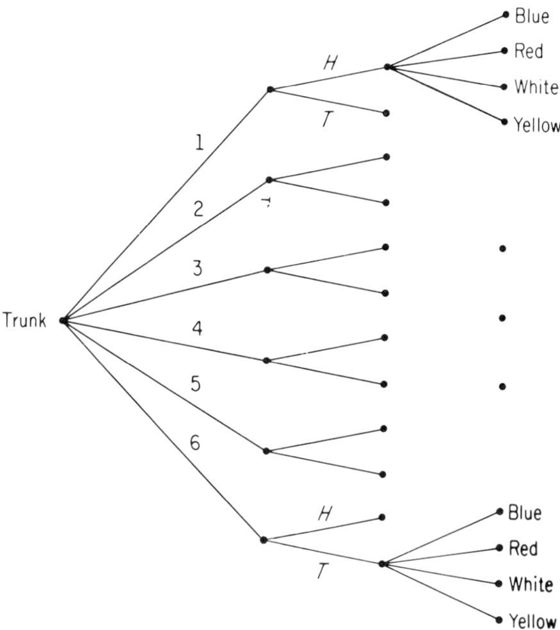

FIG. 1.5.

objects are called the *elements* of a set. An example of a set would be the outcomes in a toss of a coin. $T = \{\text{head, tail, falling on edge}\}$ where head, tail, falling on edge are the three elements in the specified set. Another example of a set would be the outcomes in the throw of two dice. This set has 36 elements and each can be represented as a 2-tuple; namely, $R = (1,1), (1,2), \ldots, (6,6)$. Sets are commonly denoted by capital letters $A, B, C, \ldots$ and elements of sets by lower case letters, $a, b, c, \ldots$. Sets can be described in two ways: (1) by enumerating the elements called the *tabular form* of sets, and (2) by describing or defining the elements referred to as the *defining form* of sets. Both methods are useful and we shall have occasion to use them.

Examples for tabular form of sets would be $X = \{1, 3, 5\}$, $Y = \{2, 4, 6, 8, 10\}$. The corresponding defining forms would be $X = \{a \mid a \text{ is positive odd integer less than } 6\}$ and $Y = \{b \mid b \text{ is positive even integer less than } 12\}$ respectively, where the vertical line is read "given" or "such that." All sets may not have both tabular and defining forms, and the most convenient form should be used. For example, the set of rational numbers between zero and one could not be put into tabular form because that set has infinitely many elements, but $Z = \{z \mid z \text{ is rational number}, 0 < z < 1\}$ would be the defining form of the set Z.

If an object a is a member of a set A, or equivalently set A contains object a as one of its elements, then we write $a \in A$ which is read "a belongs to A" or "a is a member of A" or "a is an element of A." The negation of these statements is $a \notin A$, that is, a is not an element of A. If $C = \{1, 4, 7\}$ and $D = \{3, 4, 9\}$, then $4 \in C$, $4 \in D$ $1 \notin D$, and $7 \notin D$.

If each element of set A is also an element of set B, then set A is called a *subset* of set B. This is written $A \subset B$. Set A is said to be equal to set B if $A \subset B$ and $B \subset A$. For example if $A = \{a, e, i, o, u\}$ is the set of vowels and set $B = \{u, o, a, i, e\}$, then $A = B$. That is, rearrangement of the elements in a set does not change the given set. If a set does not have any elements, the set is called the *empty* or *null* set and is denoted by ϕ. The set $W = \{x^2 \mid x \text{ an integer}, x^2 \text{ negative}\} = \phi$. If one thinks of a set as a container, the container that contains nothing is analogous to ϕ. It should be clear that $0 \neq \phi$.

In most practical applications we shall be concerned with specific subsets of a given set, say S. The set S will be referred to as the *universal* set. Without loss of generality we can consider subsets within the universal set as sets. Furthermore, since all possible subsets of a universal set could be put into one-to-one correspondence with the possible events of a sample space of an experiment, we shall use the word *event* for "subset" and *sample space* for "universal set" interchangeably, and accept their usage as synonymous.

If two sets A and B do not have common elements, then sets A and B are said to be *disjoint*. Consider the outcomes in a throw of a die. Let $A = \{1, 3, 5\}$ and $B = \{2, 4, 6\}$. Clearly these sets are disjoint. Or consider the throw of two dice and let set $X = \{(x,y) \mid x = y\}$ and $Y = \{(x,y) \mid x + y = 9\}$, where x is the face value of the first die and y represents the face value of the second die. Again, X and Y are disjoint sets. If two events E_1 and E_2 are disjoint subsets of the sample space, E_1 and E_2 are said to be *mutually exclusive* events. Intuitively, two events are mutually exclusive if both events could not occur simultaneously. The event of head and the event of tail showing simultaneously in a toss of a fair coin is an example of mutually exclusive events.

We shall next consider the use of Venn diagrams (named after the English logician, John Venn, 1834–1923) in describing events in the sample space S. S is commonly represented by a rectangle and any event or subset is designated by a circle. For example, if the outcomes of a random experiment are a, b, c and d, the sample space S would be represented in a Venn diagram as in Fig. 1.6. If $E_1 = \{a, b, c\}$ and $E_2 = \{c, d\}$, then the Venn diagram represented by Fig. 1.7 describes the two events. If $E_3 = \{d\}$, the Venn diagram representing the three events would be as described in Fig. 1.8.

It is clear from the Venn diagrams on page 7 that E_1 and E_3 are mu-

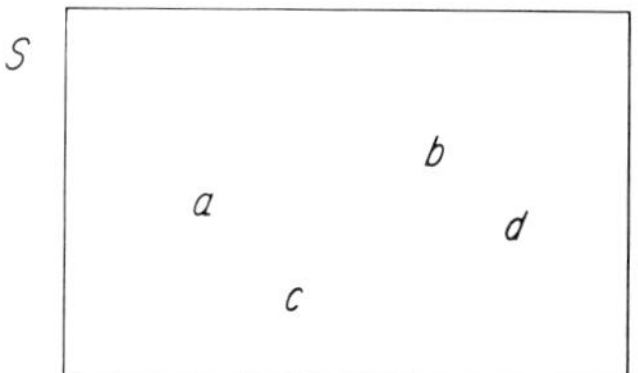

FIG. 1.6.

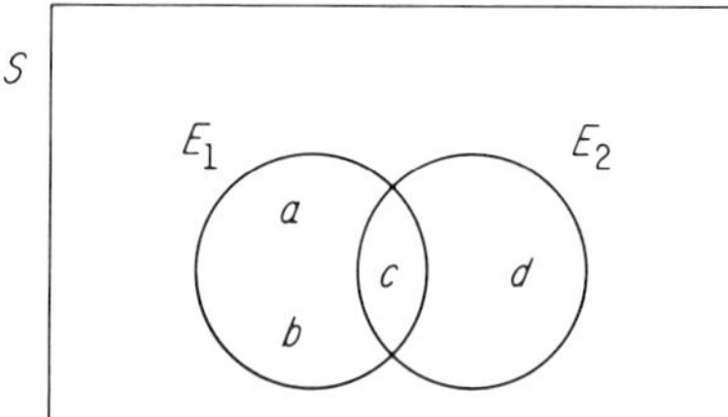

FIG. 1.7.

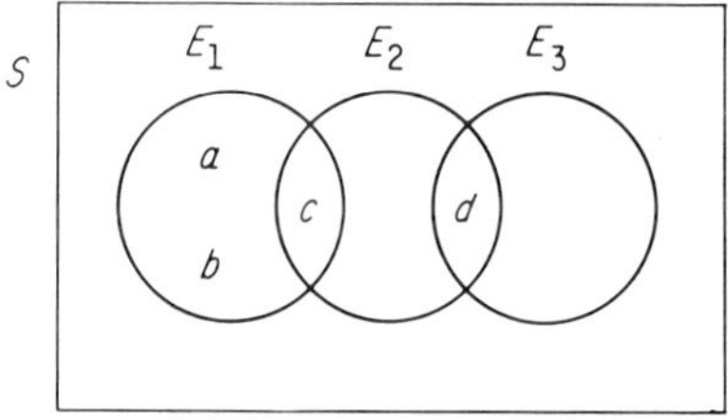

FIG. 1.8.

tually exclusive events, while E_1 and E_2, as well as E_2 and E_3, are not mutually exclusive events.

There are a few basic operations that we can perform on sets, analogous to the operations in ordinary arithmetic, which yield interesting and very useful results. The operations we shall consider are those of union, difference, and intersection of sets.

The *union* of two sets A_1 and A_2 is defined as the set of all elements that belong to A_1 or A_2 or both, and is denoted by the symbol $A_1 \cup A_2$, which reads "A_1 union A_2." Since $A_1 \cup A_2$ is a set, it could more precisely be stated as $A_1 \cup A_2 = \{a \mid a \in A_1 \text{ or } a \in A_2\}$. For example, if $A_1 = \{1, 2, 3\}$ and $A_2 = \{2, 3, 9\}$, the union of A_1 and A_2, $A_1 \cup A_2 = \{1, 2, 3, 9\}$. The Venn diagram for the set $A_1 \cup A_2$ would be represented by the shaded area in Fig. 1.9.

The *difference* of two sets A_1 and A_2 is defined as the set of elements

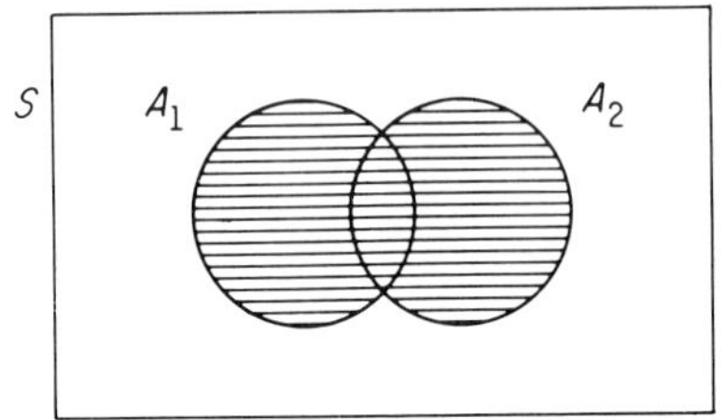

FIG. 1.9.

that belong to A_1 but not to A_2, and is denoted by the symbol $A_1 - A_2$, which reads "A_1 difference A_2" or more commonly "A_1 minus A_2." Then, $A_1 - A_2 = \{a \mid a \in A_1, a \notin A_2\}$. In the example given above $A_1 - A_2 = \{1\}$ while $A_2 - A_1 = \{9\}$. The Venn diagrams for $A_1 - A_2$ and $A_2 - A_1$ are given in Fig. 1.10*a* and Fig. 1.10*b* respectively.

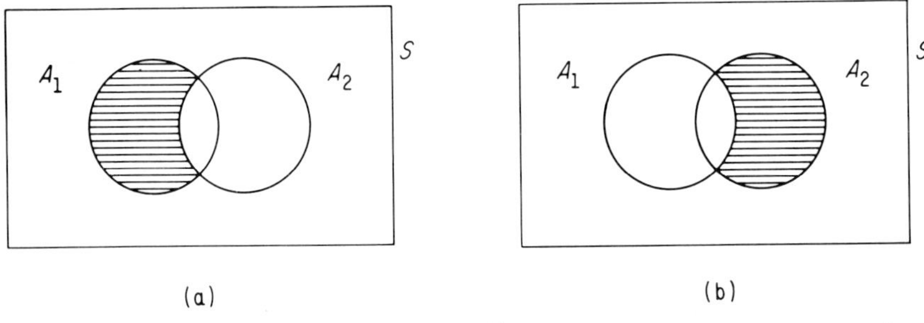

FIG. 1.10.

Consider next the set $S - A_1$, that is, the sample space minus A_1. $S - A_1$ is defined as the *complement* of set A_1 and is denoted by A_1^c, which reads "A_1 complement." In symbols, $S - A_1 = A_1^c = \{a \mid a \in S,\ a \notin A_1\}$. The Venn diagram for A_1^c is given in Fig. 1.11. The operation *intersection* of two sets A_1 and A_2 is defined as the set of elements that belong

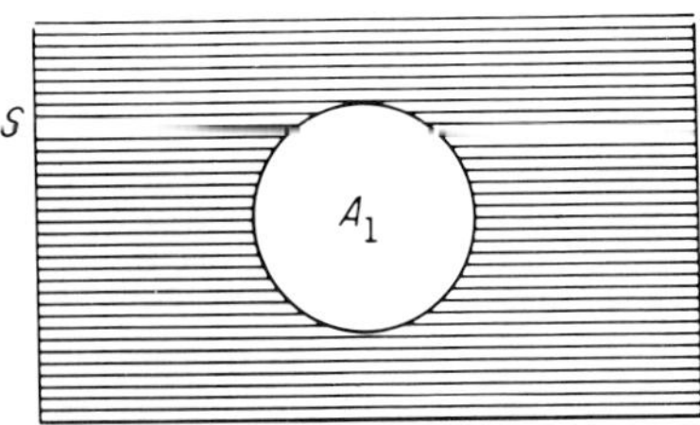

FIG. 1.11.

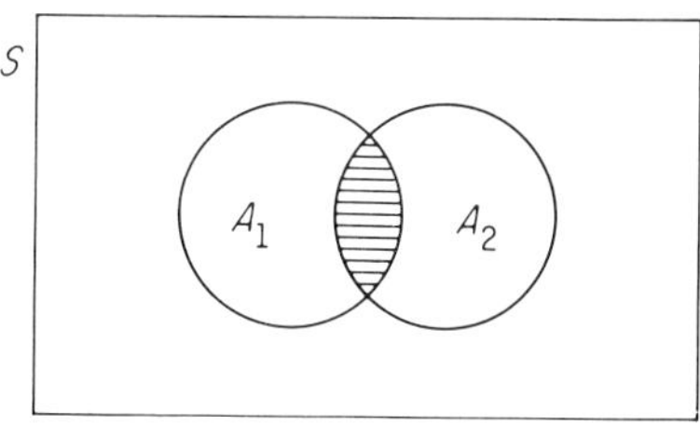

FIG. 1.12.

to A_1 and A_2, and is denoted by the symbol $A_1 \cap A_2$ which reads "A_1 intersection A_2." More precisely, $A_1 \cap A_2 = \{a \mid a \in A_1 \text{ and } a \in A_2\}$. The Venn diagram for the intersection of two sets is given in Fig. 1.12. The following example illustrates the notions presented above.

Example 1.1.1 Suppose the set $X = \{1,2,3,4\}$ and the set $Y = \{3,4,5,6\}$ are two subsets in the same sample space S, where $S = \{1,2,3,4,5,6\}$. Find: (a) X^c, (b) Y^c (c) $X^c \cup Y^c$, (d) $X^c \cap Y^c$, (e) $(X - Y) \cap (X \cap Y)$.

Solution: (a) $X^c = \{x \mid x \in S, x \notin X\} = \{5,6\}$
(b) $Y^c = \{y \mid y \in S, y \notin Y\} = \{1,2\}$
(c) $X^c \cup Y^c = \{5,6\} \cup \{1,2\} = \{1,2,5,6\}$
(d) $X^c \cap Y^c = \{5,6\} \cap \{1,2\} = \phi$
(e) $(X - Y) \cap (X \cap Y) = \{1,2\} \cap \{3,4\} = \phi$

Venn diagrams can be used conveniently to verify set theoretic equalities. The following example illustrates this notion.

Example 1.1.2 Show by use of Venn diagrams that $A = (A \cap B) \cup (A \cap B^c)$.

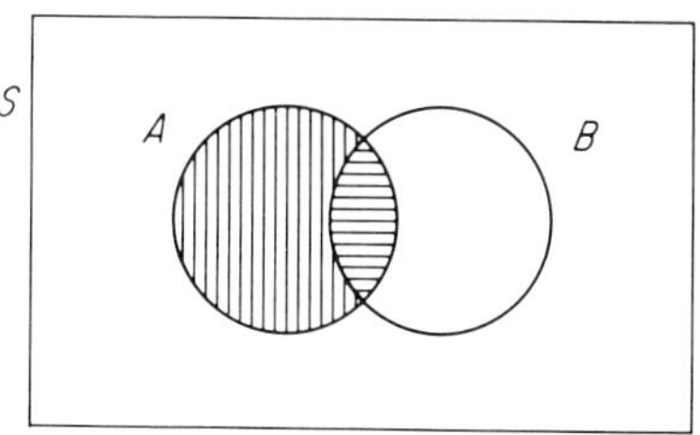

FIG. 1.13.

Solution: Let $X = A \cap B$ which is shaded with horizontal lines in Fig. 1.13. Let $Y = A \cap B^c$ which is shaded with vertical lines in Fig. 1.13. Clearly, then, $A = X \cup Y$, which verifies the given equality.

PROBLEMS

1.1.1 A penny, a nickel, a dime, a quarter, and a half dollar are tossed simultaneously on a table top. How many sample points are there in this composite experiment? Write out four different 5-tuples which represent sample points from this experiment.

1.1.2 A club has 7 engineers and 13 mathematicians. A committee of 2 is to be selected to spend a vacation in Paris. How many different groups can be chosen if it is agreed to have equal representation?

1.1.3 A transistor manufacturer wishes to test 5 transistors with 6 different voltages at 4 different temperatures. How many distinct tests are necessary?

1.1.4 During registration a student can choose among 3 mathematics, 4 physics, and 2 English professors. How many choices does this student have for his instructors in these three disciplines of study?

1.1.5 If all questions on a multiple-choice test have four possible choices, in how many different ways can a student answer an examination of 25 questions?

1.1.6 Verify your answer to Prob. 1.1.4 above by the use of a tree diagram.

1.1.7 Consider the random experiment of throwing two regular tetrahedrons one of which has its sides labeled 1, 2, 3, 4 and the other A, B, C, D. Write out all the possible sample points of this experiment in a two-dimensional cartesian coordinate system.

1.1.8 Using Venn diagrams, verify the following relationships where A and B are two subsets in the sample space S.

(a) $A \cup B = B \cup A$
(b) $A \subset (A \cup B)$
(c) $B \subset (B \cup A)$
(d) $A \cap B = B \cap A$
(e) $A \cap B \subset A$
(f) $A \cap B \subset B$
(g) $A \cap B = \phi$ implies A and B are disjoint
(h) $(A - B) \cap (B - A) = \phi$
(i) $(A - B) \cap (A \cap B) = \phi$
(j) $A \cup A^c = S$
(k) $A \cap A^c = \phi$
(l) $S^c = \phi$
(m) $\phi^c = S$

1.1.9 Let throwing of a regular six-sided die constitute a random experiment in which A is the event that odd integers show, B is the event that even integers show, C is the event that numbers showing are less than or equal to 3, and D is the event that the numbers showing are greater than or equal to 4. Write out the following in the tabular form.

(a) $A \cup B$ (b) $A \cup C$ (c) $B \cap C$ (d) $C \cap D$
(e) $A \cup D$ (f) $A^c \cup B^c$ (g) $(A \cup B)^c$ (h) $D \cup B^c$
(i) $(A \cap C)^c$ (j) $(A \cap B) \cup (A \cap C)$ (k) $A \cap (B \cup C)$

1.1.10 Using Venn diagrams, verify the results of Prob. 1.1.9.

1.1.11 Repeat Prob. 1.1.7 with two ordinary dice.

1.1.12 Using Venn diagrams, show that

(a) $(A \cup B)^c = A^c \cap B^c$ (b) $(A \cap B)^c = A^c \cup B^c$

1.1.13 In Prob. 1.1.11, let $A = \{(a,b) \mid a + b \geq 7\}$ and $B = \{(a,b) \mid a + b \leq 8\}$ where a is the face value of the first die and b the face value of the second die. Write out in the tabular form of sets the following:

(a) $A \cup B$ (d) $A^c \cup B$
(b) $A \cap B$ (e) $B^c \cap A^c$
(c) $A^c \cap B$ (f) $A^c \cup B^c$

1.2 SUMMATION NOTATION

In subsequent sections the sums of large numbers of terms occur during many discussions and it is very desirable to have a simplified notation for the summation of the terms. For example, consider your statistics class and the scores that each student obtains on the final examination. If the scores on this examination are denoted by y, the score of the first person on the alphabetical class list may be denoted by y_1, read "y sub one," the score of the second person on the list by y_2, and so on. In general, the score of the ith person on the list will be designated by y_i and if there are n persons in the class, the score of the last person will be denoted by y_n. If the instructor prefers to do so, he may arrange the scores in descending order and let y_1 represent the highest score in the class, y_2 the next highest, y_3 the next, and so on such that in this situation y_n would represent the lowest score in the class. The criteria of ordering according to alphabetical list and highest to lowest are only two of many other possibilities.

The Greek letter capital Σ is used to designate the sum of the n values —that is,

$$\sum_{i=1}^{n} y_i = y_1 + y_2 + \cdots + y_n \tag{1.2.1}$$

The left-hand side of this expression is read, "the summation of y sub i where i assumes the values 1 to n inclusive," or more commonly, "the summation of y sub i where i goes from 1 to n."

The letter i in expression (1.2.1) is called the summation index. The summation index is a *dummy variable* which simply means that the choice of letter for the summation index is irrelevant. $\sum_{i=1}^{n} y_i$ or $\sum_{j=1}^{n} y_j$ or $\sum_{k=1}^{n} y_k$ are equally valid representations for the summation of the n final examination scores. Also, the summation index need not start with the number 1. For instance, $\sum_{j=4}^{6} y_j = y_4 + y_5 + y_6$ represents the sum of the final examination scores for the 4th, 5th, and 6th students in a given arrangement. The following example illustrates the notions considered.

Example 1.2.1 Let x represent the ages of students in your class. If $x_1 = 20$, $x_2 = 18$, $x_3 = 24$, $x_4 = 21$ and $x_5 = 22$ are the ages of five students, calculate:

(a) $\sum_{k=1}^{5} (x_k - 20)^2$ (b) $\sum_{k=1}^{5} (x_k - 21)$ (c) $\left[\sum_{k=3}^{4} x_k\right]^2$

Solution: (a) $\sum_{k=1}^{5} (x_k - 20)^2 = (0)^2 + (-2)^2 + (4)^2 + (1)^2 + (2)^2$
$= 0 + 4 + 16 + 1 + 4 = 25$

(b) $\sum_{k=1}^{5} (x_k - 21) = -1 - 3 + 3 + 0 + 1 = 0$

(c) $\left[\sum_{k=3}^{4} x_k\right]^2 = [24 + 21]^2 = [45]^2 = 2025.$

Let us remark at this juncture for emphasis that $\sum_{i=1}^{n} x_i^2$ and $\left[\sum_{i=1}^{n} x_i\right]^2$ are not equivalent expressions. The former is $x_1^2 + x_2^2 + x_3^2 + \cdots + x_n^2$, whereas the latter represents $[x_1 + x_2 + x_3 + \cdots + x_n]^2$, which clearly is a different quantity.

Next we shall present some basic rules concerning summations which will be referred to repeatedly in later discussions.

Rule 1. The summation of an expression which consists of the sums of two or more terms is equal to the sum of the summations of the separate terms; that is,

$$\sum_{i=1}^{n} (x_i + y_i + \cdots + z_i) = \sum_{i=1}^{n} x_i + \sum_{i=1}^{n} y_i + \cdots + \sum_{i=1}^{n} z_i \qquad (1.2.2)$$

Rule 2. The summation of a constant multiplied by a variable is equal to the constant multiplied by the summation of the variable, that is,

$$\sum_{i=1}^{n} kx_i = k \sum_{i=1}^{n} x_i \qquad (1.2.3)$$

Rule 3. The summation of a constant is equal to n multiplied by the constant; that is,

$$\sum_{i=1}^{n} k = n \cdot k \qquad (1.2.4)$$

A specific example of Rule 1 is as follows.

$$\begin{aligned}\sum_{i=1}^{6} (x_i + y_i + \cdots + z_i) &= x_1 + y_1 + \cdots + z_1 \\ &\quad + x_2 + y_2 + \cdots + z_2 \\ &\quad + \cdots \\ &\quad + x_6 + y_6 + \cdots + z_6 \\ &= \sum_{i=1}^{6} x_i + \sum_{i=1}^{6} y_i + \cdots + \sum_{i=1}^{6} z_i\end{aligned}$$

In the following example these three rules will be used to prove an identity.

Example 1.2.2 Using the above rules of summation, show that

$$\sum_{i=1}^{n} (ax_i - b) = a\sum_{i=1}^{n} x_i - n \cdot b$$

Solution: Applying Rule 1 to the left-hand side expression, we obtain

$\sum_{i=1}^{n} ax_i - \sum_{i=1}^{n} b$. According to Rule 2, $\sum_{i=1}^{n} ax_i = a\sum_{i=1}^{n} x_i$,

and according to Rule 3, $\sum_{i=1}^{n} b = n \cdot b$ and hence

$$\sum_{i=1}^{n} (ax_i - b) = a\sum_{i=1}^{n} x_i - n \cdot b$$

Let us suppose next that the instructor in your class wishes to order the final examination scores according to the seating arrangement. Let us further suppose that there are 5 rows of seats in the classroom and 7 seats per row. If he lets i represent the rows and j the seats in a row, y_{ij} would represent the final examination score of the student sitting in the ith row and the jth seat in that row. Specifically, y_{12} represents the score of the student sitting at the first row and the second seat and y_{36} represents the score of the student sitting at the 3rd row and the 6th seat of that row. This method of arrangement of scores compared with the previous two methods—namely, alphabetical and descending order of scores—requires two summation indices and therefore two summation signs. For instance, $\sum_{i=1}^{5} \sum_{j=1}^{7} y_{ij}$ represents the total of the scores in that class. Double summations have many advantages, one of which is the formation of subgroups within a classification. Note that in this particular illustration if $i = 1$, $\sum_{j=1}^{7} y_{1j}$ represents the total score for the students in the first row and if $i = 5$, $\sum_{j=1}^{7} y_{5j}$ represents the total score for the students in the fifth row.

In general, if a variable x can be grouped into n_1 classifications and each of the n_1 classifications can be further subdivided into n_2 classifications, then the double summation $\sum_{i=1}^{n_1} \sum_{j=1}^{n_2} x_{ij}$ represents the sum of all the values. The following rules can be shown to hold true for double summations. (Refer to Prob. 1.2.7 through 1.2.10.)

Rule 4.

$$\sum_{i=1}^{n_1}\sum_{j=1}^{n_2}(x_{ij} + y_{ij} + \cdots + z_{ij})$$

$$= \sum_{i=1}^{n_1}\sum_{j=1}^{n_2} x_{ij} + \sum_{i=1}^{n_1}\sum_{j=1}^{n_2} y_{ij} + \cdots + \sum_{i=1}^{n_1}\sum_{j=1}^{n_2} z_{ij} \qquad (1.2.5)$$

Rule 5.

$$\sum_{i=1}^{n_1}\sum_{j=1}^{n_2} kx_{ij} = k\sum_{i=1}^{n_1}\sum_{j=1}^{n_2} x_{ij} \qquad (1.2.6)$$

Rule 6.

$$\sum_{i=1}^{n_1}\sum_{j=1}^{n_2} k = n_1 \cdot n_2 \cdot k \qquad (1.2.7)$$

Rules 4, 5, and 6 are analogous to Rules 1, 2, and 3 respectively.

Rule 7. The order of summation in a double summation is irrelevant —that is,

$$\sum_{i=1}^{n_1}\sum_{j=1}^{n_2} x_{ij} = \sum_{j=1}^{n_2}\sum_{i=1}^{n_1} x_{ij} \qquad (1.2.8)$$

The application of double summation is demonstrated in the following example.

Example 1.2.3 Given

$$x_{11} = 6, x_{12} = 7, x_{13} = 5$$
$$x_{21} = 8, x_{22} = 10, x_{23} = 12$$

Find:

$$\text{(a)} \sum_{i=1}^{2}\sum_{j=1}^{3} x_{ij}$$

$$\text{(b)} \sum_{i=1}^{2}\sum_{j=1}^{3} (x_{ij} - 6)^2$$

$$\text{(c)} \sum_{i=1}^{2}\sum_{j=1}^{2} (x_{ij} - 5)$$

Solution: (a) The expression for the double summation is the sum of all the terms; hence $6 + 7 + 5 + 8 + 10 + 12 = 48$.

(b) The given expression is equivalent to

$$(6 - 6)^2 + (7 - 6)^2 + (5 - 6)^2 + (8 - 6)^2 + (10 - 6)^2 + (12 - 6)^2$$

$$= (0)^2 + 1^2 + (-1)^2 + 2^2 + 4^2 + 6^2$$
$$= 0 + 1 + 1 + 4 + 16 + 36 = 58$$

(c) This expression is equal to

$$(6 - 5) + (7 - 5) + (8 - 5) + (10 - 5) = 1 + 2 + 3 + 5 = 11$$

It should be apparent to the student that the concept of double summation can be readily extended to multiple summations. We shall have occasion to use multiple summations in subsequent chapters. Let us remark also that the summation indices may be omitted from the summation signs whenever the summation includes all the terms. The absence of summation indices implies to sum over all terms and should present no confusion.

PROBLEMS

1.2.1 Write out in full the sums represented by the following summations:

(a) $\sum_{i=1}^{4} x_i$ (d) $\sum_{i=1}^{7} (x_i - 3)^2$

(b) $\sum_{j=3}^{6} y_j^2$ (e) $\sum_{j=2}^{5} 21\, w_j^3$

(c) $\left[\sum_{j=3}^{5} z_j\right]^2$ (f) $\sum_{j=1}^{5} (x_j - 1)x_j$

1.2.2 Evaluate parts (a), (d) and (f) of Prob. 1.2.1 if $x_1 = 7, x_2 = 0, x_3 = 5, x_4 = 9, x_5 = 9, x_6 = 8$ and $x_7 = 11$.

1.2.3 Express the following sums in summation notation with appropriate summation index limits:

(a) $y_5 + y_6 + y_7 + y_8 + y_9$

(b) $x_1^3 + x_2^3 + x_3^3 + x_4^3 + x_5^3 + x_6^3$

(c) $x_2^2 + x_3^3 + x_4^4 + x_5^5 + x_6^6$

(d) $(x_{14} - 25)^2 + (x_{15} - 25)^2 + \cdots + (x_{34} - 25)^2$

1.2.4 Prove Rule 2, Eq. (1.2.3).

1.2.5 Prove Rule 3, Eq. (1.2.4).

1.2.6 Prove the following relationships:

(a) $\sum_{i=1}^{n} [(x_i - \mu)^2 + (\mu - 1)x_i] = \sum_{i=1}^{n} x_i^2 - n \cdot \mu$

(b) $\sum_{i=1}^{n} [x_i(x_i - \mu) + \mu - x_i] = \sum_{i=1}^{n} x_i^2 - n \cdot \mu^2$, where $\mu = \dfrac{\sum_{i=1}^{n} x_i}{n}$

1.2.7 Prove Rule 4, Eq. (1.2.5).
1.2.8 Prove Rule 5, Eq. (1.2.6).
1.2.9 Prove Rule 6, Eq. (1.2.7).
1.2.10 Prove Rule 7, Eq. (1.2.8).
1.2.11 Express the following sums in summation notation with appropriate summation index limits:
(a) $a_{11} + a_{12} + a_{13} + a_{21} + a_{22} + a_{23}$
(b) $x_{11}^2 + x_{12}^2 + x_{13}^2 + x_{21}^2 + x_{22}^2 + x_{23}^2 + x_{31}^2 + x_{32}^2 + x_{33}^2$
1.2.12 Write out in full the sums represented by the following summations:

(a) $\sum_{i=1}^{4} \sum_{j=1}^{5} y_{ij}$

(b) $\sum_{i=1}^{3} \sum_{j=1}^{4} (y_{ij} - 4)^2$

(c) $\sum_{i=1}^{3} \sum_{j=1}^{4} x_i(y_j - 1)$

1.3 CONCEPTS OF PROBABILITY

We shall not attempt to present a detailed discussion on concepts of probability in this section, but rather introduce three approaches to the topic each of which lends itself to practical applications in different fields. Although these approaches may seem unrelated at first, a careful examination will show the similarities and the interrelations among them. These approaches are (1) the classical, (2) the relative-frequency, and (3) the axiomatic. They will be presented in the stated order.

(a) The Classical Definition. Consider the sample space S, and let E_i represent any event in S. The events $E_1, E_2, \ldots, E_k$ are said to form a partition of S if they satisfy the following two conditions:

1. $S = E_1 \cup E_2 \cup E_3 \cup \cdots \cup E_k$ (1.3.1)
2. $E_i \cap E_j = \phi$ for all $i \neq j$ (1.3.2)

For instance, let the outcomes in the toss of 3 coins represent the sample space of a random experiment. The sample space will have $2^3 = 8$ sample points corresponding to the eight 3-tuples (1,1,1), (1,1,0), (1,0,1), (0,1,1), (0,0,1), (0,1,0), (1,0,0) and (0,0,0) where 1 represents the showing of a head and 0 of a tail. Clearly then, the events E_1 = {3 heads}, E_2 = {2 heads}, E_3 = {1 head}, and E_4 = {0 head} show, form a partition of the sample space. In this illustration as well as in the discussion that follows, it is assumed that each sample point is equally likely. Let $N(S)$ represent the total number of sample points in the sample space and $n(E_i)$ the number of sample points associated with event E_i. Then

$$0 \leq n(E_i) \leq N(S) \qquad (1.3.3a)$$

and

$$\sum_{i=1}^{k} n(E_i) = N(S) \tag{1.3.3b}$$

In the above example $n(E_1) = 1$, $n(E_2) = 3$, $n(E_3) = 3$, and $n(E_4) = 1$; hence $N(S) = 8$. If E_1 and E_2 are any two events in the partition of S, it follows that

$$n(E_1 \cup E_2) = n(E_1) + n(E_2) \tag{1.3.4}$$

If E_1 and E_2 are not events in the partition of S, that is, E_1 and E_2 are not mutually exclusive, we proceed as follows to determine $n(E_1 \cup E_2)$. Consider the equality

$$(E_1 \cap E_2^c) \cup (E_1 \cap E_2) = E_1 \tag{1.3.5}$$

where $(E_1 \cap E_2^c)$ and $E_1 \cap E_2$ are two mutually exclusive events. [Refer to Fig. 1.13.] Applying Eq. (1.3.4) to these two events, we obtain

$$n(E_1 \cap E_2^c) + n(E_1 \cap E_2) = n(E_1) \tag{1.3.6}$$

Similarly,

$$n(E_2 \cap E_1^c) + n(E_1 \cap E_2) = n(E_2) \tag{1.3.7}$$

Adding Eqs. (1.3.6) and (1.3.7) and subtracting from the result the quantity $n(E_1 \cap E_2)$ one gets

$$\begin{aligned} n(E_1 \cap E_2^c) &+ n(E_2 \cap E_1^c) + n(E_1 \cap E_2) \\ &= n(E_1) + n(E_2) - n(E_1 \cap E_2) \end{aligned} \tag{1.3.8}$$

Now the left-hand side of Eq. (1.3.8) is $n(E_1 \cup E_2)$ [refer to Fig. 1.13 again], and therefore

$$n(E_1 \cup E_2) = n(E_1) + n(E_2) - n(E_1 \cap E_2) \tag{1.3.9}$$

Let us observe that Eq. (1.3.4) is a special case of Eq. (1.3.9) in which $n(E_1 \cap E_2) = 0$.

It can be easily shown (see Prob. 1.3.9) that for any three events E_1, E_2, and E_3,

$$\begin{aligned} n(E_1 \cup E_2 \cup E_3) &= n(E_1) + n(E_2) + n(E_3) \\ &\quad - n(E_1 \cap E_2) - n(E_1 \cap E_3) - n(E_2 \cap E_3) \\ &\quad + n(E_1 \cap E_2 \cap E_3) \end{aligned} \tag{1.3.10a}$$

If events E_1, E_2, and E_3 are mutually exclusive, Eq. (1.3.10a) becomes

$$n(E_1 \cup E_2 \cup E_3) = n(E_1) + n(E_2) + n(E_3) \tag{1.3.10b}$$

In general, for events that form a partition

$$n(E_1 \cup E_2 \cup \cdots \cup E_k) = n(E_1) + n(E_2) + \cdots + n(E_k) \tag{1.3.10c}$$

Next we define $P(E_i)$, the probability of an event E_i, as the ratio of $n(E_i)$ to $N(S)$; that is,

$$P(E_i) = \frac{n(E_i)}{N(S)} \tag{1.3.11}$$

If the expression Eq. (1.3.3a) is divided by $N(S)$ one obtains

$$0 \leq \frac{n(E_i)}{N(S)} \leq 1 \tag{1.3.12}$$

which is to say

$$0 \leq P(E_i) \leq 1 \tag{1.3.13}$$

In words, Eq. (1.3.13) states that the probability of any event E_i in the sample space is a number between 0 and 1. $P(E_i) = 0$ implies $n(E_i) = 0$ and $P(E_i) = 1$ implies $n(E_i) = N(S)$. If $n(E_i) = 0$, the event E_i is impossible because $n(E_i)$ represents the number of sample points associated with the event E_i. Likewise, if $n(E_i) = N(S)$ event E_i is certain because $N(S)$ represents the totality of sample points. Furthermore, dividing expression (1.3.3b) by $N(S)$ one gets

$$\sum_{i=1}^{k} \frac{n(E_i)}{N(S)} = 1 \tag{1.3.14a}$$

or equivalently,

$$\sum_{i=1}^{k} P(E_i) = 1 \tag{1.3.14b}$$

which indicates that the probability assigned to the sample space is 1.

Similarly, if Eq. (1.3.9) is divided by $N(S)$ the result becomes

$$\frac{n(E_1 \cup E_2)}{N(S)} = \frac{n(E_1)}{N(S)} + \frac{n(E_2)}{N(S)} - \frac{n(E_1 \cap E_2)}{N(S)}$$

which implies

$$P(E_1 \cup E_2) = P(E_1) + P(E_2) - P(E_1 \cap E_2) \tag{1.3.15a}$$

If E_1 and E_2 are mutually exclusive events Eq. (1.3.15a) becomes

$$P(E_1 \cup E_2) = P(E_1) + P(E_2) \tag{1.3.15b}$$

In the general case, expression (1.3.10c) is divided by $N(S)$ to yield

$$P(E_1 \cup E_2 \cup \cdots \cup E_k) = P(E_1) + P(E_2) + \cdots + P(E_k) \tag{1.3.15c}$$

Expressions (1.3.13), (1.3.14b), (1.3.15a), and (1.3.15b) represent the basic equations to find the probabilities associated with events.

We shall next present several random experiments and assign probabilities to the various events in the specified sample space.

Example 1.3.1 A president is to be elected from the membership of a political organization which has 200 members. If the ratio of male to female is 3:1 and half of both men and women are married, what is the probability that (a) the president is a man, (b) the president is a married woman and (c) the president is a married man or married woman.

Solution: Let E_1 be the event that the president is a man.
Let E_2 be the event that the president is a woman.
Let E_3 be the event that the president is a married man.

Let E_4 be the event that the president is a married woman. Then

$$P(E_1) = \frac{n(E_1)}{N(S)} = \frac{150}{200} = \frac{3}{4}$$

$$P(E_2) = \frac{n(E_2)}{N(S)} = \frac{50}{200} = \frac{1}{4}$$

$$P(E_3) = \frac{n(E_3)}{N(S)} = \frac{75}{200} = \frac{3}{8}$$

$$P(E_4) = \frac{n(E_4)}{N(S)} = \frac{25}{200} = \frac{1}{8}$$

Hence the answer to (a) is 3/4, the answer to (b) is 1/8, and we arrive at the result for (c) by applying Eq. (1.3.15a):

$$P(E_3 \cup E_4) = P(E_3) + P(E_4) - P(E_3 \cap E_4) = 3/8 + 1/8 - 0 = 1/2$$

Example 1.3.2 Consider the random experiment of throwing an ordinary six-sided die. Let $E_1 = \{1,3,5\}$, $E_2 = \{2,4,6\}$, $E_3 = \{1,2\}$ and $E_4 = \{4,5,6\}$. Find:

(a) $P(E_1)$ (b) $P(E_1 \cup E_2)$ (c) $P(E_3 \cup E_4)$ (d) $P(E_2 \cup E_3)$

Solution: $N(S) = 6$, $n(E_1) = 3$, $n(E_2) = 3$, $n(E_3) = 2$ and $n(E_4) = 3$. Hence (a) $P(E_1) = 3/6 = 1/2$.

(b) $P(E_1 \cup E_2) = 3/6 + 3/6 - 0 = 1$
(c) $P(E_3 \cup E_4) = 2/6 + 3/6 - 0 = 5/6$
(d) $P(E_2 \cup E_3) = 3/6 + 2/6 - 1/6 = 2/3$

Example 1.3.3 Refer to Prob. 1.1.11 and consider the rolling of two ordinary dice. Let $E_1 = \{(x,y) \mid x + y = 2\}$, $E_2 = \{(x,y) \mid x + y = 3\}$, $E_3 = \{(x,y) \mid x = y\}$, and $E_4 = \{(x,y) \mid x < y\}$, where x is the face value of the first and y the face value of the second die. Find:

(a) $P(E_1)$ (b) $P(E_2)$ (c) $P(E_1 \cup E_3)$ (d) $P(E_3 \cup E_4)$

Solution (a) $P(E_1) = 1/36$ because $n(E_1) = 1$ corresponding to the 2-tuple $(1,1)$ and $N(S) = 36$ corresponding to the totality of 6^2 2-tuples in the sample space.

(b) $P(E_2) = 2/36$, because $n(E_2) = 2$ corresponding to the 2-tuples $(1,2)$ and $(2,1)$.

(c) $P(E_1 \cup E_3) = 1/36 + 6/36 - 1/36 = 6/36 = 1/6$
(d) $P(E_3 \cup E_4) = 6/36 + 15/36 = 21/36$

(b) Relative Frequency Definition. In this approach, probability is defined as the ratio of success to total number of trials in a random experiment, where success is defined as the outcome of some clearly stated event. For instance, if two regular and fair dice are rolled, the event a pair of ones may be defined as success. When the experiment of rolling two dice is repeated 1,000 times and E is defined as stated above, the following result might be obtained.

Number of Successes	Number of Trials
1	25
3	50
5	100
12	500
29	1,000

Accordingly,

$$
\begin{aligned}
P(E) &= 1/25 &= .04 \quad &\text{after } 25 \text{ trials} \\
P(E) &= 3/50 &= .06 \quad &\text{after } 50 \text{ trials} \\
P(E) &= 5/100 &= .05 \quad &\text{after } 100 \text{ trials} \\
P(E) &= 12/500 &= 0.24 \quad &\text{after } 500 \text{ trials} \\
P(E) &= 29/1000 &= 0.29 \quad &\text{after } 1{,}000 \text{ trials}
\end{aligned}
$$

More precisely,

$$P(E) = \lim_{N \to \infty} \frac{n}{N} \tag{1.3.16}$$

where n is the number of successes and N is the number of trials in the random experiment.

Let us examine this experiment more closely. All the possible outcomes of the experiment will be 36 in number. If n_i, i $= 1, 2, \ldots, 36$ represents the number of times each outcome occurs in N trials, then $0 \leq n_i \leq N$ holds true or equivalently,

$$0 \leq \frac{n_i}{N} = P[E_i] \leq 1 \tag{1.3.17}$$

Also, $\sum_{i=1}^{36} n_i = N$, which implies

$$\sum_{i=1}^{36} \frac{n_i}{N} = \sum_{i=1}^{36} P(E_i) = P(S) = 1 \tag{1.3.18}$$

and for any two events E_1 and E_2,

$$P(E_1 \cup E_2) = \frac{n_1}{N} + \frac{n_2}{N} = P(E_1) + P(E_2) \tag{1.3.19}$$

because all events are mutually exclusive in this experiment. In general, if an experiment can be partitioned into k events, and n_i represents the number of times the ith event occurs in the experiment where $i = 1, 2, \ldots, k$, then

$$0 \leq \frac{n_i}{N} = P[E_i] \leq 1 \tag{1.3.20}$$

and

$$\sum_{i=1}^{k} \frac{n_i}{N} = \sum_{i=1}^{k} P(E_i) = 1 \tag{1.3.21}$$

and

$$P(E_1 \cup E_2) = \frac{n_1}{N} + \frac{n_2}{N} = P(E_1) + P(E_2) \tag{1.3.22}$$

Hence, Eqs. (1.3.20), (1.3.21) and (1.3.22) are analogous to Eqs. (1.3.13), (1.3.14*b*), and (1.3.15*b*) respectively.

The classical definition of probability states the restrictions as to what constitutes a probability model, while Eqs. (1.3.20), (1.3.21), and (1.3.22) suggest what *realistic* values may be used in assigning probabilities to events. The former indicate *how* to assign values to probabilities, while the latter specify *what* values to assign to probabilities.

Let us refer to Example 1.3.2 at this juncture to clarify the uses of classical versus relative frequency definition of probability. Suppose the die were not made of homogenous material and were loaded such that $E_1 = \{1,3,5\}$ occurred twice as many times as $E_2 = \{2,4,6\}$. That is, showing of odd numbers is favorable over even numbers. The probability values assigned to E_1 and E_2 according to the classical definition remain the same as those given in the example, while the relative frequency approach suggests that $P(E_1) = 2/3$ and $P(E_2) = 1/3$.

(c) Axiomatic Definition. In this approach the probability of an event E is defined as a number $P(E)$ such that $P(E)$ obeys the following three postulates:

1. $P(E) \geq 0$ (1.3.23)

For the certain event S,

2. $P(S) = 1$ (1.3.24)

If E_1 and E_2 are mutually exclusive events, then

3. $P(E_1 \cup E_2) = P(E_1) + P(E_2)$ (1.3.25)

These three postulates will be referred to as Axiom 1, Axiom 2, and Axiom 3 respectively. These axioms are not equivalent to the classical and relative frequency definitions. Because these postulates are assumed as axioms, they need no proof. In the classical definition, Eq. (1.3.15*c*) could not be extended to an infinite number of events, whereas in the axiomatic approach it could be further postulated that if

$$E_i \cap E_j = \phi \qquad i \neq j \qquad i,j = 1, 2, \ldots, n, \ldots$$

then

4. $P[E_1 \cup E_2 \cup \cdots \cup E_n \cup \cdots]$
$= P[E_1] + P[E_2] + \cdots + P[E_n] + \cdots$ (1.3.26)

Since the axiomatic definition of probability does not leave room for ambiguities, it is the preferable approach. In most applications, however, both the classical and the relative frequency approaches have their useful role, and we shall make use of all three definitions.

Example 1.3.4 If events E_1, E_2, E_3, and E_4 are mutually exclusive outcomes of a random experiment, are

(a) $P(E_1) = \frac{1}{2}, P(E_2) = \frac{1}{3}, P(E_3) = \frac{1}{6}, P(E_4) = 0$

(b) $P(E_1) = \frac{1}{5}, P(E_2) = -\frac{1}{5}, P(E_3) = P(E_4) = \frac{1}{2}$

(c) $P(E_1) = 1.5, P(E_2) = -.5, P(E_3) = P(E_4) = 0$

proper assignments of probability values?

Solution: (a) Axiom 1 and Axiom 2 are satisfied. Axiom 3 holds true by assumption. Therefore, these values represent a proper assignment of probabilities.

(b) Since $P(E_2) = -1/5$, Axiom 1 is not satisfied and hence this situation does not represent a proper assignment of probability values.

(c) In this case $P(E_1) > 1$ and $P(E_2) < 0$; therefore these values do not represent a proper assignment of probabilities.

PROBLEMS

1.3.1 An urn contains 6 black and 4 white balls. A ball is drawn from the urn at random, its color observed and then replaced. Describe the sample space. Describe the events. Assign probabilities to the events.

1.3.2 Four coins are tossed. Describe the sample space and indicate all sample points belonging to it. Let E_1 be the event of no heads showing, E_2 be the event of 2 heads and 2 tails showing. Assign probabilities to these events.

1.3.3 If two ordinary and fair dice are rolled, what is the probability of getting: (a) a total of 7? (b) either a total of 5 or 6? (c) neither a total of 2 nor 12? (d) a total of at least 5? (e) a total of at most 10?

1.3.4 From a class of 35 students with 21 boys one student is to be selected by lot to represent the class at a meeting. What is the probability that the class representative will be (a) a boy? (b) a girl?

1.3.5 The integers 1 to 25 inclusive are written on slips of paper and placed in a container. After thoroughly mixing the slips, one slip is drawn at random. What is the probability that the number on the slip is (a) divisible by 2? (b) divisible by 5?

1.3.6 Suppose two regular tetrahedrons are tossed for a random experiment. The faces are marked 1, 2, 3, and 4. Let $P(x = y) = 1/8$ and $P(x \neq y) = 1/24$, where x is the face value of the first and y is the face value of the second tetrahedron. Find the probabilities for the events (a) $x = 2$, (b) $y = 4$, (c) $y > 2$, (d) $x = 2$ or $y = 3$, (e) $x + y < 4$, (f) $x + y > 3$. .

1.3.7 A sample space consists of four mutually exclusive events, A, B, C, and

D. Determine for each of the following whether it satisfies the conditions of a probability model:
(a) $P(A) = .1 \quad P(B) = .2 \quad P(C) = .3 \quad P(D) = .4$
(b) $P(A) = .5 \quad P(B) = .4 \quad P(C) = .2 \quad P(D) = .25$
(c) $P(A) = .3 \quad P(B) = -.3 \quad P(C) = .5 \quad P(D) = .5$

1.3.8 A card is selected from a well-shuffled standard deck of cards. Assign probability values to the following events: (a) the card selected is a club. (b) the card selected is a queen or a king. (c) the card selected is the ace of diamonds or jack of spades.

1.3.9 Verify Eq. (1.3.10*a*).

1.3.10 Suppose A, B, and C are three mutually exclusive events forming a sample space such that $P(A) = .35$ and $P(B) = .25$. Find (a) $P(A^c \cup B)$, (b) $P(B^c)$, (c) $P(A \cup B)$, and (d) $P(C \cup A^c)$.

1.3.11 A company has 125 employees of whom some are college graduates (G), some are married (M) and some are Republicans (R). Given the Venn diagram in Fig. 1.14, with the number of sample points indicated, find:

(a) $P(G)$ (b) $P(G \cup R)$ (c) $P(R^c)$
(d) $P(G \cap M)$ (e) $P(G \cup M \cup R)$ (f) $P(G^c \cup M)$
(g) $P(G \cap M \cap R)$ (h) $P(R \cap (G \cup M))$

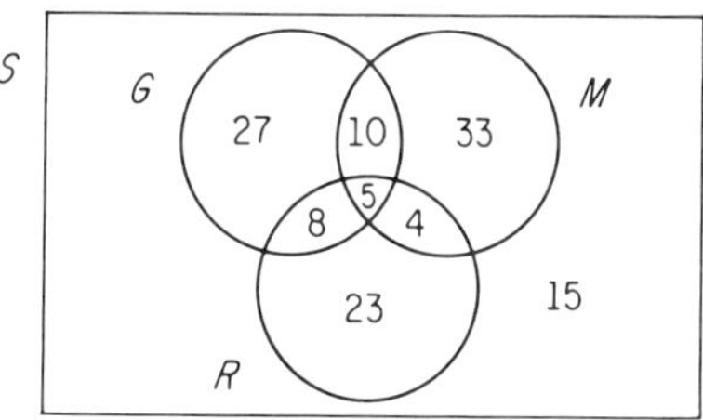

FIG. 1.14.

1.4 PERMUTATIONS COMBINATIONS, THE BINOMIAL AND MULTINOMIAL THEOREMS

In Prob. 1.3.2 the student was asked to describe the sample space of the outcomes in a random experiment where four coins were tossed. Suppose the same question were asked in the toss of 5, 6, . . . , n coins. Is there a simple and systematic way of arranging and counting the events such that the describing of the sample space becomes relatively easy? Fortunately, yes. We shall now present a brief discussion and obtain basic formulas for counting arrangements; namely, permutation and combination of objects and use them in the derivation of the binomial and multinomial theorems.

Before we give a definition and derive a formula for either permutation or combination, let us give an example.

Example 1.4.1 In how many distinct ways can 2 letters be selected from the letters a, b, c, and d?

Solution: A solution to this problem is the enumeration of all possible arrangements of the two letters; a tree diagram like that of Fig. 1.15 is very useful. Therefore, there are 12 distinct arrangements; namely,

$$ab \quad ac \quad ad \quad ba \quad bc \quad bd \quad ca \quad cb \quad cd \quad da \quad db \quad dc$$

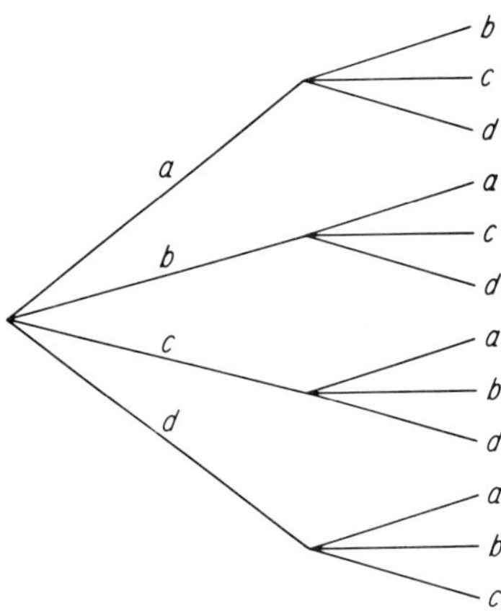

FIG. 1.15.

However, a closer examination suggests the availability of four choices to select the first letter and then three choices to select the second letter and, hence, there would be $4 \times 3 = 12$ possible arrangements or permutations.

Now we consider the general case for permutations of objects—that is, suppose we have n different objects and we wish to count the distinct arrangements of k objects taken from the n. As reasoned above, there are n choices for the selection of the first object, $(n - 1)$ choices for the selection of the second object, etc., and lastly $n - (k - 1) = n - k + 1$ choices for the selection of the kth object. Therefore the total number of distinct arrangements or permutations for k objects from the n would be

$$n(n - 1)(n - 2) \cdots (n - k + 1) = \frac{n!}{(n - k)!} \qquad (1.4.1)$$

where $n! = n(n - 1)(n - 2) \cdots (3)(2)(1)$, and is read "$n$ *factorial*."

We shall denote Eq. (1.4.1) as $P(n,k)$ which reads "the permutation of n things taken k at a time." There are several commonly used notations for permutations among which ${}_nP_k$, P^n_k, and $P(n,k)$. We prefer the last expression. Let us mention that there should be no confusion between $P(n,k)$ and $P(N)$ because the first is the permutation of n objects taken k at a time, while the latter, the probability of event N.

Suppose we wish to compute $P(n,n)$. According to Eq. (1.4.1),

$$P(n,n) = \frac{n!}{(n-n)!} = \frac{n!}{0!}$$

which holds true if $0!$ is defined to be equal to 1. Defining $0! = 1$ makes Eq. (1.4.1) valid for all $k \leq n$.

Example 1.4.2 How many permutations are possible for three cards, taken from an ordinary deck of 52 cards?

Solution: The problem is the same as the permutation of 52 objects taken 3 at a time, hence $P(52,3) = (52)(51)(50) = 132{,}600$.

Example 1.4.3 The State of California uses for car licenses three letters from the English alphabet, followed by three numbers. How many arrangements of 3 letters and 3 numbers are possible if: (a) the numbers used must be all different; (b) the letter O must not be used, and all numbers different; (c) there are no restrictions on the use of letters or numbers.

Solution: (a) Since there are no restrictions on the letters, the three letters could be the same, such as AAA. Therefore, the letters could be arranged in $(26)(26)(26) = 17{,}576$ ways. Since the numbers must be different, $P(10,3) = 10 \cdot 9 \cdot 8 = 720$ ways of permuting the numbers exist. Hence, there are $(17{,}576)(720) = 12{,}654{,}720$ ways.

(b) The three letters may be the same, except O cannot be used. Therefore, there are $(25)(25)(25) = 15{,}625$ ways of using letters. Since the numbers must be different, there are again $P(10,3) = 720$ ways of permuting them and hence $(15{,}625)(720) = 11{,}250{,}000$ different ways of preparing the license plates.

(c) Since there are no restrictions whatsoever, the total number of different arrangements of letters would be $(26)(26)(26) = 17{,}576$ and the total number of different arrangements of letters would be $10^3 = 1{,}000$. Therefore there are 17,576,000 different possible ways.

In the above discussion on permutation of objects it should have been clear that the *order* of arrangements was the important issue. In the representation of two letters, say a and b, ab is considered to be different from ba. Whenever the order of arrangement of objects is immaterial, it is referred to as the *combination* of objects. That is, combination of objects is selection regardless of order. Consider another example.

Example 1.4.4 A family is comprised of a father, a mother, two sons, and a daughter. All five members attend a concert and sit in adjacent seats. (a) How many distinct seating arrangements exist? (b) How many seating arrangements exist if order is not taken into consideration?

Solution: The answer to the first question is $P(5,5) = 5! = 120$. The answer to the second is simply 1, because regardless how they sit, the family as a unit must be selected and obviously there is only one way of doing it.

To obtain a general expression for the combination of k objects taken from n, we proceed as follows. Since there are $P(n,k) = \frac{n!}{(n-k)!}$ ways of arranging the k objects from n, and having done this, the k objects can be arranged in $k!$ ways among themselves, then $\frac{P(n,k)}{k!}$ is the expression we need for combinations. We denote this by the symbol $C(n,k)$, which reads "the combination of n objects taken k at a time" or "the combination of n things taken k at a time." There are many symbols used for combinations, among which ${}_nC_k$, C^n_k, $C(n,k)$ and $\binom{n}{k}$ are common. It should be noted that $C(n,k)$ is meaningless if $n < k$ and therefore we define $C(n,k) = 0$ for $n < k$.

Example 1.4.5 If $C(n, 12) = C(n, 8)$, find n.
Solution: Prior to solving this problem, we show that

$$C(n,k) = C(n,n-k) \tag{1.4.2}$$

Applying the definition of combinations to the left-hand side of Eq. (1.4.2) we obtain

$$C(n,k) = \frac{P(n,k)}{k!} = \frac{n!}{k!(n-k)!} \tag{1.4.3a}$$

and similarly applying the definition of combinations to the right-hand side of Eq. (1.4.2) we get

$$C(n,n-k) = \frac{P(n,n-k)}{(n-k)!} = \frac{n!}{(n-k)!k!} \tag{1.4.3b}$$

which verifies Eq. (1.4.2). Now, for the special case of $C(n, 12) = C(n, 8)$ it must be true that $12 = n - 8$ or equivalently $8 = n - 12$. Therefore $n = 20$.

Remark: If instead we proceed as follows

$$\frac{n!}{12!\,(n-12)!} = \frac{n!}{8!\,(n-8)!}$$

and hence $12!\,(n-12)! = 8!\,(n-8)!$ Expanding this last expression would yield a complicated polynomial to solve.

In our consideration of permutation of n objects taken k at a time, it was assumed that all objects were distinguishable. How will the number of distinct arrangements be affected if some of the objects were indistinguishable? For instance, the word *add* consists of 3 letters in which two d's are alike and intuitively the number of permutations of these letters is smaller than the number of permutations of 3 distinguishable letters. When there are identical objects or letters, such as d in the word *add*, we distinguish between them by using subscripts, say d_1 and d_2, and using Eq. (1.4.1) we get $P(3,3) = 6$. However, the two *d*'s can be permuted among themselves in $P(2,2) = 2! = 2$ ways with each distinct arrangement of the other letter, *a*. Thus if Q were the required number of permutations $P(2,2) \cdot Q = P(3,3)$ which yields $Q = P(3,3)/P(2,2) = 3$. Specifically, *add*, *dad*, *dda*.

In general, how many distinct permutations are possible for n objects of which some are alike? We shall derive an expression for this next. Suppose the set of n objects are partitioned and there are k subsets. Let n_1 represent the number of objects in the first subset, n_2 the number in the second, etc. then, $n_1 + n_2 + \cdots + n_k = n$. Now if all n objects were distinguishable there would be $P(n,n) = n!$ ways of permuting them. But the n_1 objects can be permuted among themselves in $P(n_1,n_1) = n_1!$ ways, the n_2 objects can be permuted among themselves in $P(n_2,n_2) = n_2!$ ways, and so on; such that if Q were the permutation required, $Q \cdot (n_1!n_2! \cdots n_k!) = n!$ Or,

$$Q = \frac{n!}{n_1!n_2! \cdots n_k!} \tag{1.4.4}$$

Equation (1.4.4) could perhaps be more readily remembered if we wrote

$$Q = \frac{P(n,n)}{P(n_1,n_1)P(n_2,n_2) \cdots P(n_k,n_k)} \tag{1.4.5}$$

Example 1.4.6 How many permutations are possible with the letters in the word *addressed* taken all together?

Solution: There are 3 *d*'s, 2 *e*'s and 2 *s*'s in the total of nine letters, hence by Eq. (1.4.4) we get $Q = \dfrac{9!}{3!2!2!} = 15{,}120$ distinct permutations.

The Binomial Theorem. We shall derive a formula, referred to as the *binomial theorem*, and use it extensively in subsequent chapters. Consider raising the binomial $a + b$ to integral powers. For $n = 1, 2, 3, 4, \ldots$ we obtain

$$
\begin{array}{ll}
n = 1 & (a+b)^1 = a + b \\
n = 2 & (a+b)^2 = a^2 + 2ab + b^2 \\
n = 3 & (a+b)^3 = a^3 + 3a^2b + 3ab^2 + b^3 \\
n = 4 & (a+b)^4 = a^4 + 4a^3b + 6a^2b^2 + 4ab^3 + b^4 \\
\cdot & \\
\cdot & \\
\cdot &
\end{array}
$$

Using mathematical induction it can be shown that certain relationships hold true for the general case, $(a + b)^n$; namely, that

1. there are $(n + 1)$ terms in the expansion of $(a + b)^n$,
2. in successive terms, the exponent of a decreases by 1 and the

exponent of b increases by 1, such that the sum of the exponents in each term remains equal to n, and

3. the coefficient of the kth term is $C(n,k-1)$.

That is,

$$(a+b)^n = C(n,0)a^n b^0 + C(n,1)a^{n-1}b^1 + C(n,2)a^{n-2}b^2 + \cdots + C(n,n)a^0 b^n = \sum_{k=0}^{n} C(n,k)a^{n-k}b^k \tag{1.4.6}$$

Equation (1.4.6) is the binomial theorem.

Example 1.4.7 Using the binomial theorem, evaluate $(1.03)^4$.

Solution: $(1.03)^4 = (1+.03)^4$

Let $1 = a$ and $.03 = b$ in the binomial expansion; then

$$\begin{aligned}(1.03)^4 &= 1 + 4(.03) + 6(.03)^2 + 4(.03)^3 + (.03)^4 \\ &= 1.12550881\end{aligned}$$

Example 1.4.8 Show that

$$\sum_{k=0}^{n} C(n,k) = 2^n$$

Solution: Since the binomial theorem holds true for all values of a and b, let $a = 1$ and $b = 1$. Then

$$(1+1)^n = 2^n = \sum_{k=0}^{n} C(n,k)$$

Example 1.4.9 In the expansion of $(x+y)^{100}$, write the 75th term assuming the terms are counted in ascending powers of y.

Solution: The coefficient of the kth term in a binomial expansion is $C(n,k-1)$, hence the 75th term is

$$C(100,74)x^{100-74}y^{74} = C(100,74)x^{26}y^{74}$$

The Multinomial Theorem. The binomial theorem derived above is a special case of the multinomial theorem, which gives the expansion of the general expression

$$(x_1 + x_2 + x_3 + \cdots + x_k)^n \tag{1.4.7}$$

When this expression is expanded, the result will have terms with the general form

$$Q\, x_1^{n_1} x_2^{n_2} \cdots x_k^{nk} \tag{1.4.8}$$

where Q is a constant, $\sum_{i=1}^{k} n_i = n$, and $n_i \geq 0$ hold true for each term.

Furthermore, a little scrutiny shows that Q is the same expression given in Eq. (1.4.4) or (1.4.5) that is to say, Q is the permutation of n objects of which n_1 are alike objects of one kind, n_2 are alike objects of another kind, etc. and hence for each term in the above expansion

$$Q = \frac{n!}{n_1!\, n_2! \cdots n_k!}$$

where $n_1, n_2, \ldots, n_k$ are the same values as the exponents in that particular term. Since we desire to obtain the totality of such terms, we sum over all terms having $\sum_{i=1}^{k} n_i = n$ and $n_i \geq 0$ and get

$$(x_1 + x_2 + \cdots + x_k)^n = \sum_{n_1, n_2, \ldots, nk} Q\, x_1^{n_1} x_2^{n_2} \ldots x_k^{nk} \qquad (1.4.9)$$

Needless to say, for large values of n, the multinomial theorem gets very laborious. However, to illustrate the theorem we give an example.

Example 1.4.10 Expand the expression $(a + b + c)^3$ by the use of multinomial theorem.

Solution: We first obtain the set of 3-tuples (n_1, n_2, n_3) which satisfy the conditions $n_1 + n_2 + n_3 = 3$ and $n \geq 0$. The set consists of 10 3-tuples, (3,0,0), (0,3,0), (0,0,3), (2,1,0), (2,0,1), (1,2,0), (1,0,2), (0,2,1), (0,1,2), (1,1,1). Therefore, the expansion will have ten terms which are

$$\frac{3!}{3!0!0!} a^3 b^0 c^0 + \frac{3!}{0!3!0!} a^0 b^3 c^0 + \cdots + \frac{3!}{1!1!1!} a^1 b^1 c^1$$

$$= a^3 + b^3 + c^3 + 3a^2 b + 3a^2 c + 3ab^2 + 3ac^2 + 3b^2 c + 3bc^2 + 6abc$$

PROBLEMS

1.4.1 Find the number of permutations that can be made from the vowels of the English alphabet if taken (a) 4 at a time, (b) 5 at a time.

1.4.2 How many distinct arrangements can be made from the letters of the word *Minnehaha* if (a) all letters are taken at a time (b) consonant letters are taken only?

1.4.3 A regular die is tossed 4 times and the face values shown were 2, 1, 6, and 4 in that order. In how many other orders could these four values have appeared?

1.4.4 Telephone numbers have seven digits grouped in the order of 3 and then 4 numbers, such as 543–1234. How many telephone numbers are possible if (a) the last group of digits could be comprised of any digit, (b) all seven digits could be comprised of any digit?

1.4.5 A college faculty council is comprised of 12 faculty members. From these members 3 committees are to be formed such that Committee A will have 5 members, Committee B will have 4 members, and Committee C will have 3 members. In how many distinct ways could the three committees be formed?

1.4.6 An article could be purchased from seven different stores in town. In how many ways can 2 stores be chosen from the seven if one wishes to inquire about the selling prices?

1.4.7 Six regular dice are colored differently. If all six dice were thrown once, (a) how many different arrangements of 3 ⚂ 's, 2 ⚃ 's, and 1 ⚀ would there be? (b) Repeat for part (a), but assume the dice all alike.

1.4.8 Show that $k\binom{n}{k} = n\binom{n-1}{k-1}$.

1.4.9 Show that $\binom{n}{k} + \binom{n}{k-1} = \binom{n+1}{k}$ for all integer values of $n > 0$ and $k \leq n$. (Pascal's rule.)

1.4.10 Using the relation in Prob. 1.4.9, extend to $n = 8$ the table of combinatorial values, (Pascal's triangle.)

n	$k = 0$	1	2	3	4 ···
0	1				
1	1	1			
2	1	2	1		
3	1	3	3	1	
·					
·					
·					

1.4.11 Show that $\sum_{k=1}^{n} k\binom{n}{k} = n2^{n-1}$. (*Hint:* Use the result of Prob. 1.4.8.)

1.5 CONDITIONAL PROBABILITY

In our discussion on probability the notation $n(E_i)$ was used to represent the number of sample points associated with the event E_i and $P(E_i)$ with its probability. It was tacitly implied in these notations that the event E_i was in a given or clearly specified sample space S, and it would have been meaningless to think of another sample space, say S_1. If there were any question about this notation, a better symbolism should have been provided stressing that the event E_i is in S. To achieve this goal, we introduce the notation $P(E_i \mid S)$ which reads, "the probability of E_i given S," and is called the *conditional probability* of E_i given S.

In most applications of probability we seldom are concerned with the event E_i alone, but rather consider some *joint* characteristics of E_i and the probability associated with it. Conditional probability plays an important role in cases of this type and before we derive general formulas we need to present the notion of marginal probability.

In Sec. 1.1 while considering sample spaces and their description we introduced the convenient notation of n-tuples, where n represents the number of experiments being carried out simultaneously. Figure 1.4, for instance, represents a 2-tuple and the two experiments are the simultaneous tossing of a coin and a die. Suppose we represent the sample space of Fig. 1.4 in tabular form.

TABLE 1.5.1

DIE	1	2	3	4	5	6	
COIN H = head	$H,1$	$H,2$	$H,3$	$H,4$	$H,5$	$H,6$	6
T = tail	$T,1$	$T,2$	$T,3$	$T,4$	$T,5$	$T,6$	6
	2	2	2	2	2	2	12

It is obvious from Table 1.5.1 that if we disregard the outcomes of the die, the sample space represents 12 sample points of which 6 correspond to the showing of a "head" on the coin and 6 to the showing of a "tail." These numbers with their sum of 12 are represented on the horizontal margins of the Table 1.5.1. Similarly, if the outcomes of the coin were disregarded, the sample space would consist of 12 sample points of which 2 correspond to the outcomes of the die showing a "1," 2 correspond to the outcomes of the die showing a "2," and so on. These numbers with their sum of 12 are represented on the vertical margins. If we are interested only in the outcomes of the coin in this composite experiment, and used the definition of probability given in Eq. (1.3.11), we obtain

$$P(H) = \frac{n(H)}{N(S)} = \frac{6}{12} = \frac{1}{2}$$

and similarly $P(T) = 1/2$. Likewise, if we were interested in the outcomes of the die only, we would get

$$P(1) = \frac{n(1)}{N(S)} = \frac{2}{12} = \frac{1}{6}$$

and similarly the rest would have a probability of $1/6$ each.

In general, for any composite experiment, if we let $n(M_i)$ represent the marginal values, we define the *marginal probability* by

$$P(M_i) = \frac{n(M_i)}{N(S)} \tag{1.5.1}$$

Equation (1.5.1) as it stands is very broad and we wish to modify it further. Let us refer to Table 1.5.1 once again and note that the marginal numbers are the totality of the sample points in a particular horizontal or vertical margin. If the horizontal and the vertical margins are looked upon as *mesh* or *net*, there would be 12 *cells* and the marginal numbers are the sums of the sample points in the cells comprising the margins. These cells can be identified by double-subscript notation. Let c_{ij} represent the cell that corresponds to the ith row and jth column. Then Table 1.5.1 represented in this notation becomes

TABLE 1.5.2

	1	2	3	4	5	6	
H = head	c_{11}	c_{12}	c_{13}	c_{14}	c_{15}	c_{16}	$c_{1.}$
T = tail	c_{21}	c_{22}	c_{23}	c_{24}	c_{25}	c_{26}	$c_{2.}$
	$c_{.1}$	$c_{.2}$	$c_{.3}$	$c_{.4}$	$c_{.5}$	$c_{.6}$	$c_{..}$

From Table 1.5.2 it is clear that the marginal for the coin at the 1st row, for instance, is

$$\sum_{j=1}^{6} c_{1j} = c_{11} + c_{12} + \cdots + c_{16}$$

and the marginal for the die at the 3rd column is

$$\sum_{i=1}^{2} c_{i3} = c_{13} + c_{23}$$

To simplify our representations further, we introduce the *dot* notation for summations and define

$$\sum_{i=1}^{r} c_{ij} = c_{.j}$$

$$\sum_{j=1}^{t} c_{ij} = c_{i.}$$

$$\sum_{i=1}^{r} \sum_{j=1}^{t} c_{ij} = c_{..}$$

It should be clear from this definition that a dot in the missing subscript position indicates summation over that subscript.

In Table 1.5.2 the various marginals of a coin and a die are represented in the dot notation. Now we define the marginal probabilities as follows:

$$P(c_{.j}) = \frac{n(c_{.j})}{n(c_{..})} \tag{1.5.2a}$$

$$P(c_{i.}) = \frac{n(c_{i.})}{n(c_{..})} \tag{1.5.2b}$$

In this section our representation will include only meshes of two dimensions but in later chapters we will have occasion to refer to cells in n-dimensions corresponding to n-tuples.

Next, let us suppose that in this composite experiment the outcome of one of the experiments is known. Say, for instance, the die has come up "4." Now if we were to ask what the probability is of getting a "head" in the toss of the coin, the answer intuitively (if not obviously) would be a greater probability value than we might have proposed, not having known the outcome of the die. To show this to be true, let E_1 represent the event of head showing in the toss of the coin and E_2 represent the event of "4" showing in the throw of the die. The probability we want is $P(E_1 \mid E_2)$; namely, the probability of E_1 given E_2. Since we know the outcome of the die is "4," the ratio needed is

$$\frac{n(E_1 \cap E_2)}{n(E_2)} = \frac{1/12}{1/6} = \frac{1}{2}$$

which is the desired probability.

To clarify the concept of conditional probability further, let us consider another example.

Example 1.5.1 There are 80 students in a graduating class of whom 50 are men and 30 women. If 1/5 of the men and 1/15 of the women are mathematics majors, find the probability that:

(a) a student selected at random from the class will be a male mathematics major

(b) a student selected from this class will be a female who is a non-mathematics major

(c) a student selected will be a mathematics major, *given* the student is a male

(d) a student selected will be female, given she is a math major

Solution: Let A represent the event of being a male in this class. Let B represent the event of being a female in this class, and let M represent the event of being a mathematics major. Then

$n(S) = 80$, $n(A) = 50$, $n(B) = 30$, $n(B \cap M) = 2$, $n(A \cap M) = 10$, $n(M) = 12$, $n(A - (A \cap M)) = 40$, and $n(B - (B \cap M)) = 28$ (See Fig. 1.16) and hence

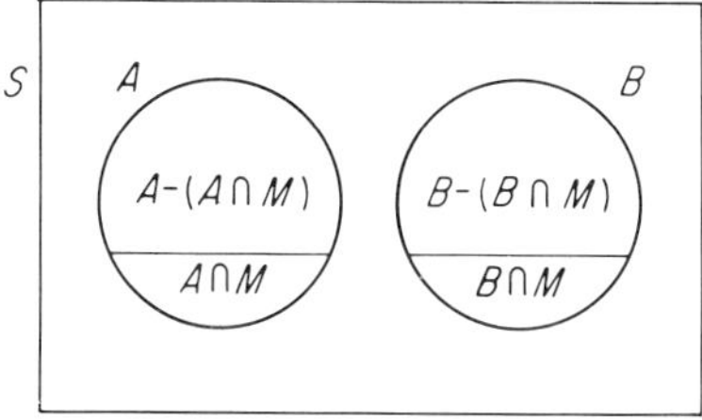

FIG. 1.16.

(a) $P(A \cap M) = 10/80 = 1/8$

(b) $P(B - (B \cap M)) = 28/80 = 7/20$

(c) $P(M \mid A) = P(M \cap A)/P(A) = 10/50 = 1/5$
(d) $P(B \mid M) = P(B \cap M)/P(M) = 2/12 = 1/6$

Now let us examine parts (c) and (d) of the above example more closely. Notice in (c) the phrase "mathematics major *given* the student is male" in effect represents a new sample space—namely, the sample space of all male graduating students. Therefore, the given information has reduced the sample space from the original 80 sample points to a new number of 50. Similarly, in (d) the phrase "female *given* the student is a mathematics major" represents a very restricted sample space consisting of 12 sample points.

It becomes apparent from these observations that a new definition is in order for the sample space if there are further specifications or restrictions on the event under question. We therefore define the conditional probability of an event E_1 *given* E_2, where both events are in the same sample space, by

$$P(E_1 \mid E_2) = \frac{P(E_1 \cap E_2)}{P(E_2)} \qquad P(E_2) \neq 0 \tag{1.5.3}$$

or

$$P(E_1 \mid E_2) \cdot P(E_2) = P(E_1 \cap E_2) \tag{1.5.4}$$

Equation (1.5.4) is preferable for the definition of conditional probability because there is no restriction on $P(E_2)$. In case $P(E_2) = 0$, then $P(E_1 \mid E_2)$ is agreed to be interpreted as having a value of zero and not representing an indeterminate form. Similarly, the conditional probability of E_2 given E_1 is

$$P(E_2 \mid E_1)\, P(E_1) = P(E_1 \cap E_2) \tag{1.5.5}$$

Next let us consider the case where $P(E_1 \mid E_2) = P(E_1)$ and $P(E_1)\, P(E_2) > 0$; that is, the given condition does not alter the sample space and E_1 and E_2 are not empty subsets of S. Then Eq. (1.5.4) becomes

$$P(E_1)\, P(E_2) = P(E_1 \cap E_2) \tag{1.5.6}$$

Whenever this situation exists for two events, the events are said to be "*stochastically independent.*" Stochastically independent is synonymous to "independent in the probability sense" and is used to distinguish it from various other concepts of independence used in mathematics.

In the following example we shall illustrate the concepts discussed above.

Example 1.5.2 A professor has 100 books in his library which are written in either English, French, or Russian. All of his books are in the field of mathematics, physics, and chemistry and Table 1.5.3 gives the breakdown of categories.

TABLE 1.5.3

	English	French	Russian
Mathematics	30	10	5
Physics	20	5	5
Chemistry	20	5	0

Determine:

(a) the probability of selecting a physics book at random

(b) the probability of selecting a book written in Russian

(c) the probability of selecting a mathematics book given that it is written in French

(d) whether the events of selecting a physics book and selecting a book written in English are stochastically independent

Solution:

(a) The marginal number for physics books is 20 + 5 + 5 = 30 and hence P(physics book) = n (physics)/$n(S)$ = 30/100 = .3

(b) The marginal number for Russian books is 5 + 5 + 0 = 10 and hence P (Russian book) = n (Russian)/$n(S)$ = 10/100 = 0.1

(c) P(mathematics | French) = P (mathematics and French)/ P (French) according to Eq. (1.5.3). Therefore, the required probability is .1/.2 = .5

(d) Two events are considered to be stochastically independent if Eq. (1.5.6) is satisfied. Let E_1 be the event of selecting a physics book and E_2 the event of selecting a book written in English. From part (a), $P(E_1) = .3$ and $P(E_2) = .7$ can be computed similarly. Now $P(E_1) \cdot P(E_2) = (.3)(.7) = .21$ but $P(E_1$ and $E_2) = .20$. Since Eq. (1.5.6) is not satisfied, the selection of a physics and an English book does not represent a stochastically independent event.

Let us comment on part (d) of the above example. Showing that Eq. (1.5.6) is not satisfied for any two events proves the experiment to be not stochastically independent. However, showing that Eq. (1.5.6) is satisfied for a particular pair of events *does not* prove stochastic independence for an experiment. The next example further demonstrates the concept.

Example 1.5.3 A company has 50 employees and the table below represents their educational attainments.

TABLE 1.5.4

	Male	Female	
B.A.	10	10	20
M.A.	11	9	20
Ph.D.	4	6	10
	25	25	50

In an experiment an employee is selected. Are the events being male and having a degree independent?

Solution: Let M represent the event of being a male, F being a female, A having a B.A. degree, B having an M.A. degree, and C having a Ph.D.

$$P(M) = P(F) = 25/50 = 1/2$$
$$P(A) = 20/50 = 2/5$$
$$P(B) = 20/50 = 2/5$$
$$P(C) = 10/50 = 1/5$$
$$P(M)\,P(A) = 1/5$$

and

$$P(M \cap A) = 10/50 = 1/5$$

Therefore Eq. (1.5.6) is satisfied for these two events, but consider another case:

$$P(M)\,P(B) = 1/2 \times 2/5 = 1/5 \text{ and } P(M \cap B) = 11/50$$

Hence Eq. (1.5.6) is not satisfied. Therefore, the events in this experiment are not stochastically independent.

In general, a random experiment is said to be stochastically independent if for all events in the partitioned sample space, $P(E_i) \cdot P(E_j) = P(E_i \cap E_j)$. The following is another example illustrating the notions discussed above.

Example 1.5.4 An urn contains 3 red, 2 white, and 1 blue balls. Another urn contains no red, 4 white, and 2 blue balls. A regular die is rolled and if the outcome is 1 or 6, a ball is to be taken from the first urn, otherwise, a ball is to be taken from the second urn. Find: (a) the conditional probability of the ball having been taken from the first urn *given* that the ball is white, and (b) the conditional probability of the ball having been taken from the second urn *given* that the ball is blue.

Solution: Let us summarize the given information in tabular form:

TABLE 1.5.5

COLOR OF BALL

	Red	White	Blue	
Urn 1	3	2	1	6
Urn 2	0	4	2	6
	3	6	3	12

Let R, W, B, represent the events of a ball being red, white, or blue respectively and U_1 and U_2 represent the events of balls having been taken from urn 1 or urn 2, respectively. The probabilities of these events are:

$$P(R) = 3/12 = 1/4$$
$$P(W) = 6/12 = 1/2$$
$$P(B) = 3/12 = 1/4$$
$$P(U_1) = 1/3$$

because in the throw of a die the probability of a 1 or 6 is $1/6 + 1/6 = 1/3$, and

$$P(U_2) = 4/6 = 2/3$$

Now the conditional probabilities are as follows:

$$P(R \mid U_1) = 3/6 = 1/2 \qquad P(R \mid U_2) = 0/6 = 0$$
$$P(W \mid U_1) = 2/6 = 1/3 \qquad P(W \mid U_2) = 4/6 = 2/3$$
$$P(B \mid U_1) = 1/6 \qquad P(B \mid U_2) = 2/6 = 1/3$$

Applying Eq. (1.5.4) we obtain,

$$P(R \cap U_1) = P(U_1) \cdot P(R \mid U_1) = 1/3 \cdot 1/2 = 1/6$$
$$P(R \cap U_2) = P(U_2) \cdot P(R \mid U_2) = 2/3 \cdot 0 = 0$$
$$P(W \cap U_1) = P(U_1) \cdot P(W \mid U_1) = 1/3 \cdot 1/3 = 1/9$$
$$P(W \cap U_2) = P(U_2) \cdot P(W \mid U_2) = 2/3 \cdot 2/3 = 4/9$$
$$P(B \cap U_1) = P(U_1) \cdot P(B \mid U_1) = 1/3 \cdot 1/6 = 1/18$$
$$P(B \cap U_2) = P(U_2) \cdot P(B \mid U_2) = 2/3 \cdot 1/3 = 2/9$$

Let us represent these probabilities in a table:

TABLE 1.5.6
COLOR OF BALL

	Red	White	Blue	
Urn 1	1/6	1/9	1/18	6/18
Urn 2	0	4/9	2/9	12/18
	3/18	10/18	5/18	18/18

Therefore

$$P(U_1 \mid W) = \frac{P(U_1 \cap W)}{P(W)} = \frac{1/9}{10/18} = \frac{1}{5}$$

and

$$P(U_2 \mid B) = \frac{P(U_2 \cap B)}{P(B)} = \frac{2/9}{5/18} = \frac{4}{5}$$

It should be evident from the above examples that the given condition or the additional information alters the probability of the event under consideration. In other words, the probability of the event *prior* to the additional relevant information given would be different from the probability of the event *after* the relevant information is given. Therefore, conditional probability is sometimes referred to as *a posteriori* probability.

Bayes' Theorem. We state next the general case for the conditional probability, which is commonly attributed to the clergyman, Thomas Bayes, and is called the *Bayes' theorem.*

If $E_1, E_2, \ldots, E_k$ form a partition of the sample space S, and A is any nonempty subset in S, then

$$P(E_j \mid A) = \frac{P(E_j \cap A)}{\sum_{i=1}^{k} P(E_i \cap A)} \tag{1.5.7}$$

But since for any two events E and A in S, $P(E \cap A) = P(A \mid E)P(E)$, then Eq. (1.5.7) can be expressed equivalently:

$$P(E_j \mid A) = \frac{P(A \mid E_j)P(E_j)}{\sum_{i=1}^{k} P(A \mid E_i)\, P(E_i)} \tag{1.5.8}$$

The proof of Bayes' theorem is very straightforward. Since $E_1, E_2, \ldots, E_k$ form a partition in S,

$$\sum_{i=1}^{k} P(E_i) = P(S) = 1$$

by Eq. (1.3.14*b*). Because A is a nonempty set in S, it could be expressed as

$$A = (A \cap E_1) \cup (A \cap E_2) \cup \cdots \cup (A \cap E_k)$$

Now as a consequence of Eq. (1.3.15*c*),

$$P(A) = \sum_{i=1}^{k} P(A \cap E_i)$$

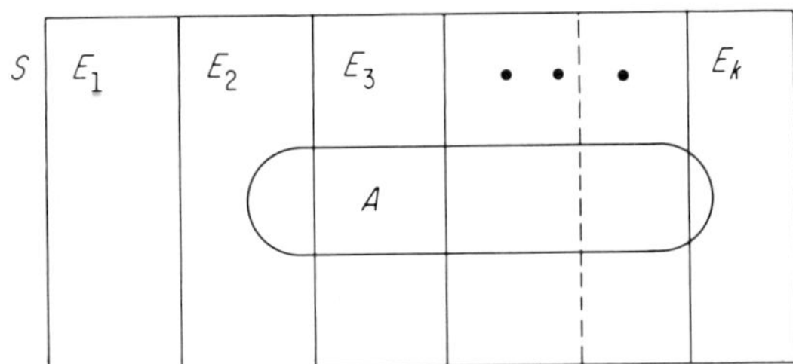

FIG. 1.17.

However, by the definition of conditional probability,

$$P(E_j \mid A) = \frac{P(E_j \cap A)}{P(A)}$$

Substituting

$$\sum_{i=1}^{k} P(A \cap E_i)$$

for $P(A)$ in the last expression, we obtain

$$P(E_j \mid A) = \frac{P(E_j \cap A)}{\sum_{i=1}^{k} P(A \cap E_i)}$$

which is Eq. (1.5.7) and Eq. (1.5.8) is an equivalent form.

To illustrate the concept of Bayes' theorem, let us consider an example.

Example 1.5.5 In a factory employees work during three shifts. It has been observed over a long period of time that the first shift produces 35 percent, the second shift 30 percent, and the third shift 35 percent of the total production. However, the number of defective items produced in each shift varies greatly and are in the proportion 1:2:3, respectively. If articles produced by the three shifts are placed in one well-mixed pile and one item randomly selected proves defective, what is the probability that the defective item was produced by (a) the first shift? (b) the third shift?

Solution: Although the first and the third shifts produce the same amount of items, the number of defective articles produced by each are different. Let E_1, E_2, and E_3 represent the events of an item being produced by the first, second, and the third shift respectively; and A represent the event of an item being defective. Then $P(E_1) = .35$, $P(E_2) = .30$, and $P(E_3) = .35$ and clearly the E_i's form a partition.

Next, $P(A \mid E_1) = 1/6$, $P(A \mid E_2) = 2/6$ and $P(A \mid E_3) = 3/6$. Using these values in Eq. (1.5.4) we obtain

$$P(A \cap E_1) = 35/600, \; P(A \cap E_2) = 60/600 \quad \text{and} \quad P(A \cap E_3) = \\ = 105/600$$

However, $A \cap E_1$, $A \cap E_2$, and $A \cap E_3$ are mutually exclusive events and their union is A. Therefore

$$P(A) = \frac{35}{600} + \frac{60}{600} + \frac{105}{600} = \frac{1}{3}$$

Applying Bayes' theorem we obtain:

(a) $P(E_1 \mid A) = 35/200 = .175$
(b) $P(E_3 \mid A) = 105/200 = .525$

PROBLEMS

1.5.1 If two ordinary and honest dice are rolled, what is the probability of getting (a) a total of 7 given that one die shows 3, (b) either a total of 5 or 6 given that one die shows 4?

1.5.2 Refer to Prob. 1.3.5. Find the probability that the number on the slip is (a) divisible by 2, given that it is even, (b) divisible by 5, given that the number is odd.

1.5.3 Refer to Prob. 1.3.6. Find the probability for the events (a) $x = 2$ given $y = 2$, (b) $y = 4$ given $x = 3$, (c) $y > 2$ given $x = 1$, (d) $x + y < 4$ given $x = 2$, and (e) $x + y > 3$ given $y < 3$.

1.5.4 Refer to Prob. 1.3.11. Find (a) $P[G \mid R]$, (b) $P[R \mid M]$, (c) $P[R \cap M \mid G]$.

1.5.5 An urn contains 4 white and 3 black chips while another identical urn contains 3 white and 6 black chips. One of the urns is selected and a chip taken out blindfolded. Find the probability that the chosen chip is white if (a) the probability of selecting the first urn is 1/2. (b) the probability of selecting the first urn is 3/4.

1.5.6 In an English class there are 30 freshmen, 25 sophomores, 25 juniors, and 20 seniors. If 50 percent of the freshmen, 40 percent of the sophomores, 60 percent of the juniors, and 25 percent of the seniors are coeds, find the probability that a student selected from this class is a

(a) male and freshman
(b) coed and senior
(c) male given that he is junior
(d) coed given that she is sophomore
(e) sophomore given that the student is a coed
(f) junior given that the student is a male.

1.5.7 In a college the proportion of full, associate, assistant professors, and instructors are 20 percent, 25 percent, 40 percent, and 15 percent respectively. If in each professorial rank 60 percent of the faculty have Ph.D. degrees, find the probability of selecting a professor at random who is (a) an associate and does not have the Ph.D. degree, (b) an instructor and has the Ph.D. degree, (c) he has the Ph.D. degree given that he is an assistant professor.

1.5.8 Suppose that nationally 60 percent of the voters are registered Democrats and the rest registered Republicans. During a presidential election it is assumed that 75 percent of the Democrats and 90 percent of the Republicans vote for the candidate of their parties while the remaining vote for the candidate of the opposite party. What is the probability that a voter selected at random has voted for the Republican candidate?

1.5.9 The probability that one million persons will visit the Minnesota State Fair is .90. The probability that it will rain during the fair week is .75. The probability that one million persons visit the fair given it does not rain is .96. (a) What is the probability that one million persons visit the fair and it does not rain during that week? (b) What is the probability that it did not rain given that one million persons did attend the fair?

1.5.10 The probability that three thousand students will be admitted to a certain college is .85. The probability that a proposed master's degree program will be initiated at this college is .65. The probability that three thousand students will attend given that the proposed program is initiated is .95. (a) What is the probability that the master's degree program is not initiated and 3,000 students attend this college? (b) What is the probability that 3,000 students will attend and the program will be initiated? (c) What is the probability that the program is initiated given that 3,000 students attend?

1.5.11 Consider the hypothetical problem that 60 percent of the cars used in a county are made in the United States, 20 percent in England, 15 percent in Germany, and 5 percent in France. If 80 percent of the American made cars are General Motors products, 50 percent of the English cars are Rootes, 60 percent of the German made cars are Volkswagen and 75 percent of the French made cars are Simca, what is the probability that a car selected at random from the car registration list is (a) a Rootes product, (b) a French car that is not a Simca?

2

Univariate Probability Functions —Discrete Case

2.0 INTRODUCTION

In Chapter 1 concepts of probability were discussed and related topics such as conditional probability and stochastic independence introduced. In this chapter we shall extend the notion of probability and define probability functions. The importance of this concept will be illustrated by some specific cases among which are (1) the binomial, (2) the Poisson, and (3) the hypergeometric probability functions.

2.1 RANDOM VARIABLES AND PROBABILITY FUNCTIONS

In all the random experiments that we discussed in the previous chapter it might have been evident to the student that we were simply considering some specific numerical description of outcomes. For instance, we were interested in the sum of the face values in the throw of two ordinary dice, and the possible values for this particular situation were the numbers 2, 3, 4, . . . , 12. The sum of the face values of the two dice is not the only numerical description that we could have considered. We mention some other possibilities: (a) the square of the sum of the face values, (b) the sum of the squares of the face values, (c) the product of the face values, (d) the difference of the face values, and *ad infinitum.*

To encompass all possible numerical description of outcomes of an experiment, we define a *random variable* as any real-valued function defined over the sample space. Let us illustrate this definition by a specific example.

Example 2.1.1 Let $w = (x - y)^2$, where x represents the face value touching the table surface of the first and y the face value of the second, in the throw of two regular tetrahedrons whose sides are labeled 1, 2, 3, and 4. Find the range of the random variable w.

Solution: There are $4^2 = 16$ sample points corresponding to the 16 2-tuples in the sample space. They are, (1,1), (1,2), . . . , (4,4). Therefore, $w = 0, 1, 4$, and 9 represent the range of the random variable.

Related to the notion of a random variable is the concept of *probability function.** A probability function of a random variable v is a function, say f, such that for each particular value of x in the range of v, $f(x)$ is the probability that v will assume this value of x. It should be evident from this definition that $f(x) \geq 0$ because $f(x)$ represents a probability which according to Axiom 1 [Eq. (1.3.23)] represents a nonnegative quantity. Also, $\sum_{\text{all } x} f(x) = 1$ because this expression represents the certain event. Therefore, an alternate definition for a probability function may be stated as follows: A probability function is any real-valued function, say f, for which the two conditions

$$f(x) \geq 0 \tag{2.1.1}$$

$$\sum_{\text{all } x} f(x) = 1 \tag{2.1.2}$$

are satisfied.

Example 2.1.2 Find the probability function for the Example 2.1.1 and graph the result.

Solution: Referring to Example 2.1.1 we note that $n(w = 0) = 4$, $n(w = 1) = 6$, $n(w = 4) = 4$ and $n(w = 9) = 2$. Clearly, $N(S) = 16$ and hence $P(w = 0) = 4/16$, $P(w = 1) = 6/16$, $P(w = 4) = 4/16$, and $P(w = 9) = 2/16$. Figure 2.1 represents the graph of the probability function associated with the random variable w.

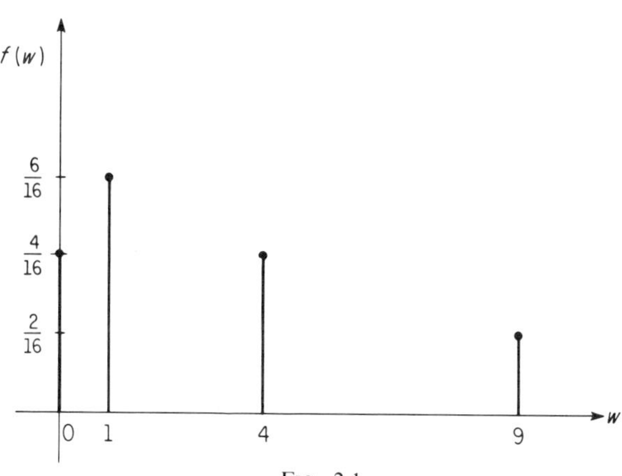

FIG. 2.1.

Example 2.1.3 Do the following satisfy the conditions of a probability function?

(a) $f(x) = 1/4$ $\qquad x = 3, 4, 5, \text{ and } 6$

(b) $f(x) = C(4,x)/2^5$ $\qquad x = 0, 1, 2, 3, \text{ and } 4$

*Often referred to as probability density function.

(c) $f(x) = \dfrac{C(3,x)C(2,3-x)}{C(5,3)}$ $\quad x = 1, 2, \text{and } 3$

Solution:
(a) $f(3) = f(4) = f(5) = f(6) = 1/4$

Therefore (2.1.1) is satisfied and clearly $\sum_{x=3}^{6} f(x) = 1$, hence $f(x)$ thus defined is a probability function.

(b) $C(4,0)/2^5 = 1/32$
$C(4,1)/2^5 = 4/32$
$C(4,2)/2^5 = 6/32$
$C(4,3)/2^5 = 4/32$
$C(4,4)/2^5 = 1/32$

Hence Eq. (2.1.1) is satisfied. However, the sum of all $f(x)$ is $1/2$ and Eq. (2.1.2) is not satisfied. Thus $f(x)$ is not a probability function.

(c) $f(1) = 3/10$
$f(2) = 6/10$
$f(3) = 1/10$

Since both conditions of a probability function are satisfied, $f(x)$ constitutes a probability function.

We remarked in Sec. 1.3 about the various approaches to the definition of probability; namely (1) the ratio of number representing an event to the total number of points in the sample space—the classical approach, (2) relative frequency of outcomes method of assigning probabilities, and (3) the axiomatic definition. Because of these approaches to the definition of probability, many authors use *frequency function* and *probability function* to define the same thing. We prefer the expression probability function because the very term indicates that probability is under consideration. In subsequent discussions we shall designate a probability function by the abbreviation p.f.

PROBLEMS

In the following functions the values of the random variable x has been indicated. (a) Find the value of the constant k which will make the given function a probability function. (b) Draw a line graph representing the probability function.

2.1.1 $f(x) = kx/3 \quad x = 2, 3, 4, 5, 6$
2.1.2 $f(x) = kx^2/2 \quad x = 1, 2, 3, 4, 5$
2.1.3 $f(x) = k(x - 5)/2 \quad x = 8, 9, 10$
2.1.4 $f(x) = k(1/5)^x \quad x = 0, 1, 2, \ldots$
2.1.5 $f(x) = k(x - 2)^2 \quad x = 1, 2, 3$
2.1.6 In the throw of two honest dice let x and y represent the outcome of the first and the second die. Let $z = x + y$

(a) Find the probability function associated with the random variable z.
(b) Draw the line graph of the probability function.

2.1.7 Repeat Prob. 2.1.6 for the random variable $z = x - y$.

2.1.8 A typist is known to type with a maximum of 5 errors per page. Write the probability function of the random variable, errors per page, assuming that $f(0) = 1/2$ and $f(x)$ is inversely proportional to the square of the number of errors.

2.2 HISTOGRAM AND FREQUENCY GRAPH

Often if not always the intent of graphic representations is to give the reader a visual aid and our objective in considering various types of graphs in this Section is to achieve this very purpose. In discussing the solution of Example 2.1.2, we gave a graphic representation of the probability function in Fig. 2.1, which was simply the line graph of the probability function under consideration.

If instead of a line graph with heights equal to the probabilities of the various outcomes, we present a bar graph with areas equal to the respective probabilities, the objective of visual aid will also be accomplished. We present such a graph in Fig. 2.2.

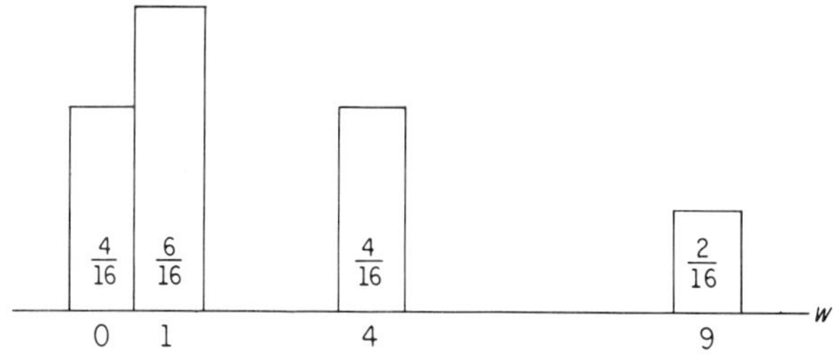

FIG. 2.2.

If the further restriction that there be no gaps between the respective bars in the bar graph were stipulated, the resulting graph is commonly referred to as a *histogram*. The word histogram literally suggests being synonymous to "written history" and perhaps the person who defined histogram in this manner might have remembered the saying, "A picture is worth a thousand words." The histogram corresponding to Fig. 2.2 is given in Fig. 2.3.

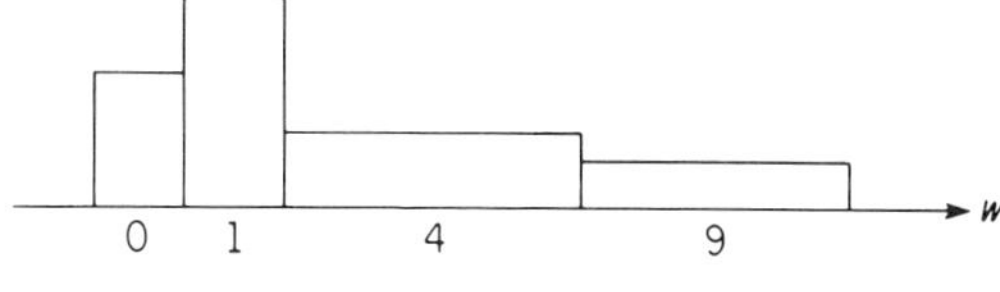

FIG. 2.3.

It should be evident from the above definition of a histogram that the total area represented is equal to unity. If instead of taking areas equal to the respective probabilities of the various outcomes, we consider areas to be proportional to the probabilities, the figure thus obtained would be similar to a histogram but it would not have an area equal to unity. Many authors refer to the latter as well as to the former as a histogram.

The reader might well inquire as to the advantages and perhaps even the necessity for defining a histogram. However, a little thought and illustration should suffice to show the need for such a definition. Let us refer to Fig. 2.1 and notice that the graph represents the probability values of the random variable $w = 0, 1, 4,$ and 9. This graphic representation of the probability function $f(w)$ has meaning only if the random variable w assumes the defined values; otherwise, the assigned probability is zero. For instance, if we wish to find the probability of $w = 1.75$ or $w = 2.38$, the answer will be zero in both cases. Whenever a probability is defined for some finite set or *denumerable set*, we say the probability function is *noncontinuous* or *discrete*, where denumerable set is any set whose elements could be put into a one-to-one correspondence with the natural numbers 1, 2, 3, 4, ... Example 2.1.3, parts (a) and (c), illustrate noncontinuous or discrete probability functions.

Next let us consider Fig. 2.3, a representation of the same probability function, namely the corresponding histogram. Here also the probability of $w = 1.75$ or $w = 2.38$ is zero; but furthermore, the probabilities of $w = 0, 1, 4,$ and 9 are zero as well because the respective areas corresponding to any point on the random variable axis, w-axis, are zero. In this situation we can find probabilities for intervals only. That is, we can find the areas between any two specified values of the random variable w, say w_1 and w_2.

Obviously, the latter type of probability assignment becomes meaningful if the random variables associated with the probability function represent finite or infinite intervals. Hence we refer to a probability function to be *continuous* if the random variable associated with it assumes values that represent continua.

In the remaining sections of this chapter we shall consider only noncontinuous type of probability functions and postpone the discussion of the continuous probability functions until Chapter 4.

PROBLEMS

In the following problems draw the histogram representation of the given probability function.

2.2.1 The probability function of Prob. 2.1.1.

2.2.2 The probability function of Prob. 2.1.2.

2.2.3 The probability function of Prob. 2.1.4.

2.2.4 $f(x) = \frac{1}{5}, \quad x = 1, 2, 3, 4, 5$

$= 0 \quad$ otherwise

2.2.5 $f(x) = \frac{C(4,x)}{2^4}, \quad x = 0, 1, 2, 3, 4$

2.2.6 $f(x) = \frac{C(3,x)C(2,3-x)}{C(5,3)}, \quad x = 1, 2, 3$

2.2.7 The probability function of Prob. 2.1.8.

2.3 DISTRIBUTION FUNCTION

Associated with each probability function $f(x)$ we consider the related function $F(x_0)$ defined as

$$F(x_0) = \sum_{x \leq x_0} f(x) \tag{2.3.1}$$

$F(x_0)$ is called the *distribution function* and as the definition indicates, the summation is over all values of the random variable that are less than or equal to the specified value of x_0. $F(x_0)$ gives the probability that the random variable x will assume values less than or equal to x_0, whereas $f(x_0)$ gives the probability that the random variable x assumes the value x_0 only. $F(x_0)$ is an expression for cumulative probabilities, and hence referred to by some authors as *cumulative frequency function.* Let us illustrate by an example the notion of distribution function.

Example 2.3.1 Given the function $f(x)$ defined as follows:

$$f(x) = \frac{x^2}{30} \quad \text{for } x = 0, 1, 2, 3, \text{and } 4$$
$$0 \quad \text{otherwise}$$

(a) Is $f(x)$ a probability function?

(b) If a probability function, find $F(0)$, $F(1)$, $F(2)$, $F(3)$, and $F(4)$.

Solution: (a) $f(0) = 0$, $f(1) = 1/30$, $f(2) = 4/30$, $f(3) = 9/30$ and $f(4) = 16/30$. Since each of these values are greater than or equal to zero and the sum of all is equal to 1, the two conditions for a probability function are satisfied. Therefore $f(x)$ thus defined is a probability function.

(b) $F(0) = f(0) = 0$
$F(1) = f(0) + f(1) = 0 + 1/30 = 1/30$
$F(2) = f(0) + f(1) + f(2) = 1/30 + 4/30 = 5/30$
$F(3) = f(0) + f(1) + f(2) + f(3) = 5/30 + 9/30 = 14/30$
$F(4) = f(0) + f(1) + f(2) + f(3) + f(4) = 14/30 + 16/30 = 1$

Next we show two basic relationships which are applicable to a distribution function, namely

1. $P(x > x_0) = 1 - F(x_0)$ (2.3.2)

2. $P(x_0 < x \leq x_1) = F(x_1) - F(x_0)$ (2.3.3)

These two relationships are very useful in computation of probabilities and may have been evident to the reader in the solution of Example 2.3.1. To show the validity of these relationships we proceed as follows:

$$P(x \leq x_0) = F(x_0) = \sum_{x \leq x_0} f(x)$$

by definition of $F(x_0)$. Subtracting these expressions from $P(S)$ we obtain

$$P(S) - P(x \leq x_0) = P(S) - F(x_0) = P(S) - \sum_{x \leq x_0} f(x) \qquad (2.3.4)$$

However, the events $x \leq x_0$ and $x > x_0$ form a partition of the sample space and, therefore

$$P(S) - P(x \leq x_0) = P(x > x_0) \qquad (2.3.5)$$

and hence

$$1 - P(x \leq x_0) = P(x > x_0)$$

or equivalently

$$1 - F(x_0) = P(x > x_0)$$

Next,

$$F(x_1) = \sum_{x \leq x_1} f(x) = P(x \leq x_1)$$

and

$$F(x_0) = \sum_{x \leq x_0} f(x) = P(x \leq x_0)$$

Subtracting the latter expression from the former we obtain

$$F(x_1) - F(x_0) = P(x \leq x_1) - P(x \leq x_0) \qquad (2.3.6)$$

Assuming that $x_0 \leq x_1$, the right-hand side of Eq. (2.3.6) represents $P(x_0 < x \leq x_1)$ and therefore, $F(x_1) - F(x_0) = P(x_0 < x \leq x_1)$.

Let us emphasize, however, that $F(x_1) - F(x_0)$ represents the probability of the interval including only the *right* end point x_1, but excluding the *left* end point x_0. For the especial case where x_0 and x_1 are two successive values, Eq. (2.3.3) becomes

$$f(x_1) = P(x = x_1) = F(x_1) - F(x_0) \qquad (2.3.7)$$

Equation (2.3.7) implies that the probability of a given term can be obtained from the distribution function.

Example 2.3.2 Verify Eqs. (2.3.2) and (2.3.3) for (a) $x_0 = 1, x_1 = 2$ and (b) $x_0 = 2, x_1 = 4$ in Example 2.3.1.

Solution: (a) $P(x > 1) = 1 - F(1) = 1 - 1/30 = 29/30$, which is equivalent to $f(2) + f(3) + f(4) = 29/30$.

$$P(1 < x \leq 2) = F(2) - F(1) = 5/30 - 1/30 = 4/30$$

which is equivalent to $P(x = 2) = f(2) = 4/30$.

(b) $P(x > 2) = 1 - F(2) = 1 - 5/30 = 25/30$

$$P(2 < x \leq 4) = F(4) - F(2) = 1 - 5/30 = 25/30$$

Note that $f(x)$ was defined for $x = 0, 1, 2, 3, 4$. Hence, the events $x > 2$ and $2 < x \leq 4$ are the same and the probabilities are identical as expected.

Example 2.3.3 A fair coin is tossed until a "tail" appears for the first time. (a) Find the probability function representing this random experiment. (b) Find the distribution function corresponding to the probability function and evaluate for $x_0 = 4$.

Solution: (a) Since the outcome of the nth toss is stochastically independent of the $(n - 1)$th toss, the probability of a "tail" showing in the first toss is $1/2$, in the second is $(1/2)^2, \ldots,$ in the nth toss is $(1/2)^n$. Therefore, $f(x) = (1/2)^x$, $x = 1, 2, 3, \ldots, n, \ldots$ is the required function. To show that this is indeed a probability function, we note that each value of $f(x)$ is positive and

$$\sum_{\text{all } x} f(x) = 1/2 + 1/4 + 1/8 + \cdots + (1/2)^n + \cdots = 1$$

since it is the sum of a geometric series with ratio $1/2$. That is,

$$\sum_{n=1}^{\infty} \left(\frac{1}{2}\right)^n = \frac{1}{2}\left(\frac{1}{1 - 1/2}\right) = 1$$

Hence Eqs. (2.1.1) and (2.1.2) are satisfied.

(b)

$$F(x_0) = \sum_{x \leq x_0} f(x) = \sum_{x \leq x_0} (1/2)^x$$

$$F(4) = 1/2 + (1/2)^2 + (1/2)^3 + (1/2)^4 = 15/16$$

PROBLEMS

In the following problems find $F(2)$ and $F(3)$ for the given probability functions. Verify in each case $f(3) = F(3) - F(2)$.

2.3.1 The probability function of Prob. 2.1.1

2.3.2 The probability function of Prob. 2.1.4.

2.3.3 The probability function of Prob. 2.2.5.

2.3.4 The probability function of Prob. 2.2.6.

2.3.5 The probability function of Prob. 2.1.6.

SOME COMMON PROBABILITY FUNCTIONS

So far in our discussions the ground rules for building probability models have been considered. Now we are ready to construct probability models, and needless to say, any function which satisfies the two conditions set forth in Eqs. (2.1.1) and (2.1.2) will represent a well-defined probability function. However, there are some probability functions that are more useful than others, just as in a men's clothing factory there may be more demand for one particular suit size than another. We shall consider in the following sections the binomial, the Poisson, the hypergeometric, the generalized hypergeometric, the multinomial, and the negative binomial probability functions.

2.4 THE BINOMIAL PROBABILITY FUNCTION

Let $p \geq 0$ and $q \geq 0$ be two related variables such that $p + q = 1$. Consider the function

$$f(x) = C(n,x)p^x q^{n-x} \qquad x = 0, 1, 2, \ldots, n \tag{2.4.1}$$

Because of the restrictions imposed on p and q, $f(x)$ thus defined is always greater than or equal to zero. Therefore, Eq. (2.1.1) is satisfied. Next let us consider $\sum_{x=0}^{n} f(x)$:

$$\sum_{x=0}^{n} C(n,x)p^x q^{n-x} = (p + q)^n = 1^n = 1$$

Therefore, Eq. (2.1.2) is also satisfied and the function $f(x) = C(n,x)p^x q^{n-x}$ is a probability function, and is called the *binomial probability function* or simply the *binomial function.*

It should be obvious from the definition of the binomial function that there are two independent parameters involved in the expression $C(n,x)p^x q^{n-x}$; namely, n and p (q is related to p and hence not independent). For any specified values of n and p, only x varies from term to term, (x is the random variable), therefore, we shall adopt the notation $f(x\colon n,p)$ to represent the binomial function and $F(x\colon n, p)$ to represent the corresponding distribution function. In general we shall use the notation $f(x\colon p_1, p_2, \ldots, p_k)$ to represent a univariate probability function with k parameters and $F(x\colon p_1, p_2, \ldots, p_k)$ to represent its distribution function.

Example 2.4.1 Find $f(2\colon 5, 1/2)$ and $f(3\colon 4, 1/4)$

Solution: $f(2\colon 5, 1/2) = C(5,2)(1/2)^2(1/2)^{5-2} = 10(1/2)^5 = 10/32$
$f(3\colon 4, 1/4) = C(4,3)(1/4)^3(3/4) = 4(1/64)(3/4) = 3/64$

Example 2.4.2 Suppose a fair coin is tossed 10 times. (a) What is the probability of getting 2 heads? (b) What is the probability of getting more than 2 heads?

Solution:

(a) $P(x = 2) = f(2\colon 10, 1/2) = C(10,2)(1/2)^2(1/2)^8 = 45/1024$
$= 0.0439$

(b) $P(x > 2) = F(10\colon 10, 1/2) - F(2\colon 10, 1/2)$

$$= 1 - \sum_{x=0}^{2} f(x\colon 10, 1/2)$$

$$= 1 - (1/1024 + 10/1024 + 45/1024)$$
$$= 968/1024 = 0.9453$$

Example 2.4.3 A regular and "unloaded" die is thrown 5 times. What is the probability of getting exactly 1 "ace"?

Solution: This is a random experiment of the same type as tossing a coin in which the parameter $p = 1/6$ and hence $q = 5/6$. Therefore, $P(x = 1) = f(1{:}\ 5, 1/6) = C(5,1)(1/6)^1(5/6)^4 = (5/6)^5$.

Next let us derive three fundamental relationships which hold true for the binomial probability function and serve a very useful purpose in computations associated with the binomial function. These relationships are:

1. $f(x{:}\ n,p) = f(n - x{:}\ n,q)$ (2.4.2)
2. $f(x{:}n,p) = F(x{:}n,p) - F(x - 1{:}n,p)$ (2.4.3)
3. $F(x{:}n,p) = 1 - F(n - x - 1{:}n,q)$ (2.4.4)

In order to show the first, we make use of the definition of the binomial function as stated in Eq. (2.4.1). The left-hand side of Eq. (2.4.2) then becomes $C(n,x)p^xq^{n-x}$. The right-hand side similarly becomes $C(n,n - x)q^{n-x}p^{n-(n-x)} = C(n,n-x)p^xq^{n-x}$. However, we have shown [refer to Eq. (1.4.2)] that $C(n,k) = C(n,n - k)$ and hence Eq. (2.4.2) holds.

The second relation is a direct consequence of the general expression shown in Eq. (2.3.7).

To show the third relationship we proceed as follows:

$$\begin{aligned} f(x{:}n,p) &= f(n - x{:}n,q) \\ f(x - 1{:}n,p) &= f(n - x + 1{:}n,q) \\ &\vdots \\ f(x - x{:}n,p) &= f(n - x + x{:}n,q) \end{aligned}$$

each of which is obtained from Eq. 2.4.2. Now, the sum of all the terms on the left-hand side in these expressions is clearly $F(x{:}n,p)$. Also, if we add to the sum of the right-hand side expressions the quantity $F(n - x - 1{:}n,q)$ and subtract the same, the equality is not altered and we will have: $F(x{:}n,p) = F(n{:}n,q) - F(n - x - 1{:}n,q)$. However, $F(n{:}n,q) = 1$ and hence Eq. (2.4.4) holds.

Example 2.4.4 Evaluate $F(9{:}\ 10, .40)$

Solution: Using Eq. (2.4.4) we obtain

$$\begin{aligned} F(9{:}\ 10, .40) &= 1 - F(0{:}\ 10, .60) \\ &= 1 - f(0{:}\ 10, .60) \\ &= 1 - C(10,0)(.6)^0(.4)^{10} \\ &= 1 - (.4)^{10} = 1 - 0.0001 \\ &= .9999 \end{aligned}$$

Example 2.4.5 In a manufacturing process 10 percent of all items turn out defective. If, from a day's output, 4 items are selected at random, what is the probability that all would be defective.

Solution: In this problem $q = .9$ (proportion of nondefectives) $p = .1$ and $n = 4$. Therefore, the solution is $C(4,4)\,(.1)^4\,(.9)^{4-4} = (.1)^4 = 1/10{,}000$. This answer could be obtained by another approach, as follows: consider choosing four items in succession from the output of the day. The probability of getting the first defective is 1/10, the probability of the second being defective is 1/10, and so on, and, hence, all four defectives would have a probability of (1/10)(1/10)(1/10)(1/10) = 1/10,000.

Table II in the Appendix contains tabulated values for

$$\sum_{x=k}^{n} f(x:\ n,p) \qquad k = 1, 2, \ldots, n$$

with increments of .05 for p. Since $\sum_{x=0}^{n} f(x:n,p) = 1$ for all p, it is not necessary to include these values in the Table. Recalling that

$$f(k:n,p) + \sum_{x=k+1}^{n} f(x:n,p) = \sum_{x=k}^{n} f(x:n,p) \tag{2.4.5}$$

we get equivalently

$$f(k:n,p) = \sum_{x=k}^{n} f(x:n,p) - \sum_{x=k+1}^{n} f(x:n,p) \tag{2.4.6}$$

Consequently the individual terms of the binomial p.f. can be determined from the cumulative terms given in Table II. The next example illustrates the use of Table II to obtain individual terms of the binomial p.f.

Example 2.4.6 Using Table II in the Appendix, evaluate:

(a) $f(4:6,.5)$ and (b) $f(7:8,.4)$

Solution:

$$\text{(a)}\ f(4:6,.5) = \sum_{x=4}^{6} f(x:6,.5) - \sum_{x=5}^{6} f(x:6,.5)$$

$$= .3438 - .1094 = .2344$$

$$\text{(b)}\ f(7:8,.4) = \sum_{x=7}^{8} f(x:8,.4) - \sum_{x=8}^{8} f(x:8,.4)$$

$$= .0085 - .0007 = .0078$$

Next, recalling that

$$\sum_{x=0}^{k} f(x:n,p) + \sum_{x=k+1}^{n} f(x:n,p) = 1$$

one obtains

$$F(k:n,p) = 1 - \sum_{x=k+1}^{n} f(x:n,p) \tag{2.4.7}$$

which represents the distribution function of the binomial probability function.

Example 2.4.7 Evaluate $F(3:4,.1)$ by using Table II in the Appendix.
Solution:

$$\begin{aligned} F(3:4,.1) &= 1 - \sum_{x=4}^{4} f(x:4,.1) \\ &= 1 - .0001 = .9999 \end{aligned}$$

Lastly, since $F(k:n,p) = 1 - F(n - k - 1: n,q)$ [Refer to Eq. (2.4.4)], we obtain substituting it in Eq. (2.4.7)

$$1 - \sum_{x=k+1}^{n} f(x:n,p) = 1 - F(n - k - 1:n,q)$$

which reduces to

$$\sum_{x=k+1}^{n} f(x:n,p) = F(n - k - 1:n,q) \tag{2.4.8}$$

This last expression is most convenient to apply when p is greater than .5.

Example 2.4.8 Evaluate (a) $F(5:7,.8)$ and (b) $F(3:4,.6)$
Solution: (a) Here $n = 7, q = .8, p = .2$ and $n - k - 1 = 5$. Hence $k = 1$. From Eq. (2.4.8) we get

$$F(5:7,.8) = \sum_{x=2}^{7} f(x:7,.2) = .4233$$

which is read directly from Table II.

(b) In this case $n = 4, q = .6, p = .4$ and $n - k - 1 = 3$. Therefore $k = 0$, and similarly we get

$$F(3:4,.6) = \sum_{x=1}^{4} f(x:4,.4) = .8704$$

PROBLEMS

2.4.1 Using mathematical induction, show that the third assumption made in the derivation of Eq. (1.4.6) holds true for all $k \leq n, n = 1, 2, \ldots, n$. That is, the coefficient of the kth term in the expansion of $(a + b)^n$ is $C(n,k - 1)$ if the terms are counted in ascending powers of b.

In Probs. 2.4.2 to 2.4.5 obtain the required answers without the use of Table II in the Appendix.

2.4.2 From a well-shuffled regular deck of playing cards a card is drawn and the suit observed. The card is replaced and shuffled again. If this is repeated 6 times, (a) what are the probabilities for getting 0, 1, 2, 3, 4, 5, and 6 spades? (b) Represent the results in a histogram. (c) Find the distribution for part (a) and evaluate $F(3)$.

2.4.3 What is the probability of getting "6" exactly 2 times in 5 throws of a regular die?

2.4.4 In a manufacturing process 1 out of every 10 articles is defective in the long run. Find the probability that a random sample of 5 from the production line will contain (a) exactly 1 defective, (b) exactly 2 defectives, (c) not more than 3 defectives, (d) all defectives.

2.4.5 A student can solve on the average half of his statistics problems and there are 8 questions on an hourly test. What is the probability of his passing if he has to have at least 6 problems solved in order to pass?

2.4.6 Show that $f(x + 1:n,p) = \dfrac{p(n - x)}{q(x + 1)} f(x:n,p)$ is the recursion formula for the binomial p.f. and using it evaluate the individual terms for $f(x:6,.4)$. Verify your results by using Eq. (2.4.6) and Table II in the Appendix.

2.4.7 Using Table II and the relationships shown in Eqs. (2.4.6), (2.4.7), and (2.4.8) find:

(a) $f(1:12,.4)$ (b) $\sum_{x=4}^{15} f(x:15,.35)$

(c) $F(8:15,.25)$ (d) $f(3:17,.65)$

(e) $\sum_{x=5}^{14} f(x:14,.75)$ (f) $F(13:20,.6)$

(g) $\sum_{x=4}^{16} f(x:19,.4)$ (h) $\sum_{x=4}^{16} f(x:19,.6)$

2.4.8 If a pair of honest dice are rolled 5 times, what is the probability of getting 7 exactly (a) 1 time? (b) 2 times?

2.4.9 If in a manufacturing process 5 percent of the articles are assumed to be defective, what is the probability that in a box containing one dozen articles (a) exactly one article will be defective? (b) at least two articles will be defective?

2.4.10 In a screening test on the average 15 percent of the applicants fail. In a group of 10 applicants, what is the probability that (a) at least two will fail? (b) at most two will fail?

2.4.11 Four persons match pennies and odd person wins. Assuming that there must be an odd person in order to have a match, find the probability that in 16 matches player A will win (a) exactly 4 times, (b) no times.

2.4.12 Assuming that the proportion of newborn boys equals that of newborn girls, what is the probability that among 17 newborn babies exactly 10 will be boys?

2.5 THE POISSON PROBABILITY FUNCTION

In this and subsequent sections a special number designated by e is needed, which we shall introduce next. The number is obtained by the summation $\sum_{x=0}^{\infty} \frac{1}{x!}$ and has a value of 2.71828 . . . . Equivalent definitions of e are $\lim_{n \to \infty} \left(1 + \frac{1}{n}\right)^n$ and $\lim_{n \to 0} (1 + n)^{1/n}$. More generally, it can be shown that $e^k = \sum_{x=0}^{\infty} \frac{k^x}{x!}$, where k is any real number.

A very important and particularly useful probability function is the function

$$f(x:k) = e^{-k} \frac{k^x}{x!} \qquad x = 0, 1, 2, 3, \ldots \qquad \text{and } k > 0 \tag{2.5.1}$$

where k is the only parameter. This function is known as the *Poisson probability function*, named after the French mathematician Siméon Poisson (1781–1840).

We next show that Eq. (2.5.1) satisfies the two conditions of a probability function. Since k is assumed to be a positive constant and x assumes only positive integral values, then clearly $f(x:k) \geq 0$. To show that $\sum_{x=0}^{\infty} f(x:k) = 1$, we proceed as follows:

$$\sum_{x=0}^{\infty} f(x:k) = e^{-k}(1 + k + k^2/2! + k^3/3! + \cdots + k^n/n! + \cdots) \tag{2.5.2}$$

The expression in parentheses on the right-hand side of Eq. (2.5.2) represents e^k, and hence

$$\sum_{x=0}^{\infty} f(x:k) = e^{-k}e^k = 1$$

The random variable x in the Poisson probability function is an example of denumerable set and represents another discrete probability function.

Example 2.5.1 Evaluate $f(2:1)$ and $f(2:2)$ if $f(x:k)$ is the Poisson probability function.

Solution:

$$f(2:1) = e^{-1}\,1^2/2! = \tfrac{1}{2}e^{-1} = .1839$$
$$f(2:2) = e^{-2}\,2^2/2! = 2e^{-2} = .2707$$

Example 2.5.2 Find the probability that $x > 2$ in the Poisson function $f(x:1)$.

Solution: Using Eq. (2.3.2) we get

$$P(x > 2) = 1 - F(2:1) = 1 - \sum_{x=0}^{2} e^{-1}\,1^x/x! = 1 - e^{-1}(1 + 1 + \tfrac{1}{2})$$
$$= .080$$

Example 2.5.3 The number of typographical errors per page prior to proofreading is assumed to obey the Poisson probability function with $k = 3$. Find the probability of finding at most one error, if a page is examined at random.

Solution: The expression wanted is $F(1:3)$, and therefore applying the definition of the Poisson distribution function we obtain for $x = 1$:

$$\begin{aligned} F(1:3) &= e^{-3}(3^0/0! + 3^1/1!) \\ &= e^{-3}(1 + 3) = 4e^{-3} = .1991 \end{aligned}$$

RELATIONSHIP BETWEEN THE BINOMIAL AND THE POISSON PROBABILITY FUNCTIONS

The binomial probability function approximately obeys the Poisson function if $n \to \infty$ and $p \to 0$, such that np remains constant, say k. That is,

$$\lim_{\substack{n\to\infty \\ p\to 0}} C(n,x)p^x q^{n-x} = \frac{e^{-k}k^x}{x!} \tag{2.5.3}$$

To show this, we make use of the definition of the binomial probability function and take the limit after representing the function in a simplified form.

The left-hand side of Eq. (2.5.3) can be expressed thus:

$$\begin{aligned} &\lim_{n\to\infty} \frac{n!}{x!(n-x)!}\left(\frac{k}{n}\right)^x\left(1 - \frac{k}{n}\right)^{n-x} \\ &\qquad = \frac{k^x}{x!}\lim_{n\to\infty}\frac{n!}{n^x(n-x)!}\left(1 - \frac{k}{n}\right)^{n-x} \end{aligned} \tag{2.5.4}$$

Let us observe that $n^x(n-x)!$ is comprised of the product of n factors; namely, x terms in n^x and $(n-x)$ terms in $(n-x)!$. Therefore, Eq. (2.5.4) can be equivalently stated:

$$\frac{k^x}{x!}\lim_{n\to\infty}\frac{n(n-1)(n-2)\cdots(n-x+1)(n-x)!}{n\cdot n\cdot n\cdots n(n-x)!}\left(1 - \frac{k}{n}\right)^{n-x} \tag{2.5.5}$$

which is the same as

$$\frac{k^x}{x!}\lim_{n\to\infty}\left[\left(\frac{n}{n}\right)\left(\frac{n-1}{n}\right)\left(\frac{n-2}{n}\right)\left(\frac{n-3}{n}\right)\cdots\left(\frac{n-x+1}{n}\right)\right]\left(1 - \frac{k}{n}\right)^{n-k} \tag{2.5.6}$$

Clearly, the limit as $n \to \infty$ of the x terms in the brackets is equal to 1; let us recall that $\lim_{n\to\infty}\left(1 - \frac{k}{n}\right)^n = e^{-k}$. Now since x is finite, $\lim_{n\to\infty}\left(1 - \frac{k}{n}\right)^{n-x} = e^{-k}$. Therefore, Eq. (2.5.6) becomes $\frac{k^x}{x!}e^{-k}$, which proves Eq. (2.5.3).

We illustrate this relationship by an example.

Example 2.5.4. In a production line it is known that 10 percent of all items are defective. What is the probability that in a sample of 10 items selected at random, no more than one defective will be found.

Solution: (a) Using the binomial probability function, we have $p = 1/10$ and $q = 9/10$ with $n = 10$. Therefore

$$F(1:10,.1) = C(10,0)(1/10)^0(9/10)^{10} + C(10,1)(1/10)^1(9/10)^9 = .7361$$

(b) Using the Poisson probability function,

$$k = np = 10(10\%) = 1$$

thus for $x = 1$ $F(x:1)$ becomes

$$F(1:1) = e^{-1}(1^0/0! + 1^1/1!) = e^{-1}(1 + 1) = 2e^{-1} = .7358$$

The difference in the answers of part (a) and (b) is negligible for all practical purposes.

Table III in the Appendix contains tabulated values for

$$\sum_{x=m}^{\infty} \frac{e^{-k}k^x}{x!} \qquad m = 0,1,2, \ldots$$

with increments of .1 for k. For instance, if $k = 1.0$ and $m = 4$, we find the tabulated value to be .0190 in Table III. Again, if $k = 2.5$ and $m = 7$, we read the tabulated value to be .0142.

Since

$$\sum_{x=m}^{\infty} f(x:k) = f(m:k) + \sum_{x=m+1}^{\infty} f(x:k) \tag{2.5.7}$$

holds true for the Poisson p.f., we get by rewriting the terms

$$f(m:k) = \sum_{x=m}^{\infty} f(x:k) - \sum_{x=m+1}^{\infty} f(x:k) \tag{2.5.8}$$

which is related to Eq. (2.3.7). Equation (2.5.8) can be used to obtain the individual terms of the Poisson probability function.

Example 2.5.5 Using Table III in the Appendix, evaluate

(a) $f(4:4)$ (b) $f(2:1)$ and (c) $f(3:0.1)$

Solution: (a) Here $m = 4$ and $k = 4$. Applying Eq. (2.5.8) we get

$$f(4:4) = \sum_{x=4}^{\infty} f(x:4) - \sum_{x=5}^{\infty} f(x:4)$$

$$= .5665 - .3712 = .1953$$

(b) Here $m = 2$ and $k = 1$ hence

$$f(2:1) = \sum_{x=2}^{\infty} f(x:1) - \sum_{x=3}^{\infty} f(x:1)$$

$$= .2642 - .0803 = .1839$$

which as expected agrees with the result of Example 2.5.1.

(c) Here $m = 3$ and $k = .1$ and therefore

$$f(3{:}.1) = .0002 - .0000 = .0002$$

Next let us recall that the distribution function for the Poisson p.f. is

$$F(m{:}k) = \sum_{x=0}^{m} f(x{:}k) = \sum_{x=0}^{\infty} f(x{:}k) - \sum_{x=m+1}^{\infty} f(x{:}k)$$

$$= 1 - \sum_{x=m+1}^{\infty} f(x{:}k) \qquad (2.5.9)$$

That is, Eq. (2.5.9) represents the expression for the computation of the distribution function associated with the Poisson probability function.

Example 2.5.6 Using Table III, evaluate (a) $F(10{:}12)$ (b) $F(7{:}.6)$

Solution: (a) In this case $m = 10$ and $k = 12$. From Table III we obtain for $\sum_{x=11}^{\infty} f(x{:}12) = .6528$ and hence $F(10{:}12) = 1 - .6528 = .3472$

(b) Here $m = 7$ and $k = .6$. From Table III we note $\sum_{x=8}^{\infty} f(x{:}.6) = 0.0000$ and therefore $F(7{:}.6) = 1 - 0.0000 = 1$.

PROBLEMS

2.5.1 In a Poisson probability function $f(x{:}k)$, $f(1{:}k) = f(2{:}k)$; find the value of the parameter k.

In the computation of Probs. 2.5.2 to 2.5.5 do not use Table III in the Appendix.

2.5.2 If the probability of an accident occurring on a certain road between two towns can be approximated by the Poisson p.f. $f(x{:}1)$, find the probability that there will be (a) exactly one accident, (b) exactly two accidents, and (c) more than three accidents between the two cities.

2.5.3 In a manufacturing process the number of defective items produced follow a Poisson function. If the management has been using $f(x{:}2)$ for their estimation of number of defective articles, what is the probability that there will be four or more defective items?

2.5.4 An insurance company calculates the probabilities for the number of fatal accidents during a given year by the Poisson p.f. $f(x{:}.5)$. Find the probability that (a) two or less fatal accidents will occur, and (b) more than 2 such accidents will take place.

2.5.5 If the misfired missiles are assumed to obey the Poisson p.f. $f(x{:}.2)$, calculate the probabilities for misfired missiles on a testing center for values of $x = 1, 2, 3$, and 4.

2.5.6 Show that $f(x + 1:k) = \dfrac{k}{x + 1} f(x:k)$ is the recursion formula for the individual terms of the Poisson p.f. $f(x:k)$. Using the given relationship, evaluate $f(x:0.2)$ for $x = 1, 2, 3$, and 4. Compare your result with that of Prob. 2.5.5.

2.5.7 Using Table III in the Appendix, evaluate:

(a) $f(5:2)$ (b) $f(3:1.4)$

(c) $F(3:1.6)$ (d) $\sum_{x=4}^{\infty} f(x:3.3)$

(e) $\sum_{x=4}^{8} f(x:3.5)$ (f) $\sum_{x=1}^{10} f(x:5)$

2.5.8 Repeat Prob. 2.4.9 if the box contained six dozen articles.

2.5.9 Repeat Prob. 2.4.10 for 40 applicants.

2.6 THE HYPERGEOMETRIC PROBABILITY FUNCTION

Another discrete probability function which plays an important role in applications is the hypergeometric probability function. Before we give the formula representing this probability function, however, let us consider an example.

Example 2.6.1 A manager of an apartment building has 3 dozen light bulbs in his possession of which 5 are known to be defective. If he selects three lamps at random, what is the probability that

(a) all three will be defective?
(b) two will be defective?
(c) one will be defective?
(d) none will be defective?

Solution: (a) The probability of obtaining 3 defectives is the same as the first lamp selected being defective, the second lamp being defective and the third lamp also being defective. Let

$A = P$ (first lamp defective)
$B = P$ (second lamp defective | first defective)
$C = P$ (third lamp defective | first and second defective)

Clearly then $P(A) = 5/36$, $P(B) = 4/35$ and $P(C) = 3/34$. Since A, B, and C are stochastically independent, $P(A \text{ and } B \text{ and } C) = P(A)\, P(B)\, P(C) = (3/34)(4/35)(5/36)$ which is identical to the expression

$$\frac{\binom{5}{3}\binom{31}{0}}{\binom{36}{3}}$$

(b) In order to compute the probability of having two defectives we analyze as follows: there are three arrangements of two defectives and one nondefective; namely, DDN, DND, and NDD where D represents a defective and N a nondefective. By a similar argument as in part (a) we get

$$P(2 \text{ defective and } 1 \text{ nondefective}) = (5/36)(4/35)(31/34)(3)$$

where the factor 3 is because of the 3 different possible arrangements. This probability is equivalent to the expression

$$\frac{\binom{5}{2}\binom{31}{1}}{\binom{36}{3}}$$

(c) Following a similar pattern of analysis we obtain for the required probability the value of (5/36)(31/35)(30/34)(3), which is equivalent to the expression

$$\frac{\binom{5}{1}\binom{31}{2}}{\binom{36}{3}}$$

(d) To obtain three lamps of which none are defective is the same as obtaining 3 nondefectives and this event has a probability of (31/36)(30/35)(29/34), which is equivalent to the expression

$$\frac{\binom{5}{0}\binom{31}{3}}{\binom{36}{3}}$$

The foregoing example is a special case of a general, random experiment of selecting n objects from N of which M are of one kind and $N - M$ another. For instance, in the example just considered $M = 5$ is the number of defectives and $N - M = 36 - 5 = 31$ is the number of nondefectives. Now, if x represents the number of the first kind of objects among the n selected, then of necessity $n - x$ represents the number of the second kind. Since there are $\binom{N}{n}$ possible ways of selecting n objects from among N, $\binom{N}{n}$ corresponds to the total number of sample points in the sample space of n-tuples describing the random experiment. Furthermore, since x represents the number of objects of first kind among the n selected, then $\binom{M}{x}\binom{N-M}{n-x}$ represents the number of sample points corresponding to exactly x objects of the first kind and $n - x$ objects of the second kind. Therefore, applying the definition of probability of an event, Eq. (1.3.11), we obtain:

$$f(x) = \frac{\binom{M}{x}\binom{N-M}{n-x}}{\binom{N}{n}} \qquad x = 0, 1, \ldots, n \tag{2.6.1}$$

Clearly Eq. (2.6.1) contains 3 independent parameters, namely N, M, and n. Hence we shall express Eq. (2.6.1) as

$$f(x:N, M, n) = \frac{\binom{M}{x}\binom{N-M}{n-x}}{\binom{N}{n}} \qquad x = 0, 1, \ldots, n \tag{2.6.2}$$

Equation (2.6.2) is defined as the *hypergeometric* probability function.

Answers to Example 2.6.1 then may readily be obtained from the expression

$$f(x:36, 5, 3) = \frac{\binom{5}{x}\binom{31}{3-x}}{\binom{36}{3}} \qquad x = 0, 1, 2, \text{and } 3$$

in which $x = 3$ gives the answer to part (a), $x = 2$ to part (b) $x = 1$ to part (c) and $x = 0$ to part (d).

Let us next show that the hypergeometric probability function defined by Eq. (2.6.2) satisfies the two conditions for a probability function. To show this we present the following analysis.

Because the numerator and the denominator of Eq. (2.6.2) are composed of combinatorials, then the fraction is always positive, which satisfies the condition of Eq. (2.1.1). In order to show that the condition

$$\sum_{x=0}^{n} f(x:N, M, n) = 1$$

is satisfied, we prove the equivalent relation

$$\sum_{x=0}^{N} \binom{M}{x}\binom{N-M}{n-x} = \binom{N}{n} \tag{2.6.3}$$

and to prove Eq. (2.6.3) we consider the binomial identity

$$\sum_{n=0}^{N} \binom{N}{n} p^n = (1+p)^N = (1+p)^M (1+p)^{N-M}$$

$$= \left[\sum_{y=0}^{M} \binom{M}{y} p^y\right]\left[\sum_{z=0}^{N-M} \binom{N-M}{z} p^z\right] \quad (2.6.4)$$

Recalling that $\binom{n}{k} = 0$ for either $n < k$ or $k < 0$, Eq. (2.6.4) may be equivalently written

$$\sum_{n=0}^{N} \binom{N}{n} p^n = \left[\sum_{y=0}^{N} \binom{M}{y} p^y\right]\left[\sum_{z=0}^{N} \binom{N-M}{z} p^z\right] \quad (2.6.5)$$

$$= \left[\sum_{y=0}^{N} \binom{M}{y} p^y\right]\left[\sum_{z=n}^{N} \binom{N-M}{n-z} p^{n-z}\right] \quad (2.6.6)$$

$$= \left[\sum_{x=0}^{N} \binom{M}{x} p^x\right]\left[\sum_{n=x}^{N} \binom{N-M}{n-x} p^{n-x}\right] \quad (2.6.7)$$

$$= \sum_{n=0}^{N} \left[\sum_{x=0}^{N} \binom{M}{x}\binom{N-M}{n-x}\right] p^n \quad (2.6.8)$$

Comparing this last expression with the left-hand side of Eq. (2.6.4) we observe that

$$\sum_{x=0}^{N} \binom{M}{x}\binom{N-M}{n-x} = \binom{N}{n}$$

which verifies Eq. (2.6.3) and, therefore, the hypergeometric probability function indeed represents a probability function.

Example 2.6.2 A radio manufacturing company puts into a container 24 items for shipment. An inspector at the plant will test two items at random prior to shipping and if any of the tested items prove defective he will return the whole container for examination. What is the probability that a box will not be shipped, if there are actually two defectives in each container?

Solution: This is a situation where fixed numbers of items of two types are being considered and therefore the hypergeometric probability function would be the proper function to use. The expression needed is $f(1:24,2,2) + f(2:24,2,2)$ since the box will not be shipped if either one or two prove defective. When evaluated this yields the value 15/92.

Example 2.6.3 Another radio manufacturing company puts into a container 24 items for shipment but has the following scheme of inspection. An inspector takes an item at random for examination and after inspecting replaces it in the box. Then a second inspector takes an item at random from this box and inspects it. The box is not shipped if either inspector finds a defective. What is the probability that a box will not be shipped if there are actually two defectives in each box?

Solution: It should be evident to the reader that the schemes of inspection in Example 2.6.2 and in the example at hand are unlike. In the former, the first item is not replaced in the container prior to the second item being tested, while in this case the first item tested is *replaced* and hence the second inspector may retest the *same* item which the first inspector tested. Here the box is not shipped if the first inspector or the second inspector or both find a defective. Now, this corresponds to a random experiment in which there are 24^2 sample points representing all possible 2-tuples (x_1,x_2) where x_1 is either a defective or a nondefective item selected by the first inspector, and x_2 is either defective or nondefective item selected by the second inspector. Clearly, there are $2 \times 22 = 44$ sample points corresponding to x_1 nondefective and x_2 defectives. Similarly, there are 44 sample points corresponding to x_1 defective and x_2 nondefective. Also, $2 \times 2 = 4$ sample points represent x_1 defective and x_2 defective. (There are $22 \times 22 = 484$ sample points corresponding to x_1 nondefective and x_2 nondefective, such that the total sample points $484 + 44 + 44 + 4 = 576$ is indeed 24^2.) Therefore, the probability sought is

$$(44 + 44 + 4)/576 = 92/576 = 23/144$$

Out of curiosity let us calculate the binomial expression

$$\begin{aligned} f(1{:}2, 2/24) + f(2{:}2, 2/24) &= \tbinom{2}{1}(1/12)(11/12) + \tbinom{2}{2}(1/12)^2(11/12)^0 \\ &= 22/144 + 1/144 = 23/144 \end{aligned}$$

Now let us remark about the above results. In discussing the binomial probability function we assumed p and q to be positive constants such that $p + q = 1$. In applying the binomial probability function, we tacitly assumed that p and q did not change during the experiment; that is, if p and q represent the proportions of objects of two types in a collection, the respective proportions remain fixed throughout all trials. This implies the sample space is comprised of n^2 2-tuples representing all possible points of a random experiment *with* replacement. If however, the experiment were carried out *without replacement*, the corresponding sample space would have $n(n - 1)$ 2-tuples representing all possible points in the experiment. Hence the binomial probability function is apropos to situations where the random experiment is with replacement, (proportions of two types of objects do not alter from trial to trial) and the hypergeometric probability function to situations where the random experiment is without replacement.

RELATIONSHIP BETWEEN THE HYPERGEOMETRIC AND THE BINOMIAL FUNCTIONS

Intuitively if not obviously it should be evident from the remarks on the result of Example 2.6.3 that if N were large, the binomial and the hypergeometric probability functions would become indistinguishable. We show next that as $N \to \infty$ and p remains fixed, the hypergeometric probability function equals the binomial probability function—that is,

$$\lim_{N\to\infty} \frac{\binom{M}{x}\binom{N-M}{n-x}}{\binom{N}{n}} = \binom{n}{x} p^x q^{n-x} \tag{2.6.9}$$

To show the validity of Eq. (2.6.9) we proceed as follows. Since p and q are the proportions of sample points corresponding to objects of type 1 and type 2 in a sample space of size N, Np and Nq then are the exact numbers of objects of type 1 and type 2. Hence $M = Np$ and $N - M = Nq$, which when substituted in Eq. (2.6.9) yield

$$\lim_{N\to\infty} \frac{\binom{Np}{x}\binom{Nq}{n-x}}{\binom{N}{n}} \tag{2.6.10}$$

If we express the combinatorials in factorial form, Eq. (2.6.10) can be written equivalently

$$\lim_{N\to\infty} \frac{(Np)!}{(Np-x)!x!} \frac{(Nq)!}{(n-x)!(Nq-n+x)!} \frac{n!(N-n)!}{N!} \tag{2.6.11}$$

When Eq. (2.6.11) is expanded it yields

$$\lim_{N\to\infty} \left\{\frac{n!}{x!(n-x)!}\right\} \cdot$$

$$\left\{\frac{[Np(Np-1)\cdots(Np-x+1)][(Nq)(Nq-1)\cdots(Nq-n+x+1)]}{N(N-1)(N-2)\cdots(N-n+1)}\right\} \tag{2.6.12}$$

Now the expression in the first braces is clearly $\binom{n}{x}$. The numerator of the expression in the second braces has n factors (x factors in the first brackets and $n - x$ factors in the second brackets), and the denominator has n factors also. Therefore if we divide numerator and denominator of the expression in the second braces by N^n, Eq. (2.6.12) becomes

$$\lim_{N\to\infty}\binom{n}{x}\cdot\frac{p\left(p-\frac{1}{N}\right)\cdots\left(p-\frac{x-1}{N}\right)q\left(q-\frac{1}{N}\right)\cdots\left(q-\frac{n-x-1}{N}\right)}{1\left(1-\frac{1}{N}\right)\cdots\left(1-\frac{n-1}{N}\right)} \tag{2.6.13}$$

Clearly the limit as $N \to \infty$ Eq. (2.6.13) becomes $\binom{n}{x}p^x q^{n-x}$ which is the binomial probability function.

In most applications of the hypergeometric probability function N is relatively large compared to n and hence the binomial probability function may conveniently be used instead. If $N \geq 10n$, the difference between the two probability functions is very small for all practical purposes, and hence $N \geq 10n$ becomes the basic rule of choice between the two functions.

For instance, if we refer to Examples 2.6.2 and 2.6.3 we note that the binomial probability function can be used in both cases since $N = 24$ and $n = 2$. In these examples, the difference between the results obtained is .003.

PROBLEMS

2.6.1 Mr. Probability places 10 pieces of money in an envelope of which four are \$20 bills and six are \$10 bills. He tells Mrs. Probability that for her birthday present she may draw three bills from the envelope blindfolded. If the bills are in random order, find the probability that (a) she will draw all \$20 bills, (b) one \$20 and two \$10 bills, (c) all \$10 bills.

2.6.2 A lot of one dozen contains three defective articles. If a sample of four is taken without replacement, calculate the probabilities for 0, 1, 2, 3, and 4 defectives in the sample.

2.6.3 Given that a lot contains 5 dozen articles and it is known that there are six defectives. What is the probability that in a random sample of four there will be two defectives?

2.6.4 Repeat Prob. 2.6.3 if the sample of four is selected with replacement. Compare your answer and effort involved in the two methods.

2.6.5 In a manufacturing process two types of insulating materials are used. 75 percent of the items are manufactured with type A and the rest with type B insulators. From a lot of 24 a sample of 3 is taken at random. What is the probability of getting two items of type A and 1 item of type B? Assume the lot to contain the same proportions as the general production.

2.7 THE GENERALIZED HYPERGEOMETRIC PROBABILITY FUNCTION

If instead of having two kinds of objects among N, we consider three or more kinds, the hypergeometric and the binomial functions would not be the proper functions to utilize. For instance, if a box contains 20 objects of which 10 are white, 5 are green, 3 are blue, and 2 are red, we could not use any of the probability functions discussed so far to find the probability of obtaining one object of each color if 4 were drawn from the box. Furthermore, the probability required would be different in case the objects are not replaced in the box after each drawing than if they were replaced. However, the required probability can be calculated. We suggest the following analysis.

Case 1. Drawings are without replacement. The probability of obtaining a white object first, a green object second, a blue object third, and a red object last is (10/20)(5/19)(3/18)(2/17). However, we are not concerned in the order of the colors and there are 24 permutations; that is, $P(4,4) = 4! = 24$, and therefore the required probability is (24)(10/20)(5/19)(3/18)(2/17), which is equivalent to the expression

$$\frac{\binom{10}{1}\binom{5}{1}\binom{3}{1}\binom{2}{1}}{\binom{20}{4}}$$

Case 2. Drawings are with replacement. Following a similar argument as in case 1, we obtain (24)(10/20)(5/20)(3/20)(2/20), which is equivalent to

$$\frac{4!}{1!1!1!1!}(1/2)(1/4)(3/20)(1/10)$$

Case 1 discussed above is an example of the generalized hypergeometric probability function whose sample space is composed of all possible 4-tuples corresponding to the four colors of the objects. The hypergeometric probability function considered in Sec. 2.6 is a special case of the generalized case whose sample space is comprised of 2-tuples corresponding to the two types of objects. In general, if the sample space is comprised of k-tuples, the proper probability function is the generalized hypergeometric function which we state next.

If $M_1, M_2, \ldots, M_k$ represent the number of objects of type 1, type 2, ..., type k among N such that $\sum_{i=1}^{k} M_i = N$, and n objects were selected from the N such that $x_1, x_2, \ldots, x_k$ represent the number of type 1, type 2, ..., type k among n, then

$$f(x_1, x_2, \ldots, x_k : N, M_1, M_2, \ldots, M_k, n) = \frac{\binom{M_1}{x_1}\binom{M_2}{x_2}\binom{M_3}{x_3}\cdots\binom{M_k}{x_k}}{\binom{N}{n}} \tag{2.7.1}$$

which is the *generalized hypergeometric probability function.* If instead of the number of objects of type 1, type 2, . . . , type k the proportions of objects of type 1, type 2, . . . , type k were given Eq. (2.7.1) can equivalently be expressed as

$$f(x_1,x_2,\ldots,x_k:N,p_1,p_2,\ldots,p_k,n) = \frac{\binom{Np_1}{x_1}\binom{Np_2}{x_2}\cdots\binom{Np_k}{x_k}}{\binom{N}{n}} \tag{2.7.2}$$

where $p_1,p_2,\ldots,p_k$ represent the proportions of type 1, type 2, . . . , type k objects among N.

Next let us consider an example where we can apply the generalized hypergeometric probability function.

Example 2.7.1 From a well-shuffled regular deck of playing cards 6 cards are drawn without replacement. What is the probability that (a) there will be 2 clubs, 2 spades, and 2 hearts? (b) there will be 3 clubs and 3 spades? (c) all will be clubs?

Solution: $N = 52$ in a regular deck of playing cards, $M_1 = M_2 = M_3 = M_4 = 13$ corresponding to the number of each suit, $n = 6$, and hence

(a)

$$f(2,2,2,0:52,13,13,13,13,6) = \frac{\binom{13}{2}\binom{13}{2}\binom{13}{2}\binom{13}{0}}{\binom{52}{6}} = \frac{\left[\binom{13}{2}\right]^3}{\binom{52}{6}}$$

(b)

$$f(3,3,0,0:52,13,13,13,13,6) = \frac{\binom{13}{3}\binom{13}{3}\binom{13}{0}\binom{13}{0}}{\binom{52}{6}} = \frac{\left[\binom{13}{3}\right]^2}{\binom{52}{6}}$$

(c)

$$f(6,0,0,0:52,13,13,13,13,6) = \frac{\binom{13}{6}\binom{13}{0}\binom{13}{0}\binom{13}{0}}{\binom{52}{6}} = \frac{\binom{13}{6}}{\binom{52}{6}}$$

RELATIONSHIP BETWEEN THE GENERALIZED HYPERGEOMETRIC AND THE MULTINOMIAL PROBABILITY FUNCTIONS

In our discussion of the relationship between the hypergeometric and the binomial probability functions we considered the situation in which N approached infinity while p and q remained fixed. Here we shall consider the limiting case of the generalized hypergeometric function; that is, examine Eq. (2.7.2) as N approaches infinity and the proportions $p_1, p_2, \ldots, p_k$ remain fixed. Following a similar argument as before, we can readily show that

$$\lim_{N \to \infty} \frac{\binom{Np_1}{x_1}\binom{Np_2}{x_2}\cdots\binom{Np_k}{x_k}}{\binom{N}{n}} = \frac{n!}{x_1!x_2!\cdots x_k!}(p_1)^{x_1}(p_2)^{x_2}\cdots(p_k)^{xk} \quad (2.7.3)$$

where

$$\sum_{i=1}^{k} x_i = n \quad \text{and} \quad \sum_{i=i}^{k} p_i = 1$$

Equation (2.7.3) is called the *multinomial probability function*, which is a special case of the multinomial theorem presented in Sec. 1.4.

Example 2.7.2 Repeat Example 2.7.1 if the cards are replaced before each drawing.

Solution: Since the cards are replaced before each drawing, the proportions remain fixed and $p_1 = p_2 = p_3 = p_4 = \frac{1}{4}$.

(a) $\dfrac{6!}{2!2!2!0!}\left(\dfrac{1}{4}\right)^6$ (c) $\dfrac{6!}{6!0!0!0!}\left(\dfrac{1}{4}\right)^6$

(b) $\dfrac{6!}{3!3!0!0!}\left(\dfrac{1}{4}\right)^6$

PROBLEMS

2.7.1 What is the probability of selecting 3 cards from a deck of well-shuffled regular playing cards without replacement such that (a) all three will be spades, (b) one will be spade and two clubs, (c) one of each of hearts, clubs, and spades?

2.7.2 Refer to Prob. 2.6.1 and suppose Mr. Probability adds to the contents of the envelope three \$5 bills and two \$1 bills. Answer the same questions.

2.7.3 In a community there are three major recreational clubs A, B, and C. If there are 30 members in A, 50 in B, and 20 in C, find the probability of selecting three persons at random from the combined membership such that there will be one person from each club.

2.7.4 Four different sizes of glass bottles are manufactured by a company with 4:3:2:1 ratios, and placed in identical boxes by mistake. What is the probability that four boxes opened at random will represent one of each size of bottle?

2.7.5 An insurance company has settled 1,000 claims during the past year, of which 500 were for car damage, 400 for property damage, and 100 for medical. If the same proportions are anticipated this coming year, what is the probability that the first four claims of the new year were all (a) property damage, (b) car damage, (c) medical?

2.7.6 An urn contains 1 red and 4 white balls which are identical except for color. Balls are selected without replacement, singly, until the red ball is drawn. Find the probability that the red ball is drawn on the (a) first draw (b) second draw.

2.7.7 Suppose in Prob. 2.7.6 the urn contains n balls of which 1 is red and the rest white. Find the probability that the red ball will be drawn on the kth draw, assuming balls are selected without replacement.

2.7.8 An urn contains n balls of which 2 are red and the rest are white. They are identical except for color. If balls are drawn without replacement until the second red ball is selected, write an expression for the probability of this event happening on the kth draw.

2.7.9 Human blood types are classified into four mutually exclusive categories O, A, B, and AB. Assuming that the proportions of these types are .45, .40, .10, and .05, respectively, find the probability that among six randomly selected persons (a) three will have type O and three type B blood, (b) all six will have type A blood, and (c) all six will have type AB blood.

2.7.10 Repeat Prob. 2.7.6 if the balls are replaced after each draw.

2.7.11 Repeat Prob. 2.7.8 if the balls are replaced after each draw.

2.8 THE NEGATIVE BINOMIAL PROBABILITY FUNCTION

Suppose in a manufacturing situation an inspector tests the production of certain items from a specific machine until 3 defectives have appeared. What is the probability that 0, 1, 2, . . . nondefectives appear in the process of 3 defectives showing? For instance, what is the probability that the inspector had to test 10 items in order to find 3 defectives? Clearly, the 3rd defective must have occured on the 10th testing, and 2 defectives on the 9th or prior trials. Also, there must have appeared in the process 7 nondefectives if the 3rd defective occurred on the 10th examination. If we denote defectives by d and nondefectives by n, the following are possible cases: (a) *ddnnnnnnnd*, (b) *nnnnndndnd*, (c) *nnnnnnnddd*, etc. Recalling our discussion of permutations of N objects of which n_1 are of one type and n_2 of another, we obtain in total $N!/n_1!n_2!$ distinct arrangements. Therefore, for the problem under consideration $9!/2!7! = 36$ ways exist for 2 defectives and 7 nondefectives to appear during the first 9 testings and the 3rd defective occurring on the 10th testing.

Now if p is the proportion of defectives and q the proportion of nondefectives in the production of these items, then, $\left[\left(\frac{9!}{2!7!}\right)p^2q^7\right]p$ is the required probability, which may be equivalently expressed $\binom{9}{7}p^3q^7$. Let us observe, however, that the 3rd defective could have appeared on any testing *on or after* the 3rd. Therefore, following a similar argument we can show that $\binom{2}{0}p^3q^0, \binom{3}{1}p^3q^1, \ldots, \binom{3+x-1}{x}p^3q^x$ are the probabilities that the 3rd defective would appear on the 3rd, 4th, . . . , $(3+x)$th testing.

In general, if we desire k defectives, $\binom{k+x-1}{x}p^kq^x \qquad x = 0, 1, \ldots$

is the expression for k defectives to appear on the kth, $(k+1)$st, . . . testings, where k and p are the two independent parameters.

Let us next consider

$$f(x:k,p) = \binom{k+x-1}{x} p^k q^x \quad x = 0, 1, 2, \ldots \tag{2.8.1}$$

to see if $f(x:k,p)$ thus defined represents a probability function. Since p and q represent proportions, $p \geq 0$, $q \geq 0$ and $p + q = 1$. Also $\binom{k+x-1}{x}$ is positive for $x = 0, 1, 2, \ldots$ and hence $f(x:k,p) \geq 0$. To show that $\sum_{x=0}^{\infty} f(x:k,p) = 1$, we proceed as follows:

$$\sum_{x=0}^{\infty} \binom{k+x-1}{x} p^k q^x = p^k \left[\binom{k-1}{0} + \binom{k}{1} q + \binom{k+1}{2} q^2 + \cdots \right] \tag{2.8.2}$$

The expression in the brackets on the right-hand side of Eq. (2.8.2) is the series expansion of $(1-q)^{-k}$, which is equivalent to p^{-k}. Consequently, the right-hand side of Eq. (2.8.2) may be written as $p^k p^{-k}$ which is equal to 1.

We next show that

$$\binom{k+x-1}{x} = (-1)^x (-k)_x / x! \tag{2.8.3}$$

where $(-k)_x$ is defined as the product of x successively decreasing integers starting with $-k$. For instance, $(-3)_4 = (-3)(-4)(-5)(-6) = 360$ and $(-4)_3 = (-4)(-5)(-6) = -120$.

Now the left-hand side of Eq. (2.8.3) when expressed in factorials becomes

$$\begin{aligned} \frac{(k+x-1)!}{x!(k-1)!} &= \frac{(k+x-1)(k+x-2)\cdots(k+1)(k)(k-1)!}{x!(k-1)!} \\ &= \frac{k(k+1)(k+2)\cdots(k+x-1)}{x!} \end{aligned} \tag{2.8.4}$$

The numerator of the right-hand side of Eq. (2.8.4) has x terms and hence Eq. (2.8.4) may equivalently be expressed as

$$\frac{(-1)^x(-k)(-k-1)(-k-2)\cdots(-k-x+1)}{x!} = \frac{(-1)^x(-k)_x}{x!}$$

Consequently, the probability function defined by Eq. (2.8.1) is equivalent to

$$f(x:k,p) = \frac{(-1)^x(-k)_x}{x!} p^k q^x \qquad x = 0, 1, 2, \ldots \tag{2.8.5}$$

Because of the close similarities between the binomial probability function and this one, the probability function represented by either Eq. (2.8.1) or Eq. (2.8.5) is called the *negative binomial probability function.*

Example 2.8.1 A fair coin is tossed until (a) 1 head, (b) 2 heads show. What is the probability that these events will occur on or before the 4th toss?

Solution: For this problem the negative binomial function can be utilized. Therefore,

(a)

$$\sum_{x=0}^{3} f(x:1, 1/2) = \sum_{x=0}^{3} \binom{x}{x} (1/2)^1 (1/2)^x$$

$$= 1/2\,[(1/2)^0 + (1/2)^1 + (1/2)^2 + (1/2)^3] = 15/16$$

Let us refer to Example 2.3.3*b* and observe that the two problems are identical and the results agree.

(b)

$$\sum_{x=0}^{3} f(x:2, 1/2) = \sum_{x=0}^{3} \binom{x+1}{x} (1/2)^2 (1/2)^x$$

$$= \frac{1}{4}\left[\binom{1}{0} + \binom{2}{1} 1/2 + \binom{3}{2} (1/2)^2 + \binom{4}{3} (1/2)^3\right] = \frac{13}{16}$$

RELATIONSHIP BETWEEN THE NEGATIVE BINOMIAL AND THE POISSON PROBABILITY FUNCTIONS

Let us consider a special limiting case of the negative binomial probability function in which $kq = pc$, where c is a constant, p and q are proportions of the two types of objects, and $k \longrightarrow \infty$. That is,

$$\lim_{k\to\infty} \frac{(k)(k+1)(k+2)\cdots(k+x-1)}{x!} p^k q^x \tag{2.8.6}$$

If we express $kq = pc$ equivalently as $qk = (1 - q)c$ and solve for q, we get $q = c/(k + c)$. Therefore Eq. (2.8.6) becomes

$$\lim_{k\to\infty} \frac{(k)(k+1)(k+2)\cdots(k+x-1)}{x!} \left(1 - \frac{c}{c+k}\right)^k \left(\frac{c}{c+k}\right)^x \tag{2.8.7}$$

$$= \frac{c^x}{x!} \lim_{k\to\infty} \frac{k(k+1)(k+2)\cdots(k+x-1)}{(c+k)^x} \left(1 - \frac{c}{c+k}\right)^k \tag{2.8.8}$$

$$= \frac{c^x}{x!} \lim_{k\to\infty} \left(\frac{k}{k+c}\right)\left(\frac{k+1}{k+c}\right)\left(\frac{k+2}{k+c}\right)\cdots\left(\frac{k+x-1}{k+c}\right)\left(1 - \frac{c}{k+c}\right)^k$$

$$= \frac{c^x}{x!} e^{-c} \tag{2.8.9}$$

Clearly, Eq. (2.8.9) is the Poisson probability function with parameter c.

Example 2.8.2 A fair die is thrown until 6 "aces" show. What is the probability that the 6th ace will occur on the 8th throw?

Solution: The desired probability is the same as $x = 2$ in the negative binomial $f(x{:}6, 1/6) = \binom{x+5}{x}\left(\frac{1}{6}\right)^6\left(\frac{5}{6}\right)^x$ which yields $\frac{525}{6^8}$.

We conclude this chapter by presenting the accompanying large flow diagram, Fig. 2.4, which summarizes the interrelationships between the important discrete probability functions discussed in the chapter. The student may find this diagram useful for reference.

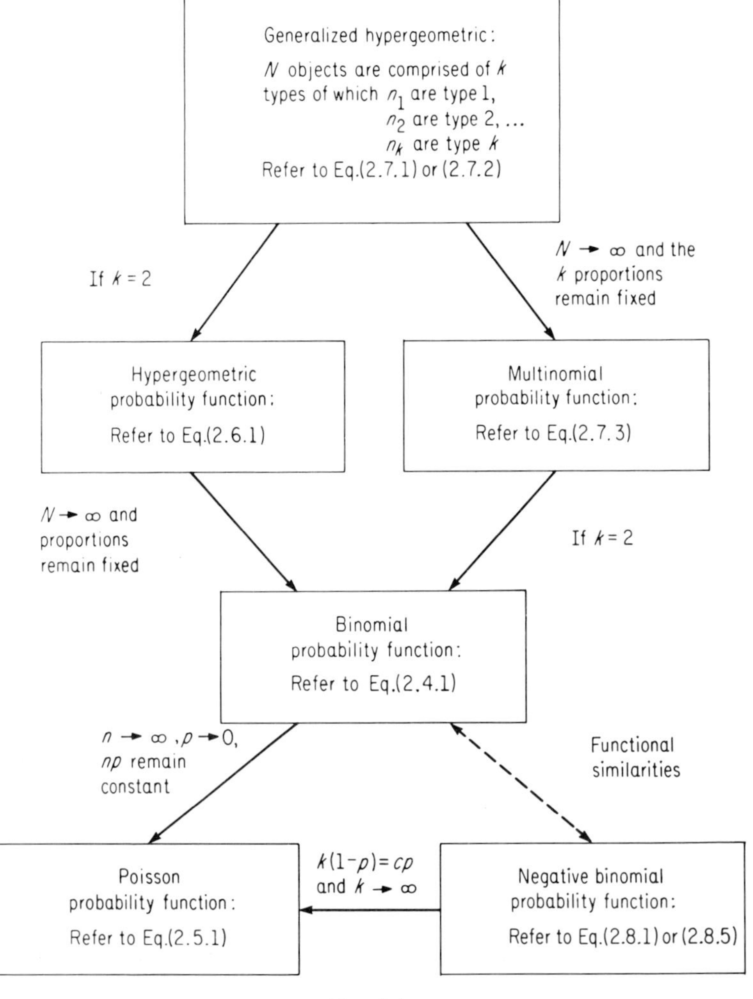

FIG. 2.4.

PROBLEMS

2.8.1 Evaluate the negative binomial of Example 2.8.2 for $x = 0, 1, 2, 3$, and 4 and graph the result.

2.8.2 In a manufacturing process 10 percent of the articles produced are defective. What is the probability that in a sampling process the 4th defective will show on the seventh examination?

2.8.3 If 60 percent of registered voters in a community are Republicans and the rest Democrats, what is the probability that if six persons are interviewed at random the first Democrat will be the fourth individual?

3

Moments of Probability Functions —Discrete Case

3.0 INTRODUCTION

In the opening remarks of Chapter 1 we made the analogy between suit models for men and probability models. In Chapter 2 we presented the rules for constructing a probability model and gave several examples of such. However, without some further detailed information available both the suit and the probability models would be of no practical use. For instance, suppose the suit buyer sees a coat in a display rack and wants to buy it. He notices the tag on the coat indicating "size 44 long." This represents detailed information—namely, the chest measurement is 44 inches and the sleeve length longer than average. If the suit buyer wanted a suit with chest measurement of 38 inches and sleeve lengths shorter than average, this particular coat would be unsatisfactory. Similarly, detailed information for a probability model assists the statistician to determine the applicability, fitness and usefulness of it to a given situation. In this chapter some measurable characteristics of probability functions will be considered, but prior to doing that let us mention some intuitive measurable characteristics that may seem plausible to the beginning student.

The student was requested in Probs. 2.2.1 through 2.2.7 to draw the histograms of the indicated probability functions. If we refer to those histograms we notice that some of them are symmetric while others are asymmetric. The line upon which the center of gravity is located is readily observed for the symmetric cases, whereas a certain amount of computation is necessary for the asymmetric histograms.

Furthermore, all but the probability function specified in Prob. 2.2.3 were defined for finite values of the random variable, while the probability function given in Prob. 2.2.3 was defined for countably infinite values. Consequently, this probability function is "spread out" relatively more than the others. It is these types of characteristics: (1) symmetry, (2) line upon which the center of gravity is located, (3) how much the probability function is "spread out" from the line upon which the center of gravity is located that we shall examine. These concepts will become meaningful when precise definitions of terminology are given. To do this we shall first present a related topic.

3.1 EXPECTATION

If $f(x)$ is a probability function and $g(x)$ any function of the random variable x, the expectation of $g(x)$ is defined as the sum of the products of $g(x)$ and $f(x)$ for all x. In symbols,

$$E[g(x)] = \Sigma g(x) \cdot f(x) \tag{3.1.1}$$

which reads "the expectation of $g(x)$."

Example 3.1.1 Given $g(x) = x^2 + 2x + 1$ and $f(x) = x^2/30$ defined for $x = 0, 1, 2, 3$, and 4. Find $E[g(x)]$.

Solution: Referring to Example 2.3.1, we note that $f(x)$ is a probability function and hence

$$E[(x + 1)^2] = \sum_{x=0}^{4} (x + 1)^2 \cdot x^2/30 = 292/15$$

Example 3.1.2 Find the $E[x]$ if $f(x) = (.5)(x - 2)^2 \quad x = 1, 2, 3$
$= 0 \quad \text{otherwise}$

Solution: From the result of Prob. 2.1.5 we know that $f(x)$ is a probability function and hence

$$E[x] = 1(.5)(1 - 2)^2 + 2(.5)(2 - 2)^2 + 3(.5)(3 - 2)^2 = 2.$$

Example 3.1.3 Find the expectation for x^2 if $f(x)$ is the binomial probability function $f(x{:}4,.5)$.

Solution: $E[x^2] = \dfrac{1}{16}[0 \cdot 1 + 1 \cdot 4 + 2^2 \cdot 6 + 3^2 \cdot 4 + 4^2 \cdot 1]$

$$= \frac{80}{16} = 5$$

Next we shall derive some theorems concerning expectations which will be used in subsequent considerations. The three rules that we shall present are not the only theorems concerning expectation. In later chapters some other rules also will be presented. However, the following rules simplify computations of expectations greatly.

Rule 1. The expectation of a constant k is equal to k. That is,

$$E[k] = k \tag{3.1.2}$$

Proof: $E[k] = \Sigma k \cdot f(x) = k \Sigma f(x) = k \cdot 1 = k$

Rule 2. The expectation of $k \cdot g(x)$, where k is a constant and $g(x)$ a random variable, is k times the expectation of $g(x)$. That is,

$$E[k \cdot g(x)] = k \cdot E[g(x)] \tag{3.1.3}$$

Proof:

$$E[k \cdot g(x)] = \Sigma\, k \cdot g(x) \cdot f(x)$$
$$= k\, \Sigma\, g(x) \cdot f(x) = k \cdot E[g(x)]$$

Rule 3. The expectation of the sum of two or more random variables is the sum of the respective expectations. That is,

$$E\left[\sum_{i=1}^{n} g_i(x)\right] = \sum_{i=1}^{n} E[g_i(x)] \tag{3.1.4}$$

Proof:

$$E\left[\sum_{i=1}^{n} g_i(x)\right] = \Sigma\left(\sum_{i=1}^{n} g_i(x) \cdot f(x)\right)$$
$$= \sum_{i=1}^{n} (\Sigma\, g_i(x) \cdot f(x))$$
$$= \sum_{i=1}^{n} E[g_i(x)]$$

Example 3.1.4 Find the expectation for $g(x) = x(x-1)(x-2)$.
Solution: $E[x(x-1)(x-2)] = E[x^3 - 3x^2 + 2x]$
$= E[x^3] - E[3x^2] + E[2x]$ by Rule 3
$= E[x^3] - 3\,E[x^2] + 2\,E[x]$ by Rule 2

PROBLEMS

In Probs. 3.1.1 to 3.1.5 (a) verify that the given function is a probability function, (b) find $E[x]$, (c) find $E[x^2]$, (d) using the results of parts (b) and (c), calculate the quantity VAR $= E[x^2] - (E[x])^2$.

3.1.1 $f(x) = \dfrac{x}{25} \qquad x = 3, 4, 5, 6, 7$
$= 0 \qquad$ otherwise

3.1.2 $f(x) = x^2/55 \qquad x = 0, 1, 2, 3, 4, 5$
$= 0 \qquad$ otherwise

3.1.3 $f(x) = \dbinom{3}{x}(1/3)^x(2/3)^{3-x} \qquad x = 0, 1, 2, 3$
$= 0 \qquad$ otherwise

3.1.4 $f(x) = \dfrac{\dbinom{5}{x}\dbinom{5}{3-x}}{\dbinom{10}{3}} \qquad x = 0, 1, 2, 3$
$= 0 \qquad$ otherwise

3.1.5 $f(x) = \frac{(x-1)^2}{5} \qquad x = -1, 0, 1$

$= 0 \qquad$ otherwise

3.1.6 Using Rules 1, 2, and 3 of expectations, evaluate $E[4x^2 - 7x + 3]$ if the probability function is the expression given in Prob. 3.1.3. (*Hint:* Use the answers obtained in parts (b) and (c) in Prob. 3.1.3.)

3.1.7 Repeat Prob. 3.1.6 with the probability function of Prob. 3.1.4.

3.1.8 $f(x) = \frac{1}{n} \qquad x = 1, 2, \cdots, n$

$= 0 \qquad$ otherwise

is called the *uniform probability function.* Show that $E[x]$ and $E[x^2]$ for this probability function are $\frac{n+1}{2}$ and $\frac{1}{6}(n+1)(2n+1)$ respectively.

3.2 MOMENTS

Suppose a particular family of expectations is considered next, namely, $E[(x-a)^k]$, where a is any constant and $k = 0, 1, 2, \ldots$. We define this family of expectations by the notation $m_{k,a}$, which reads, "the kth moment about the point a." In symbols,

$$m_{k,a} = E[(x-a)^k] = \Sigma (x-a)^k \cdot f(x) \qquad (3.2.1)$$

A few specific examples should suffice to illustrate the concept of moments about a given point.

Example 3.2.1 Find $m_{3,2}$ for the binomial probability function $f(x{:}4, .5)$.

Solution: $m_{3,2} = E[(x-2)^3]$

$$= \sum_{x=0}^{4} (x-2)^3 \tbinom{4}{x} (.5)^x (.5)^{4-x} = 0$$

In words, the third moment about the point 2 for the binomial probability function $f(x{:}4, .5)$ is equal to zero.

Example 3.2.2 Find $m_{2,0}$ if $f(x)$ is the same probability function as given in Example 3.2.1.

Solution: $m_{2,0} = E[(x-0)^2] = E[x^2]$

$$= \sum_{x=0}^{4} x^2 \cdot f(x) = 5$$

Note that this result is identical with the expression in Example 3.1.3.

Let us remark in passing that if $k = 0$, $m_{0,a}$ is equal to one for all probability functions. That is, $E[(x-a)^0] = 1$, which can be readily verified by Rule 1 of expectations.

Now that $m_{k,a}$ has been defined in general, let us consider two cases in particular which are of vital importance in statistical considerations. The two cases are (1) $a = 0$ and (2) $a = E[x]$.

(1) Moments about the Origin. If $a = 0$, Eq. (3.2.1) becomes $m_{k,0} = E[x^k]$ which according to adopted convention reads "the kth moment about the point zero," or "the kth moment about the origin." Specifically, $m_{1,0} = E[x]$, $m_{2,0} = E[x^2]$, $m_{3,0} = E[x^3]$. This notation can be further simplified if we define*

$$m_{k,0} = m_k \tag{3.2.2}$$

Computations of m_1 and m_2 have already been given in Examples 3.1.2 and 3.1.3 respectively; next an example of m_3 bill be presented.

Example 3.2.3 Find the third moment about the origin for the binomial probability function $f(x:4,.5)$.

Solution:
$$m_3 = E[x^3] = \sum_{x=0}^{4} x^3 \binom{4}{x}(.5)^x(.5)^{4-x}$$
$$= 14$$

Let us remark that the kth moment about any point has some constant value, since x and $(x - a)^k$ are both random variables which by definition represent real numbers.

In the following pages we shall scrutinize the meaning and implications of the first, second, and third moments about the origin. We shall compare the numeric values obtained for these moments to some observable characteristic of a given probability function. Let us pursue our discussion with an example.

Example 3.2.4 Draw the histogram for the binomial probability function $f(x:6,.5)$ and calculate m_1.

Solution: Figure 3.1 represents the requested histogram and

$$m_1 = \sum_{x=0}^{6} x \cdot f(x) = \frac{1}{64}[0 \cdot 1 + 1 \cdot 6 + 2 \cdot 15 + \cdots + 6 \cdot 1]$$
$$= 3$$

Referring to Fig. 3.1, we note that the center of gravity of the histogram lies on the line $x = 3$. The symmetry of the graph permits us to obtain the value of 3 without computation, which is equal to the calculated value of m_1.

The student may recall from his studies in physics that the x coordinate of the center of gravity is defined as

*Many authors define the kth moment about the origin by the symbol μ'_k. We feel, however, that our symbolism is less confusing to the beginning student.

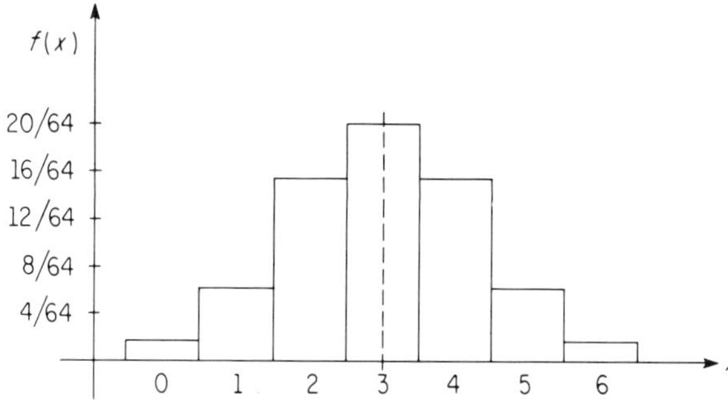

FIG. 3.1. Histogram of binomial $f(x: 6, 0.5)$.

$$\bar{x} = \frac{\sum_{i=1}^{n} x_i \cdot f(x_i)}{\sum_{i=1}^{n} f(x_i)} \tag{3.2.3}$$

where $f(x_i)$ represents the mass located at x_i and $\sum_{i=1}^{n} f(x_i)$ is equal to the total mass. In the special case where the total mass is equal to unity, which is analogous to a probability model, Eq. (3.2.3) reduces to $\bar{x} = \sum_{i=1}^{n} x_i \cdot f(x_i)$. In other words, if a unit mass were distributed homogenously over the given histogram (probability model) then m_1 and $\bar{x}$ would represent the same line.

In physics, $\bar{x}$ is often referred to as "the first moment" and hence the carryover to the present definition of m_1. For this reason many authors use m_1 and $\bar{x}$ synonymously. Clearly then, $m_1 = \bar{x}$ represents the line upon which the center of gravity of a probability function is located. In an intuitive sense, $m_1 = \bar{x}$ is the line around which a given probability function is centered, hence the reference to m_1 as the *average, arithmetic mean*, or simply the *mean*. Therefore, first moment about the origin, m_1, arithmetic mean, mean, and $\bar{x}$ are all synonymous terms.

At this juncture let us interject that because a probability function is analogous to a unit mass being homogenously distributed over a unit area, many authors use the term *probability density function* synonymously with probability function. Furthermore, it is commonly referred to a random variable as being distributed according to a certain probability function. For instance, the expressions "a random variable x is distributed according to the binomial p.f. with parameters n and p" and "a random variable

x obeys the binomial p.f. with parameters n and p." are interpreted to be equivalent.

The mean of a probability function represents a significant characteristic and practically always is considered a descriptive measure of the model. Because repeated reference will be made to the mean of most probability models, we define $\mu = m_1$ to further simplify notation. It will be shown in a later chapter that the mean of a given data is the "best" estimate of the mean of the underlying probability model. The word "best" is used in a technical sense and its precise definition will be presented also in the same chapter.

Next let us proceed to calculate the means of various probability functions that were discussed in Chapter 2.

(a) The Mean of the Binomial Probability Function $f(x:n,p)$

$$\begin{aligned}\mu = E[x] &= \sum_{x=0}^{n} x \cdot \frac{n!}{x!\,(n-x)!} p^x q^{n-x} \\ &= n \cdot p \sum_{x=1}^{n} \frac{(n-1)!}{(x-1)!(n-x)!} p^{x-1} q^{n-x} \\ &= n \cdot p(p+q)^{n-1} = n \cdot p(1) = n \cdot p \qquad (3.2.4)\end{aligned}$$

Therefore, the mean of any binomial probability function is the product of the two parameters n and p. Referring to Example 3.2.4, we notice that $6(.5) = 3$ agrees with the previous computation.

(b) The Mean of the Poisson Probability Function $f(x:k)$

$$\begin{aligned}\mu = E[x] &= \sum_{x=0}^{\infty} x \cdot \frac{k^x e^{-k}}{x!} = k \cdot \sum_{x=1}^{\infty} \frac{k^{x-1} e^{-k}}{(x-1)!} \\ &= k \cdot e^k \cdot e^{-k} = k \qquad (3.2.5)\end{aligned}$$

Therefore the mean of any Poisson probability function is the parameter k. For instance, the mean of the Poisson $f(x:3)$ is $x = 3$.

(c) The Mean of the Hypergeometric Probability Function $f(x:N,M,n)$

$$\mu = E[x] = \sum_{x=0}^{N} x \cdot \frac{\binom{M}{x}\binom{N-M}{n-x}}{\binom{N}{n}} \qquad (3.2.6)$$

or equivalently,

$$\mu \binom{N}{n} = \sum_{x=0}^{N} x \binom{M}{x}\binom{N-M}{n-x} \qquad (3.2.7)$$

If the identity $x\binom{M}{x} = M\binom{M-1}{x-1}$ is substituted in the last expression, one gets

$$\mu\binom{N}{n} = \sum_{x=1}^{N} M\binom{M-1}{x-1}\binom{N-M}{n-x} = M\sum_{y=0}^{N}\binom{M-1}{y}\binom{(N-1)-(M-1)}{n-1-y}$$

(3.2.8)

However, the summation on the right-hand side of Eq. (3.2.8) is equal to $\binom{N-1}{n-1}$, [refer to the result of Eq. (2.6.3)] and hence it becomes

$$\mu\binom{N}{n} = M\binom{N-1}{n-1},$$

which when solved for μ yields

$$\mu = n \cdot \frac{M}{N} \tag{3.2.9}$$

The mean of the hypergeometric probability function then, is a constant related to the parameters N, M, and n. The mean for the hypergeometric function $f(x:10,5,2)$, for example, is $x = 2(5/10) = 1$.

Example 3.2.5 Find the mean of the hypergeometric probability function given in Example 2.6.1.

Solution: In Example 2.6.1 $N = 36$, $M = 5$ and $n = 3$. Therefore, from Eq. (3.2.9) we get, $\mu = 3 \cdot \frac{5}{36} = \frac{5}{12}$.

(2.) Moments about the Mean. If $a = E[x]$, that is a is equal to the mean, Eq. (3.2.1) becomes

$$m_{k,\mu} = E[(x-\mu)^k] = \Sigma\,(x-\mu)^k f(x) \tag{3.2.10}$$

Since in all subsequent applications we shall be primarily concerned with m_k and $m_{k,\mu}$ only, Eq. (3.2.10) could further be modified in its notation if we define

$$\mu_k = m_{k,\mu} \tag{3.2.11}$$

That is, μ_k represents the kth moment about the mean and this symbol is commonly used in literature.

Example 3.2.6 Find μ_1, μ_2, and μ_3 for the binomial probability function $f(x:4,.5)$.

Solution: Using the Eq. (3.2.4) one gets $m_1 = \mu = 2$. Hence

$$\mu_1 = E[(x-2)] = \sum_{x=0}^{4}(x-2)\binom{4}{x}(.5)^x(.5)^{4-x} = 0$$

$$\mu_2 = E[(x-2)^2] = \sum_{x=0}^{4}(x-2)^2\binom{4}{x}(.5)^x(.5)^{4-x} = 1$$

$$\mu_3 = E[(x-2)^3] = \sum_{x=0}^{4}(x-2)^3\binom{4}{x}(.5)^x(.5)^{4-x} = 0$$

Consider next the special case of $k = 2$. Eq. (3.2.10) becomes

$$\mu_2 = E[(x - \mu)^2] \tag{3.2.12}$$

which the student may recognize to be analogous to the "moment of inertia" just as $m_1 = \mu$ was analogous to the x coordinate of the center of gravity of a unit mass distribution. From the definition it should be evident that μ_2 is always greater than or equal to zero. Also, μ_2 would be large if for some x, $|x - \mu|$ is large.

To state it differently, if some values of the random variable x are "spread far apart" from the mean, μ_2 would be relatively large in value. Intuitively then, μ_2 is a measure of "spread" or dispersion from the mean and hence the greater the value of the second moment about the mean, the larger the spread from the mean. μ_2 is very significant in statistical studies and therefore a special name is given to it—*variance.* Variance, μ_2, and second moment about the mean are synonymous terms.

The positive square root of the variance is defined as the *standard deviation* of the probability function and conventionally the Greek letter σ is associated with it. That is,

$$\sigma = \text{standard deviation} = \sqrt{\text{variance}}$$

Consequently, the symbol σ^2 is used to represent the variance and Eq. (3.2.12) is the formula for the variance. We mention for emphasis that in order to calculate the standard deviation of a probability function, the variance must be calculated first. Referring to Example 3.2.6 the student may notice that the variance of a specific probability function was computed there and before presenting further examples let us express Eq. (3.2.12) in an equivalent form which will be used repeatedly in later considerations.

According to Eq. (3.2.12), $\sigma^2 = E[(x - \mu)^2]$, which when expanded becomes

$$\sigma^2 = E[x^2 - 2\mu x + \mu^2] \tag{3.2.13}$$

Because $\mu = E[x]$ and is a constant, one can apply Eqs. (3.1.2), (3.1.3), and (3.1.4) to obtain

$$\begin{aligned}\sigma^2 &= E[x^2] - 2\mu E[x] + E[\mu^2] \\ &= E[x^2] - 2\mu^2 + \mu^2 \\ &= E[x^2] - (E[x])^2 \end{aligned} \tag{3.2.14}$$

In words, Eq. (3.2.14) states, the variance of a probability function is equal to the second moment about the origin minus the square of the mean.

Another expression for variance which is often used will be presented next. It proves useful particularly in the derivation of the variances for the binomial, Poisson, and the hypergeometric probability functions.

Consider the identity

$$\begin{aligned} E[x^2] &= E[x^2 - x + x] \\ &= E[x(x-1) + x] \\ &= E[x(x-1)] + E[x] \end{aligned} \tag{3.2.15}$$

and if $(E[x])^2$ is subtracted from both sides, one gets

$$E[x^2] - (E[x])^2 = E[x(x-1)] + E[x] - (E[x])^2 \tag{3.2.16}$$

Since Eqs. (3.2.14) and (3.2.16) are equivalent either the left-hand or the right-hand side of Eq. (3.2.16) may be used in computing the variance. Which expression to use is often a matter of convenience.

(a) The Variance of the Binomial Probability Function. Since $E[x]$ of the binomial probability function is already known, one needs to calculate $E[x(x-1)]$ to obtain the expression for the variance. Now for the binomial $f(x{:}n,p)$

$$\begin{aligned} E[x(x-1)] &= \sum_{x=0}^{n} x(x-1)\binom{n}{x}p^x(1-p)^{n-x} \\ &= \sum_{x=2}^{n} \frac{n!p^x(1-p)^{n-x}}{(n-x)!(x-2)!} \\ &= n(n-1)p^2\sum_{x=2}^{n} \frac{(n-2)!p^{x-2}(1-p)^{n-x}}{(n-x)!(x-2)!} \\ &= n(n-1)p^2(p+q)^{n-2} \qquad \text{where } q = 1-p \\ &= n(n-1)p^2 \end{aligned} \tag{3.2.17}$$

Recalling that $E[x] = np$, for the binomial $f(x{:}\ n,\ p)$, and substituting in Eq. (3.2.16) the result of Eq. (3.2.17) one gets

$$\sigma^2 = n(n-1)p^2 + np - (np)^2 = np(1-p) = npq \tag{3.2.18}$$

Or equivalently, the standard deviation $\sigma = \sqrt{npq}$.

Example 3.2.7 Find the standard deviation for the binomial $f(x{:}64,\ .5)$.

Solution: Using Eq. (3.2.18) we obtain $\sigma^2 = 64(.5)(.5) = 16$ and hence $\sigma = 4$.

(b) The Variance of the Poisson Probability Function. Here also we shall calculate $E[x(x-1)]$ first and utilize the expression of Eq. (3.2.17) to obtain the variance.

$$\begin{aligned} E[x(x-1)] &= \sum_{x=0}^{\infty} x(x-1)\frac{e^{-k}k^x}{x!} = \sum_{x=2}^{\infty} \frac{e^{-k}(k^2)k^{x-2}}{(x-2)!} \\ &= k^2e^{-k}\sum_{y=0}^{\infty}\frac{k^y}{y!} = k^2e^{-k}e^k = k^2 \end{aligned}$$

Since $E[x]$ for the Poisson is the parameter k, Eq. (3.2.16) yields

$$\sigma^2 = k^2 + k - (k)^2 = k \tag{3.2.19}$$

That is, the variance of the Poisson probability function is the parameter k. It is interesting to observe that the mean and the variance of the Poisson probability function are the same and equal to the single parameter involved. The standard deviation for this probability function is $\sqrt{k}$. The mean and the standard deviation for the Poisson function $f(x\colon 4)$, for instance are 4 and 2 respectively.

(c) The Variance of the Hypergeometric Probability Function. As in the derivation of the variances for the binomial and Poisson probability functions, one proceeds similarly for the hypergeometric. That is,

$$E[x(x-1)] = \sum_{x=0}^{N} x(x-1)\frac{\binom{M}{x}\binom{N-M}{n-x}}{\binom{N}{n}} \tag{3.2.20}$$

which is equivalent to

$$\binom{N}{n}E[x(x-1)] = \sum_{x=0}^{N} x(x-1)\binom{M}{x}\binom{N-M}{n-x} \tag{3.2.21}$$

Substituting the identity

$$x(x-1)\binom{M}{x} = M(M-1)\binom{M-2}{x-2}$$

to the right-hand side of Eq. (3.2.21), one obtains

$$\binom{N}{n}E[x(x-1)] = \sum_{x=2}^{N} M(M-1)\binom{M-2}{x-2}\binom{N-M}{n-x} \tag{3.2.22}$$

The right-hand side of Eq. (3.2.22) is equal to

$$M(M-1)\sum_{y=0}^{N}\binom{M-2}{y}\binom{(N-2)-(M-2)}{n-2-y} \tag{3.2.23}$$

Referring to Eq. (2.6.3), however, one observes that the summation in the expression Eq. (3.2.23) is equal to $\binom{N-2}{n-2}$ and therefore

$$\binom{N}{n}E[x(x-1)] = M(M-1)\binom{N-2}{n-2} \tag{3.2.24}$$

Solving Eq. (3.2.24) for $E[x(x-1)]$ yields

$$E[x(x-1)] = \frac{M(M-1)n(n-1)}{N(N-1)} \tag{3.2.25}$$

Recalling that mean of the hypergeometric probability function is $n \cdot \dfrac{M}{N}$ and applying Eq. (3.2.16) one gets

$$\sigma^2 = \frac{M(M-1)n(n-1)}{N(N-1)} + n\frac{M}{N} - \left(n\frac{M}{N}\right)^2$$
$$= \frac{nM(N-M)(N-n)}{N^2(N-1)} \tag{3.2.26}$$

Example 3.2.8 Find the variance for the hypergeometric probability function $f(x:36,6,4)$.

Solution: Using the result of Eq. (3.2.26) and substituting $N = 36$, $M = 6$ and $n = 4$, one gets $\sigma^2 = 32/63$.

Let us remark about the result of the means and the variances of the binomial, Poisson, and the hypergeometric probability functions. In Sec. 2.5 while considering the limiting relationships between the binomial and the Poisson functions, we assumed that $n \cdot p$ remained constant and equal to k as $n \to \infty$ and $p \to 0$. However, Eq. (3.2.4) indicates that $n \cdot p$ is the mean of the binomial $f(x:n,p)$ and Eq. (3.2.5) shows that k represents the mean of the Poisson function $f(x:k)$. It becomes apparent that the assumptions of Sec. 2.5 are equivalent to holding the means of the binomial and the Poisson equal during the limiting process.

Also, Eq. (3.2.18) shows that the variance of the binomial is $n \cdot p(1-p)$ and Eq. (3.2.19) indicates the variance of the Poisson function to be equal to k. Now, under the assumptions of Sec. 2.5 one gets for the variance of the binomial

$$\lim_{\substack{n\to\infty \\ p\to 0}} n \cdot p(1-p) = k \lim_{p\to 0} (1-p) = k$$

which is the variance of the Poisson.

Likewise, in Sec. 2.6 while discussing the limiting relationship between the hypergeometric and the binomial functions it was assumed that as $N \to \infty$, $M/N = p$ remained constant, where p is the proportion of objects of one type. This implies that $n \cdot p = n \cdot M/N$ is held constant. This however, implies that the means of the binomial and the hypergeometric probability functions are held constant and equal. Furthermore, if the limiting case of Eq. (3.2.26) is considered, one gets $\sigma^2 = n \cdot p(1-p)$ which is shown to be the variance of the binomial probability function. With this additional insight, the student may find it worthwhile to review Secs. 2.5 and 2.6 at this juncture.

Chebyshev's Theorem. Now that the definition of variance is given and the variances of several probability functions computed, the reader might have recognized the significance of the variance as a measure of spread or dispersion from the mean. That is, it might have been evident that a large variance or standard deviation implies a large spread from the mean and a small variance or standard deviation indicates small spread from the mean. Next, a more precise relationship between the standard deviation and its corresponding probability function will be presented. The theorem is named after the Russian mathematician P. L. Chebyshev (1821–1894.) Chebyshev's theorem states that at least the fraction $(1 - 1/k^2)$ of the total probability of a random variable lies within k standard deviations of the mean, that is, $P(|x - \mu| \leq k\sigma) > 1 - 1/k^2$. This theorem can be stated equivalently that at most the fraction $1/k^2$ of the total probability of a random variable lies outside k standard deviations of the mean; that is, $P(|x - \mu| > k\sigma) < 1/k^2$.

Before the proof of Chebyshev's theorem is presented, let us refer to the result of an example previously considered. In Example 3.2.6, μ_2, the variance of the binomial $f(x: 4, .5)$ was computed to be equal to 1, and equivalently the standard deviation σ is 1. On the line graph of this particular binomial probability function (see Fig. 3.2) let us locate the

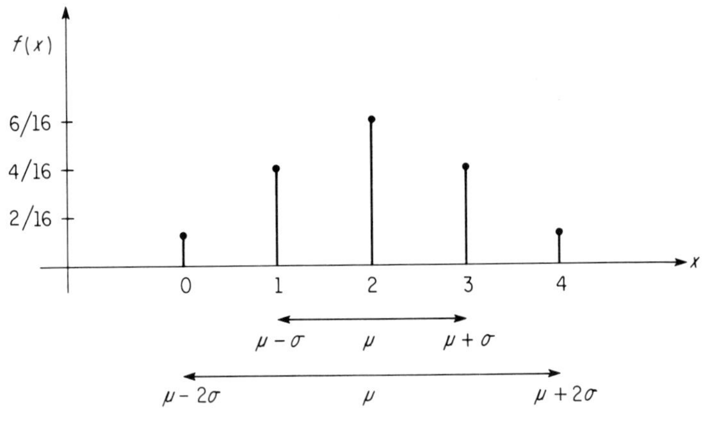

FIG. 3.2.

points corresponding to $\mu - 2\sigma$, $\mu - \sigma$, μ, $\mu + \sigma$ and $\mu + 2\sigma$, which are 0, 1, 2, 3, and 4 respectively because $\mu = 2$ and $\sigma = 1$.

Clearly all the probabilities of the random variable x, $x = 0, 1, 2, 3,$ and 4 are included within 2 standard deviations of the mean, while according to Chebyshev's theorem at least 3/4 of the total probability lies within two standard deviations of the mean. Chebyshev's theorem gives a conservative expression for the spread of the probability function but because it applies to any probability function, it proves very helpful.

The proof of the Chebyshev's theorem is presented next. Let R_1, R_2, and R_3 represent a partition of the sample space of the random variable x such that R_1, R_2, and R_3 represent regions for $x < \mu - k\sigma$, $|x - \mu| \leq k\sigma$ and $x > \mu + k\sigma$ respectively. Then,

$$\sigma^2 = \Sigma(x - \mu)^2 f(x) = \sum_{R_1} (x - \mu)^2 f(x) + \sum_{R_2} (x - \mu)^2 f(x) + \sum_{R_3} (x - \mu)^2 f(x) \tag{3.2.27}$$

Since each term of the right-hand side expression in Eq. (3.2.27) is positive or zero,

$$\sigma^2 \geq \sum_{R_1} (x - \mu)^2 f(x) + \sum_{R_3} (x - \mu)^2 f(x) \tag{3.2.28}$$

Now in regions R_1 and R_3 $(x - \mu)^2 > k^2\sigma^2$ because of the restrictions on the partition, and consequently

$$\sigma^2 > \sum_{R_1} f(x)(k\sigma)^2 + \sum_{R_3} f(x)(k\sigma)^2 \tag{3.2.29}$$

Dividing this last expression by $k^2\sigma^2$ yields

$$\frac{1}{k^2} > \sum_{R_1} f(x) + \sum_{R_3} f(x) = P(|x - \mu| > k\sigma) \tag{3.2.30}$$

Because the three regions form a partition by assumption,

$$P(|x - \mu| > k\sigma) + P(|x - \mu| \leq k\sigma) = 1 \tag{3.2.31}$$

and therefore

$$P(|x - \mu| \leq k\sigma) > 1 - \frac{1}{k^2} \tag{3.2.32}$$

Equations (3.2.30) and (3.2.32) are equivalent relationships and represent Chebyshev's theorem.

Example 3.2.9 Given the binomial $f(x\colon 16, .5)$, use Chebyshev's theorem to find the proportion of the probability between $x = 2$ and $x = 14$. What is the corresponding exact probability?

Solution: For the binomial $f(x\colon n, p)$, $\mu = n \cdot p$ and $\sigma^2 = npq$ which for this case yields $\mu = 8$ and $\sigma = 2$. $(2 - 8)/2 = -3$ and $(14 - 8)/2 = 3$ and hence $k = 3$. Applying Eq. (3.2.32) one gets

$$P(2 \leq x \leq 14) > 1 - \frac{1}{3^2} = \frac{8}{9} = 0.8888$$

Equivalently, $P(x \leq 1 \text{ or } x \geq 15) < \frac{1}{9} = 0.1111$. This last expression has an exact probability equal to $2[F(1\colon 16, .5)]$ because the function is symmetric. Therefore,

$$\begin{aligned} P(x \leq 1 \text{ or } x \geq 15) &= 2[\tbinom{16}{0}(.5)^{16} + \tbinom{16}{1}(.5)^{16}] \\ &= (.5)^{15} + (.5)^{11} \\ &= .0006 \end{aligned}$$

The probability that x lies between 2 and 14 inclusive is equal to $1 - .0006 = .9994$.

PROBLEMS

In Probs. 3.2.1 to 3.2.3 calculate the variance and the standard deviation for the probability function referred to.

3.2.1 The probability function of Prob. 3.1.1.

3.2.2 The probability function of Prob. 3.1.3.

3.2.3 The probability function of Prob. 3.1.5.

3.2.4 Evaluate the inequality given in Eq. (3.2.32) for $k = 1.5, 2.0, 2.5$, and 3.0 and tabulate the result.

3.2.5 Apply Chebyshev's theorem to find the upper limits for the probability that fewer than 20 and more than 44 heads will show in the toss of 64 unbiased coins.

3.2.6 Apply Chebyshev's theorem to find the lower limits for the probability that a random variable described by the Poisson probability function $f(x:9)$ will assume a value between $x = 5$ and $x = 13$.

3.2.7 Show that the first moment about the mean, μ_1, is equal to zero for all probability functions.

3.2.8 Show that the kth moment about any point $m_{k,a}$ is greater than or equal to zero if k is a positive even integer.

3.3 SKEWNESS

In the previous section two specific moments were considered: (1) the first moment about the origin—the mean, and (2) the second moment about the mean—the variance. In Prob. 3.2.7 it was requested to show that μ_1 is equal to zero for all probability functions. Also, as a special case of Prob. 3.2.8 one can show that μ_k is greater than or equal to zero if k is an even integer. If k is an odd integer greater than one, however, μ_k may have values which are either negative or positive or equal to zero. The characteristic of odd moments of a probability function having negative, positive or zero values can be utilized to broadly classify probability functions into three major types. The most convenient number to achieve this objective is $k = 3$ and it will be shown next that μ_3 is either negative, positive or equal to zero.

Let R_1, R_2, and R_3 constitute a partition of the sample space of the random variable x such that R_1, R_2, and R_3 represent the regions for $x < \mu$, $x = \mu$ and $x > \mu$ respectively. Therefore

$$\mu_3 = \sum_{R_1} (x - \mu)^3 f(x) + \sum_{R_2} (x - \mu)^3 f(x) + \sum_{R_3} (x - \mu)^3 f(x) \quad (3.3.1)$$

The assumptions on the regions R_1, R_2, and R_3 imply

$$\sum_{R_1} (x - \mu)^3 f(x) < 0 \quad (3.3.2)$$

$$\sum_{R_3} (x - \mu)^3 f(x) > 0 \quad (3.3.3)$$

$$(x - \mu)^3 f(x) = 0 \quad \text{for } R_2 \quad (3.3.4)$$

and hence

$$\mu_3 = \sum_{R_1} (x - \mu)^3 f(x) + \sum_{R_3} (x - \mu)^3 f(x) \quad (3.3.5)$$

Now if

$$\left| \sum_{R_1} (x - \mu)^3 f(x) \right| < \sum_{R_3} (x - \mu)^3 f(x)$$

Equation (3.3.5) yields $\mu_3 > 0$ otherwise $\mu_3 \leq 0$. This implies μ_3 is either positive, or negative, or equal to zero.

Next let us consider a special case. If the random variable x is symmetric about its mean, for each value of $x - \mu$ in R_3 there exists a corresponding equivalent value in R_1 but negative in numeric sign. When these negative values are cubed, they retain their signs and the expression

$$-\sum_{R_1} (x - \mu)^3 f(x) = \sum_{R_3} (x - \mu)^3 f(x) \tag{3.3.6}$$

holds true and can be expressed equivalently

$$\sum_{R_1} (x - \mu)^3 f(x) + \sum_{R_3} (x - \mu)^3 f(x) = 0 \tag{3.3.7}$$

Therefore, if the probability function is symmetric about its mean, $\mu_3 = 0$. The converse of this statement, however, is not necessarily true. That is, if $\mu_3 = 0$ the probability function may be asymmetric as shown in the following example.

Example 3.3.1 Find the m_1, μ_2 and μ_3 for the probability function defined as follows:

$$\begin{aligned} f(x) &= .4 \quad \text{if } x = -2 \\ &= .5 \qquad x = 1 \\ &= .1 \qquad x = 3 \\ &= 0 \quad \text{otherwise} \end{aligned}$$

Solution: $m_1 = x \cdot f(x) = -2(.4) + 1(.5) + 3(.1) = 0$ which states that the mean is zero.

$$\begin{aligned} \mu_2 = \sigma^2 &= \Sigma (x - \mu)^2 f(x) \\ &= (-2)^2(.4) + 1^2(.5) + 3^2(.1) = 3 \end{aligned}$$

which states that the variance is equal to 3.

$$\begin{aligned} \mu_3 &= \Sigma (x - \mu)^3 f(x) \\ &= (-2)^3(.4) + 1^3(.5) + 3^3(.1) = 0 \end{aligned}$$

However, $f(x)$ represents an asymmetric probability function.

If a probability function is asymmetric about its mean it is said to be *skewed.* If $\mu_3 > 0$ the probability function is said to be *positively skewed* or to have *positive skewness*, and if $\mu_3 < 0$ the probability function is said to be *negatively skewed* or to have *negative skewness.* The next two examples represent skewed probability functions.

Example 3.3.2 Graph the binomial $f(x{:}4, .25)$ and calculate (a) m_1 (b) μ_2 and (c) μ_3.

Solution: $f(0) = 81/256$, $f(1) = 108/256$, $f(2) = 54/256$, $f(3) = 12/256$ and $f(4) = 1/256$

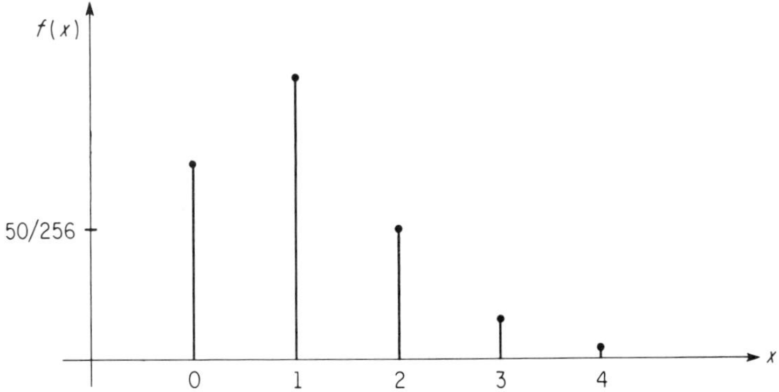

FIG. 3.3. Positively skewed binomial function.

(a) $m_1 = \mu = n \cdot p = 4(.25) = 1$ by Eq. (3.2.4)
(b) $\mu_2 = \sigma^2 = n \cdot p(1 - p) = 4(.25)(.75) = 0.75$ by Eq. (3.2.18).

(c) $\mu_3 = \sum_{x=0}^{4} (x - 1)^3 f(x) = 3/8$

Since $\mu_3 > 0$, the binomial $f(x:4,.25)$ is a positively skewed probability function.

Example 3.3.3 Graph the binomial $f(x:4,.75)$ and calculate (a) m_1, (b) μ_2, and (c) μ_3.

Solution: $f(0) = 1/256$, $f(1) = 12/256$, $f(2) = 54/256$, $f(3) = 108/256$ and $f(4) = 81/256$

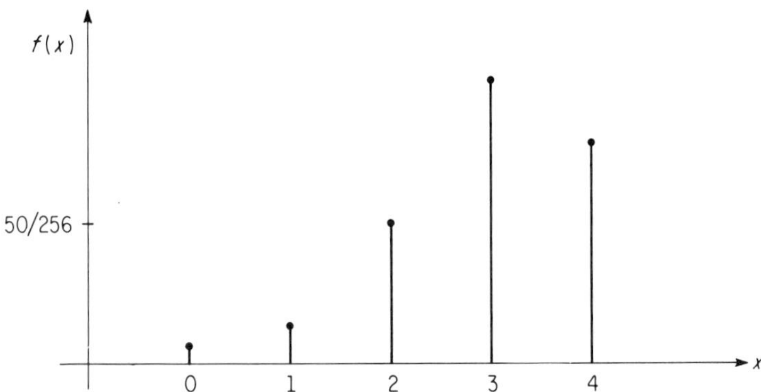

FIG. 3.4. Negatively skewed binomial function.

(a) $m_1 = \mu = n \cdot p = 4(.75) = 3$ by Eq. (3.2.4)
(b) $\mu_2 = \sigma^2 = n \cdot p(1 - p) = .75$ by Eq. (3.2.18)

(c) $\mu_3 = \sum_{x=0}^{4} (x - 3)^3 f(x) = -3/8$

Since $-3/8 < 0$, the binomial $f(x{:}4, .75)$ is a negatively skewed probability function.

Some general remarks are in order at this juncture. In Example 3.2.6 it was shown that the binomial $f(x{:}4,.5)$ has $\mu_3 = 0$ and referring to Fig. 3.2 we note its symmetry. Example 3.3.2 above shows that the binomial $f(x{:}4,.25)$ is positively skewed and Example 3.3.3 shows that the binomial $f(x{:}4,.75)$ is negatively skewed. In comparing these three binomial probability functions with the parameter $n = 4$ but different values of the parameter p, one recognizes a pattern; namely, the mean of the positively skewed binomial function is located toward the left of the corresponding symmetric probability function and the mean of the negatively skewed function is situated toward the right of the corresponding symmetric probability function. This pattern can be described differently. One observes that the positively skewed binomial probability function is extended more toward the right (long tail in the positive direction) and the negatively skewed binomial function is extended more toward the left (long tail in the negative direction.) These remarks hold true in general for most of the commonly used probability functions and one could determine from a crude sketch whether a probability function is positively or negatively skewed.

Now let us refer to Eq. (3.3.7), which was shown to apply for the symmetric probability functions. Suppose a new value of x in R_3 is added to the summation. The expression becomes greater than zero—that is, the function is positively skewed. If instead a value of x in R_1 is added to the summation, the probability function becomes negatively skewed. This verifies the above described patterns of skewed probability functions.

PROBLEMS

3.3.1 Show that $\mu_3 = m_3 - 3m_2 \cdot m_1 + 2(m_1)^3$ for all probability functions.

3.3.2 Verify that the hypergeometric $f(x{:}N,M,2)$ is symmetric if $2M = N$. Evaluate the special case where $M = 5$.

3.3.3 Verify that the binomial $f(x{:}n,p)$ is symmetric if $p = .5$ and asymmetric otherwise.

3.3.4 Construct two examples of each for probability functions that are (a) symmetric about their means, (b) positively skewed, (c) negatively skewed.

3.4 THE MOMENT-GENERATING FUNCTION

The expectation of the function $g(x) = e^{tx}$ plays an important role in the determining of the moments about the origin of a given probability function and hence is called the *moment-generating function* of the random variable. Symbolically,

$$\text{m.g.f.} = M_x(t) = E[e^{tx}] = \Sigma\, e^{tx} f(x) \tag{3.4.1}$$

The moment-generating function does not always exist because the sum in Eq. (3.4.1) may not necessarily converge. If however, the sum does converge and the Maclaurin series of e^{tx} is substituted in the expression (3.4.1), one obtains

$$M_x(t) = \Sigma\left[1 + tx + \frac{(tx)^2}{2!} + \cdots + \frac{(tx)^n}{n!} + \cdots\right] f(x) \tag{3.4.2}$$

which is equivalent to

$$M_x(t) = \Sigma\, f(x) + t\,\Sigma\, xf(x) + \frac{t^2}{2!}\,\Sigma\, x^2 f(x) + \cdots \tag{3.4.3}$$

Now each term of Eq. (3.4.3) represents the various moments about the origin multiplied by a coefficient—that is,

$$M_x(t) = 1 + t\cdot m_1 + \frac{t^2}{2!} m_2 + \cdots = \sum_{i=0}^{n} \frac{t^i}{i!} m_i \tag{3.4.4}$$

If differentiability under the summation is assumed and Eq. (3.4.4) is differentiated with respect to t, one gets

$$\frac{d}{dt} M_x(t) = m_1 + t\cdot m_2 + \cdots + \frac{t^{n-1}}{(n-1)!} m_n = \sum_{i=1}^{n} \frac{t^{i-1}}{(i-1)!} m_i \tag{3.4.5}$$

When the expression (3.4.5) is evaluated at $t = 0$, it yields

$$\frac{d}{dt} M_x(0) = m_1 \tag{3.4.6}$$

which is the first moment about the origin.

If the expression (3.4.5) is differentiated with respect to t, we obtain

$$\frac{d^2}{dt^2} M_x(t) = \sum_{i=2}^{n} \frac{t^{i-2}}{(i-2)!} m_i \tag{3.4.7}$$

which when evaluated at $t = 0$ yields

$$\frac{d^2}{dt^2} M_x(0) = m_2 \tag{3.4.8}$$

If $M_x(t)$ is differentiated k times and evaluated at $t = 0$, one gets

$$\frac{d^k}{dt^k} M_x(0) = m_k \tag{3.4.9}$$

This expression states that the kth moment about the origin of a probability function is equal to the kth derivative of the moment-generating function evaluated at $t = 0$, provided the moment-generating function exists and differentiability under the summation can be assumed. The following examples illustrate the usefulness of the moment-generating function.

Example 3.4.1 Find the moment-generating function of the binomial $f(x:n,p)$ and using the result in Eq. (3.4.9) find (a) m_1, (b) m_2, and from (a) and (b) show that $\sigma^2 = n \cdot p(1 - p)$.

Solution:
$$M_x(t) = E[e^{tx}] = \sum_{x=0}^{n} e^{tx}\binom{n}{x}p^x(1-p)^{n-x}$$
$$= \sum_{x=0}^{n}\binom{n}{x}(pe^t)^x(1-p)^{n-x}$$
$$= \sum_{x=0}^{n}\binom{n}{x}(pe^t)^x q^{n-x} = (pe^t + q)^n$$

where $q = 1 - p$.

(a) $\frac{d}{dt} M_x(t) = n(pe^t + q)^{n-1} pe^t$

which when evaluated at $t = 0$ yields $n \cdot p$. This agrees with the result of Eq. (3.2.4).

(b) $\frac{d^2}{dt^2} M_x(0) = n^2p^2 + n \cdot p(1 - p) = m_2$

and hence

$$\sigma^2 = n^2p^2 + n \cdot p(1 - p) - (n \cdot p)^2$$
$$= n \cdot p(1 - p) = n \cdot p \cdot q$$

which agrees with the result of Eq. (3.2.18).

Example 3.4.2 Find the moment-generating function of the Poisson probability function $f(x:k)$.

Solution:
$$M_x(t) = E[e^{tx}] = \sum_{x=0}^{\infty} e^{tx} e^{-k} k^x/x!$$
$$= \sum_{x=0}^{\infty} e^{-k}(ke^t)^x/x!$$
$$= e^{-k}e^{ke^t} = e^{k(e^t-1)}$$

PROBLEMS

3.4.1 Using the result of Example 3.4.2, verify that the mean and variance for the Poisson probability function are equal to the single parameter k.

3.4.2 $f(x) = pq^{x-1} \quad x = 1, 2, 3, \cdots$
$= 0 \quad$ otherwise

where $q = 1 - p$, is defined as the *geometric probability function.* Find the moment-generating function for this probability function. Applying Eq. (3.4.9) to this result, find (a) the mean, (b) m_2, (c) σ^2 from the result of parts (a) and (b).

3.4.3 $f(x) = p$ if $x = 1$
$= q$ $\quad x = 0$
$= 0$ otherwise

where $q = 1 - p$, is defined as the *Bernoulli probability function.* Using the notion of Sec. 3.4, find (a) the mean and (b) the variance of the Bernoulli probability function.

3.4.4 Find the moment-generating function of the negative binomial probability function given in Sec. 2.8. From this result obtain the mean and the variance of the negative binomial probability function.

3.5 FACTORIAL MOMENT-GENERATING FUNCTION

Referring to Eq. (3.2.14), one observes that the first and the second moments about the origin were used to obtain an identity for the variance of a probability function. Referring to Eq (3.2.16), one observes that the first moment about the origin and $E[x(x - 1)]$ were used to get another identity for the variance of a probability function. The expression $E[x(x - 1)]$ was not given a special name in Sec. 3.2 but its usefulness was demonstrated. In this section expectations of the type, $E[x(x - 1)]$, $E[x(x - 1)(x - 2)]$, etc. will be considered. $E[x(x - 1)(x - 2) \cdots (x - k + 1)]$ is defined as the kth factorial moment and hence $E[x]$ is the first factorial moment, $E[x(x - 1)]$ is the second factorial moment, $E[x(x - 1)(x - 2)]$ is the third factorial moment, and so on. The factorial moments are derived in a similar fashion as the moments about the origin were obtained from the moment-generating function, except that $E[t^x]$ is utilized instead of $E[e^{tx}]$.

The factorial moment-generating function is defined as

$$F_x(t) = E[t^x] = \Sigma\, t^x f(x) \tag{3.5.1}$$

Now, assuming differentiability under the summation and differentiating the expression (3.5.1) with respect to t, one gets

$$\frac{d}{dt} F_x(t) = \Sigma\, x \cdot t^{x-1} f(x) \tag{3.5.2}$$

When Eq. (3.5.2) is evaluated at $t = 1$, it yields

$$\frac{d}{dt} F_x(1) = \Sigma\, x f(x) = E[x] \tag{3.5.3}$$

If the expression (3.5.2) is differentiated once again and t set equal to 1, one obtains

$$\frac{d^2}{dt^2} F_x(1) = \Sigma\, x(x - 1) f(x) = E[x(x - 1)] \tag{3.5.4}$$

which is the second factorial moment. In general, the kth factorial moment is obtained by differentiating the factorial moment-generating function k times and evaluating at $t = 1$. That is,

$$\frac{d^k}{dt^k} F_x(1) = E[x(x-1)(x-2)\ldots(x-k+1)] \tag{3.5.5}$$

The following example illustrates the application of the factorial moment-generating function.

Example 3.5.1 Find the factorial moment-generating function for the Poisson probability function $f(x{:}k)$ and using the method of this section, verify that $m_1 = k$ and $\sigma^2 = k$.

Solution:

$$F_x(t) = \sum_{x=0}^{\infty} t^x \cdot e^{-k} k^x/x! = e^{-k} \sum_{x=0}^{\infty} (tk)^x/x!$$

$$= e^{-k} e^{tk} = e^{k(t-1)}$$

$$\frac{d}{dt} F_x(t) = ke^{k(t-1)}$$

and therefore

$$\frac{d}{dt} F_x(1) = k \cdot e^0 = k$$

$$\frac{d^2}{dt^2} F_x(t) = k^2 e^{k(t-1)}$$

and hence

$$\frac{d^2}{dt^2} F_x(1) = k^2$$

Using Eq. (3.2.16) for the computation of the variance one gets, $\sigma^2 = k^2 + k - (k)^2 = k$ which verifies the previously obtained values for the mean and the variance of the Poisson probability function.

It may be evident from the identity $t^x = e^{x \log t}$ that

$$F_x(t) = E[t^x] = E[e^{x \log t}] = M_x(\log t) \tag{3.5.6}$$

That is, the factorial moment-generating function and the moment-generating function are related as shown by Eq. (3.5.6). Whether $F_x(t)$, the factorial moment-generating function, or $M_x(t)$, the moment-generating function should be computed is a question of convenience. The student is requested to compare Example 3.4.2 and Example 3.5.1 to see the interrelationship expressed.

PROBLEMS

3.5.1 Find the factorial moment-generating function for the Bernoulli probability function given in Prob. 3.4.3. Apply Eq. (3.5.5) to obtain (a) $E[x]$, (b) $E[x(x-1)]$, (c) Using the results of parts (a) and (b), calculate σ^2 for the Bernoulli probability function.

3.5.2 Find the factorial moment-generating function for the geometric probability function given in Prob. 3.4.2, and answer the same questions as in Prob. 3.5.1.

TABLE 3.1.1

SUMMARY TABLE OF REFERENCE FOR SOME DISCRETE PROBABILITY FUNCTIONS

Probability Function	Moment-Generating Function $M_X(t)$	Mean	Variance	References
Binomial $f(x:n,p) = \binom{n}{x} p^x q^{n-x} \quad x = 0, 1, \ldots, n$ $= 0$ otherwise where $0 < p < 1$ $q = 1 - p$	$(pe^t + q)^n$	$n \cdot p$	$n \cdot p \cdot q$	Eq. (2.4.1) Eq. (3.2.4) Eq. (3.2.18) Example 3.4.1
Poisson $f(x:k) = \dfrac{e^{-k}k^x}{x!} \quad x = 0, 1, 2, \ldots$ $= 0$ otherwise $k > 0$	$e^{k(e^t - 1)}$	k	k	Eq. (2.5.1) Eq. (3.2.5) Eq. (3.2.19) Example 3.4.2
Bernoulli $f(x:p) = p \quad x = 1$ $= q \quad x = 0$ $= 0$ otherwise	$pe^t + q$	p	$p \cdot q$	Prob. 3.4.3
Geometric $f(x:p) = pq^{x-1} \quad x = 1, 2, \ldots$ $= 0$ otherwise	$\dfrac{pe^t}{1 - qe^t}$	$\dfrac{1}{p}$	$\dfrac{q}{p^2}$	Prob. 3.4.2
Negative Binomial $f(x:k,p) = \binom{k + x - 1}{x} p^k q^x \quad x = 0, 1, \ldots$ $= 0$ otherwise	$\left(\dfrac{p}{1 - qe^t}\right)^k$	$\dfrac{k \cdot q}{p}$	$\dfrac{k \cdot q}{p^2}$	Eq. (2.8.1) Prob. 3.4.4

4

Univariate Probability Functions —Continuous Case

4.1 DEFINITION OF CONTINUOUS PROBABILITY FUNCTION

Each probability function considered in the previous chapters has been defined for random variables that assume discrete values only. In Example 2.1.1, for instance, the random variable $w = (x - y)^2$ had the values 0, 1, 4, and 9. It was obvious that w could assume these specific values and nothing else. To ascribe the value 3.5 or 4.45 to w would have been meaningless. The line graph representing the probability associated with each values of w was given in Fig. 2.1.

Suppose now we consider the following random experiment. A transistor is selected at random from a lot of 32 and tested for breakdown voltage. For example, 71 volts might be the observed value. Is 71 volts the *exact* or the *approximate* breakdown voltage? A little scrutiny should clearly indicate that there is always an error of measurement involved in an experiment of this type and the degree of accuracy has to do with the calibration of the instrument used. Hence 71 volts is not an exact but rather an approximate value. What does it signify to state that the breakdown voltage is 71 volts approximately? Intuitively, it simply suggests that the voltage reading on the voltmeter is closer to 71 than either 70 or 72. In other words, the breakdown voltage is in the interval $70.5 < v < 71.5$ and its exact value can not be observed.

Suppose next the voltage of another transistor was measured at 70 volts, and still another at 68, and still another at 66, etc. If v represents the breakdown voltage of any transistor of a given kind, $65 < v < 72$ might be an interval that contains all breakdown voltages. Furthermore, if all 32 transistors had the breakdown voltages shown in the table, the corresponding histogram would be as in Fig. 4.1.

Voltage	*Frequency*
66	1
67	5
68	10
69	10
70	5
71	1

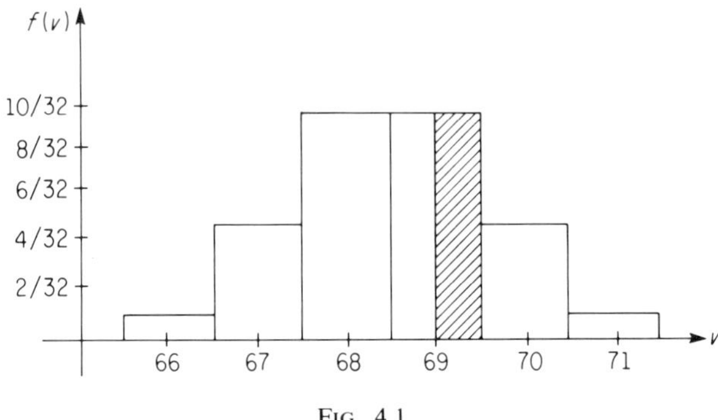

FIG. 4.1.

From Fig. 4.1 it should be evident that the area bounded by the interval $65.5 < v < 66.5$ is equal to $(1/32) \cdot 1 = 1/32$ and corresponds to the probability of breakdown voltages in that interval. Similarly, the area bounded within the interval $68.5 < v < 69.5$ is equal to $(10/32) \cdot 1 = 10/32$ which represents the probability of breakdown voltages in that interval. Next let us consider the interval $69.0 < v < 69.5$. What is the probability associated with this interval? Since areas corresponding to a given interval represent probabilities associated with that interval, $(10/32) \cdot (.5) = 5/32$ is the desired probability and the shaded area in Fig. 4.1 represents this probability.

Suppose that in the above-mentioned experiment we had a voltmeter which was calibrated in decivolts. Then a breakdown voltage of 70.6 volts, for instance, represents a reading in the interval $70.55 < v < 70.65$. If a large number of transistors were tested for breakdown voltages with a voltmeter of this type, the histogram associated with the random experiment will be similar to that of Fig. 4.1 except that the rectangles will be narrower by a factor of 1/10. If the voltmeter were calibrated in millivolts or microvolts, the rectangles in the histogram will be very narrow and approach to a line and consequently the histogram will approach to a continuous or piecewise continuous curve. That is, the random variable v, the breakdown voltage, will assume all values in a given continuum and $f(v)$ represents a continuous function associated with v.

In general and more precisely, let $f(x)$ represent a continuous function and the interval $a \leq x \leq b$ be divided into subintervals at the points $x_1, x_2, \ldots, x_n$ such that

$$a = x_1 < x_2 < x_3 < \cdots < x_n = b \text{ and if } \delta x_k = x_{k+1} - x_k$$

for $k = 1, 2, \ldots, n - 1$ such that $x_k < x'_k < x_{k+1}$; then

$$\lim_{\max \delta x_k \to 0} \sum_{k=1}^{n} f(x'_k)\delta x_k = \int_a^b f(x)\,dx$$

represents an area. (See Fig. 4.2.) If the value thus calculated is equal to one, $f(x)$ is said to be a probability function of the continuous type. Clearly, this definition is similar to the discrete case and the similarities

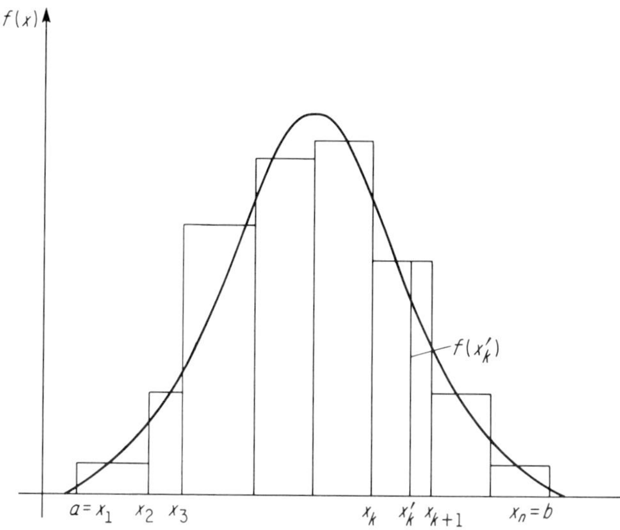

FIG. 4.2.

between them should be apparent. That is, any piecewise continuous function, say f, for which

1. $f(x) \geq 0$ (4.1.1)

2. $\int_{-\infty}^{\infty} f(x)\,dx = 1$ (4.1.2)

holds true, defines a probability function of the continuous type.* The basic difference is that the summation is replaced by integration in the second condition, which implies that $f(x)$ is an integrable function.

Let us remark here that the interval $a \leq x \leq b$ represents an infinite range because the random variable x assumes all possible values in that interval. Furthermore, a and b could assume values approaching negative and positive infinity respectively.

The following examples illustrate the notion of probability functions of the continuous type.

*Some authors associate the term *probability function* (p.f.) only with discrete random variables (r.v.) and refer by the term *probability density function* (p.d.f.) to those with continuous random variables.

Example 4.1.1 Given $f(x) = x - x^2 \quad 0 \le x \le 1$
$= 0$ otherwise

(a) Graph the given function.
(b) Is $f(x)$ a p.f.?
(c) If not, find the constant which will make it a p.f.
Solution:

(a)

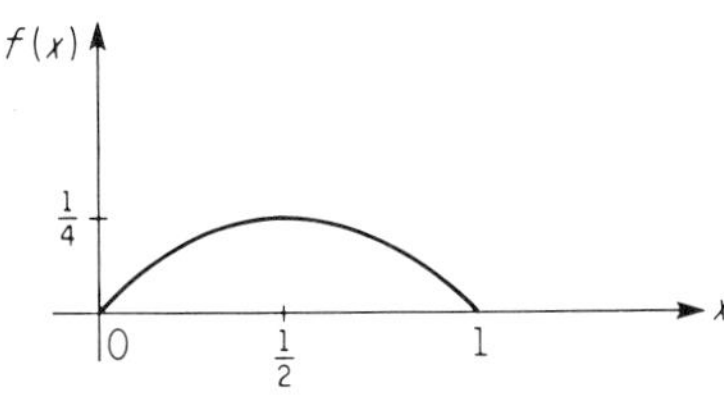

FIG. 4.3.

(b) $$\int_0^1 (x - x^2)\,dx = 1/6$$

hence $f(x)$ is not a probability function.

(c) In order to have a probability function, we solve

$$\int_0^1 k(x - x^2)\,dx = 1$$

for k and from part (b) obtain $k = 6$.
Therefore,

$$\begin{aligned} f(x) &= 6(x - x^2) \qquad 0 < x < 1 \\ &= 0 \qquad \text{otherwise} \end{aligned}$$

does represent an example of a probability function of the continuous type.

Example 4.1.2 Given

$$\begin{aligned} f(x) &= \frac{x}{2} \qquad 0 < x < 1 \\ &= \frac{1}{2} \qquad 1 < x < 2 \\ &= \frac{3 - x}{2} \qquad 2 < x < 3 \\ &= 0 \qquad \text{otherwise} \end{aligned}$$

(a) Graph the function.
(b) Verify that $f(x)$ is a probability function.

Solution:

(a)

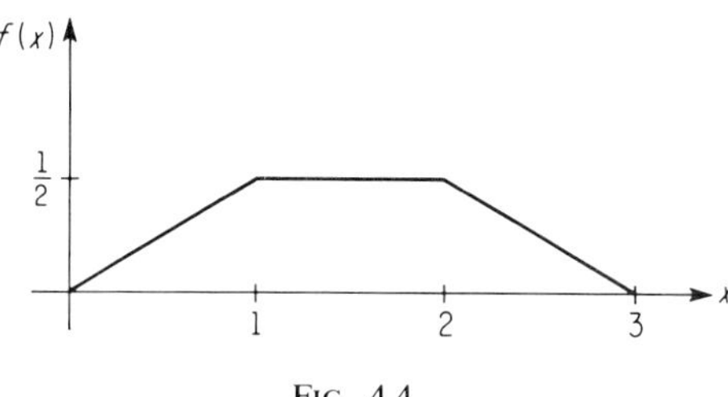

FIG. 4.4.

(b) To show $f(x)$ is a probability function, one verifies that Eqs. (4.1.1) and (4.1.2) are satisfied. Clearly $f(x) > 0$ in the specified intervals of x, and

$$\int_0^1 \frac{x}{2}\,dx + \int_1^2 \frac{1}{2}\,dx + \int_2^3 \frac{(3-x)}{2}\,dx = 1$$

Hence the given function is a probability function.

As in the discrete case, in the continuous case also, there is associated with each probability function $f(x)$ its distribution function $F(x)$ defined by the expression

$$F(x) = \int_{-\infty}^{x} f(t)\,dt \tag{4.1.3}$$

The distribution function possesses the following properties

1. $F(-\infty) = 0$ (4.1.4)
2. $F(\infty) = 1$ (4.1.5)
3. $\dfrac{d}{dx} F(x) = f(x)$ almost everywhere (4.1.6)
4. $F(x > x_0) = 1 - F(x_0)$ (4.1.7)
5. $F(b) - F(a) = P[a < x < b]$ (4.1.8)

Properties of Eqs. (4.1.7) and (4.1.8) are analogous to Eqs. (2.3.2) and (2.3.3) respectively. Furthermore, because $\int_a^a f(x)\,dx = 0$,

$$P[a \leq x \leq b] = P[a < x < b] \tag{4.1.9}$$

We illustrate with an example the concepts presented above.

Example 4.1.3 Given $f(x) = \frac{3}{2}(x-1)^2 \qquad 0 < x < 2$

$= 0 \qquad$ otherwise

(a) Graph this function.
(b) Find the distribution function.
(c) Graph the distribution function.
(d) Find $P[.5 < x < 1]$.

Solution:

(a)

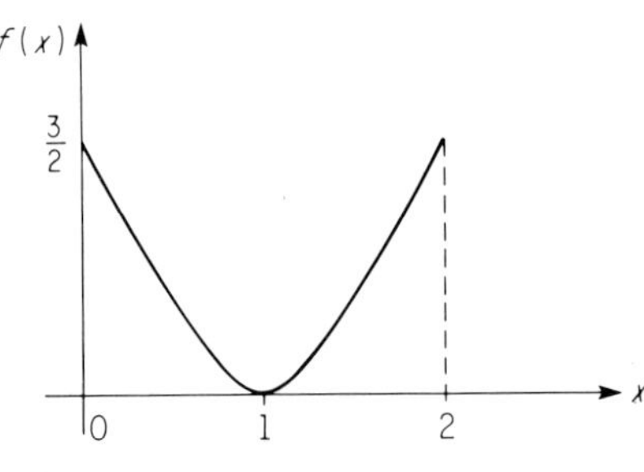

FIG. 4.5.

(b) $\int_0^x \frac{3}{2}(t-1)^2\,dt = \frac{x}{2}[x^2 - 3x + 3] = F(x)$

(c)

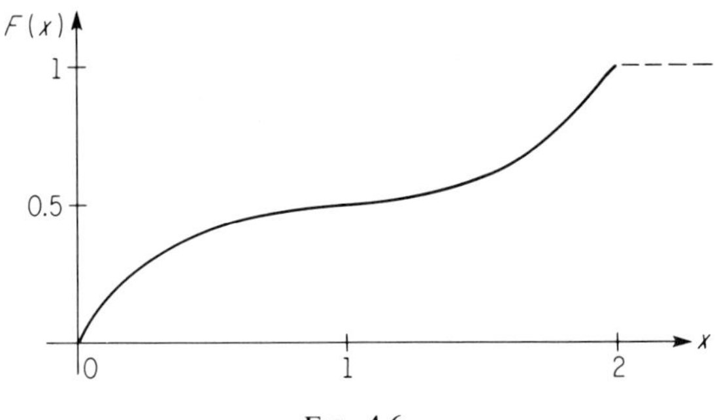

FIG. 4.6.

(d) $P[0.5 < x < 1] = \int_{.5}^{1} f(x)\,dx = F(1) - F(.5) = \frac{1}{16}$

PROBLEMS

In Probs. 4.1.1 to 4.1.6, (a) graph the given function, (b) verify that it represents a probability function, (c) find the corresponding distribution function, (d) graph the distribution function, and (e) find the indicated probabilities.

4.1.1 $f(x) = \frac{1}{b - a} \quad a \le x \le b$
$= 0 \quad \text{otherwise}$

is called the *uniform* or the *rectangular probability function.*
$P\left[\frac{a + b}{2} < x < b\right] =$ ________.

4.1.2 $f(x) = x \quad 0 \le x \le 1$
$= 2 - x \quad 1 < x \le 2$
$= 0 \quad \text{otherwise}$

is called the *triangular probability function.*
$P[x < 1.5] =$ ________.

4.1.3 $f(x) = 3x^2 \quad 0 \le x \le 1$
$= 0 \quad \text{otherwise}$

$P[x > .5] =$ ________.

4.1.4 $f(x) = \frac{1}{2} \sin x \quad 0 \le x \le \pi$
$= 0 \quad \text{otherwise}$

is called the *sine probability function.*
$P\left[\frac{\pi}{4} < x < \frac{\pi}{2}\right] =$ ________.

4.1.5 $f(x) = xe^{-x} \quad x \ge 0$
$= 0 \quad \text{otherwise}$

$P[1 < x < 3] =$ ________.

4.1.6 $f(x) = \sqrt{\frac{2}{\pi} - x^2} \quad x \le \sqrt{\frac{2}{\pi}}$
$= 0 \quad \text{otherwise}$

is called the *semicircular probability function.*
$P\left[-\frac{1}{2} < x < \frac{1}{4}\right] =$ ________.

4.2 EXPECTATION AND RELATED TOPICS

In Sec. 3.1 the expectation of $g(x)$ was defined as $\sum_{\text{all } x} g(x)f(x)$, where $g(x)$ is any function of the random variable x and $f(x)$ a specific p.f. associated with x. In Sec. 4.1 the probability function of a continuous random variable was defined and the similarities between the two types, discrete and continuous, were discussed. It was indicated that the major difference between the two cases was in the use of integration instead of summation for the continuous case. In the definition of expectation of $g(x)$ for the continuous case the same similarity holds also. That is,

$$E[g(x)] = \int_{-\infty}^{\infty} g(x)f(x)\,dx \tag{4.2.1}$$

provided x represents a continuous random variable. Therefore, all the rules of expectation given in Sec. 3.1 for the discrete case hold true for the continuous case as well. (See Prob. 4.2.1.)

Example 4.2.1 Find $E[x]$ and $E[x^2]$ for the p.f. given in Example 4.1.1.

Solution: From part (c) of Example 4.1.1 we note

$$f(x) = 6(x - x^2) \qquad 0 \le x \le 1$$
$$= 0 \quad \text{otherwise}$$

to represent the required p.f. Hence

$$E[x] = 6 \int_0^1 x(x - x^2)\,dx = \frac{1}{2}$$

$$E[x^2] = 6 \int_0^1 x^2(x - x^2)\,dx = \frac{3}{10}$$

Next let us consider moments associated with a continuous p.f. In Sec. 3.2 moments about any point a were defined as special cases of expectation. Since the rules of expectation hold true for the continuous case, we use the same symbolism to define moments for this case also. That is,

$$m_{k,a} = \int_{-\infty}^{\infty} (x - a)^k f(x)\,dx \tag{4.2.2}$$

represents the kth moment about the point a, and

$$m_k = \int_{-\infty}^{\infty} x^k f(x)\,dx \tag{4.2.3}$$

defines the kth moment about the origin.*

The moments about the mean are defined similarly:

$$\mu_k = \int_{-\infty}^{\infty} (x - \mu)^k f(x)\,dx = \int_{-\infty}^{\infty} (x - E[x])^k f(x)\,dx \tag{4.2.4}$$

represents the kth moment about the mean.

The relationships between moments about the mean and the moments about the origin verified for the discrete case hold true for the continuous type also. The proofs are identical except summation signs are replaced with integral signs. For instance, the variance, $\sigma^2 = E[x^2] - (E[x])^2$, [refer to Eq. (3.2.14)] holds true for any probability function of the continuous type.

Example 4.2.2 Using the identity given by Eq. (3.2.14), find the variance for the probability function of Example 4.2.1.

Solution: Since $E[x^2]$ and $E[x]$ are calculated in Example 4.2.1, one

*See footnote, p. 78.

simply gets

$$\sigma^2 = \frac{3}{10} - \left(\frac{1}{2}\right)^2 = \frac{1}{20}$$

The moment-generation function for the continuous case is defined as

$$\text{m.g.f.} = M_x(t) = E[e^{tx}] = \int_{-\infty}^{\infty} e^{tx} f(x)\,dx \qquad (4.2.5)$$

and the kth moment about the origin is obtained as in the discrete case by the expression

$$m_k = \frac{d^k}{dt^k} M_x(0) \qquad (4.2.6)$$

The moment-generating function, if it exists, is a powerful tool in the continuous case also. The student is requested to review Sec. 3.4 at this juncture.

Example 4.2.3 Given that $f(x) = e^{-x} \quad x > 0$
$= 0 \quad$ otherwise

is a p.f. Find (a) the moment-generating function (b) the mean and (c) the variance for the given p.f.

Solution:

(a)

$$M_x(t) = \int_0^{\infty} e^{tx} e^{-x}\,dx$$

$$= \int_0^{\infty} e^{-x(1-t)}\,dx$$

$$= \frac{1}{1-t}$$

(b) $\frac{d}{dt} M_x(t) = (1 - t)^{-2}$, which when evaluated for $t = 0$, yields $m_1 = 1$.

(c) $\frac{d^2}{dt^2} M_x(t) = 2\,(1 - t)^{-3}$, which for $t = 0$ becomes $m_2 = 2$. Therefore, $\sigma^2 = 2 - 1^2 = 1$. Let us remark that for this example, $m_k = k!$.

Often it is necessary to translate the origin of a p.f. in order to describe a random variable. For instance,

$$f(x) = e^{-x+3} \quad x > 3$$
$$= 0 \quad \text{otherwise}$$

represents the probability function given in Example 4.2.3 except that this function is shifted 3 units. If the moment-generating function of either probability function is given, can we derive the moment-generating function of the other from the known? To answer this question, we show that

$$M_{x+k}(t) = e^{kt} M_x(t)$$

By definition,

$$\begin{aligned} M_{x+k}(t) &= E[e^{(x+k)t}] \\ &= E[e^{xt} \cdot e^{kt}] \\ &= e^{kt} E[e^{xt}] \qquad \text{since } e^{kt} \text{ is a constant} \\ &= e^{kt} M_x(t) \end{aligned} \tag{4.2.7}$$

This expression states that the moment-generating function of the k unit shifted function is equal to e^{kt} multiplied by the moment-generating function of the original. Let us illustrate this by an example.

Example 4.2.4 Using the result of Example 4.2.3, find the moment-generating function of

$$\begin{aligned} f(x) &= e^{-x+3} \qquad x > 3 \\ &= 0 \qquad \text{otherwise} \end{aligned}$$

and using it to show that the mean is 4.

Solution: The moment-generating function of $f(x) = e^{-x}$ $x > 0$ is $(1 - t)^{-1}$, as shown in Example 4.2.3. Hence the required moment-generating function is

$$M_{x+3}(t) = e^{3t}(1 - t)^{-1}$$

Now

$$\frac{d}{dt} M_{x+3}(t) = e^{3t}(1 - t)^{-2} + 3e^{3t}(1 - t)^{-1}$$

which when evaluated for $t = 0$ yields 4.

There is considerable theory associated with the moment-generating function, and some interesting and very useful theorems have been shown. Such theoretical considerations are beyond the scope of this introductory volume. We shall, however, merely state one theorem, the uniqueness theorem, which will be needed in subsequent chapters.

Theorem 4.2.1 If a moment-generating function $M_x(t)$ exists in an interval containing the origin $t = 0$, then there is only one p.f. having this moment-generating function.

This is a very useful theorem. For instance, in Example 4.2.4 it was shown that the moment-generating function for the p.f. $f(x) = e^{-x+3}$ was

$$M_x(t) = \frac{e^{3t}}{1 - t}$$

The uniqueness theorem states that $f(x) = e^{-x+3}$ is the *only* p.f. whose moment-generating function is $\dfrac{e^{3t}}{1 - t}$.

PROBLEMS

4.2.1 Assuming that $f(x)$ is a p.f. of the continuous type, verify

(a) $E[k] = k$

(b) $E[kg(x)] = kE[g(x)]$

(c) $E\left[\sum_{i=1}^{n} g_i(x)\right] = \sum_{i=1}^{n} E[g_i(x)]$

4.2.2 Refer to Prob. 4.1.1. Find (a) $E[x]$, (b) $E[x^2]$ and (c) Using the result of parts (a) and (b) compute the variance of the uniform p.f.

4.2.3 Refer to Prob. 4.1.4. Find the same quantities requested in Prob. 4.2.2, for the sine p.f.

4.2.4 Given that
$$f(x) = ke^{-kx} \qquad x > 0$$
$$= 0 \qquad \text{otherwise}$$
is a p.f. Find the mean and the variance of the given p.f.

4.2.5 Given
$$f(x) = xe^{-x} \qquad x > 0$$
$$= 0 \qquad \text{otherwise}$$
Find the mean of this p.f. (*Hint:* Use the results of Prob. 4.2.4 assuming $k = 1$.)

4.2.6 (a) Find the moment-generating function of the p.f. given in Prob. 4.2.5 above.

(b) Using this m.g.f. find the value of $E[x]$ and $E[x^2]$.

(c) Evaluate $\sigma^2 = E[x^2] - (E[x])^2$ by using the values obtained in part (b).

4.2.7 The moment-generating function of a certain p.f. is $(1 - 2t)^{-n/2}$. Find the mean and the variance of this p.f.

4.2.8 Given that the moment-generating function of a p.f. is $e^{t^2/2}$. (a) Find the mean and the variance of the probability function. (b) What is the m.g.f. of the associated p.f. $f(y)$ if $y = x + 7$? (c) Using the result of part (b), find the mean and the variance of the p.f. $f(y)$. (d) Compare the results of parts (a) and (c).

SOME COMMON PROBABILITY FUNCTIONS OF THE CONTINUOUS TYPE

As in the case of noncontinuous probability functions, in the case of continuous also one could construct many probability models which satisfy the conditions expressed in Eqs. (4.1.1) and (4.1.2). However, there are a few probability functions of the continuous type which are of greater significance and of more usage than others, just as the binomial, the hypergeometric, and the Poisson probability functions were among the discrete type. In subsequent sections of this chapter we shall present the general gamma probability function and some of its special cases, the normal or the Gaussian probability model, and the log-normal probability functions. Also in this chapter we shall discuss some useful topics of calculus and their applications to probability functions.

4.3 THE GAMMA PROBABILITY FUNCTION

The function expressed by

$$f(x:a,n) = \frac{1}{\Gamma(n)a^n} x^{n-1}e^{-x/a} \qquad x > 0, a > 0, n > 0$$

$$= 0 \quad \text{otherwise} \tag{4.3.1}$$

is called the *gamma probability function*, where $\Gamma(n)$ is a constant and is defined as

$$\int_0^\infty x^{n-1}e^{-x}\,dx \tag{4.3.2}$$

Prior to showing that Eq. (4.3.1) satisfies the two conditions for a probability function, let us verify that

$$\Gamma(n) = (n-1)\Gamma(n-1) = (n-1)! \tag{4.3.3}$$

To show this, one integrates Eq. (4.3.2) by parts. Letting

$$x^{n-1} = u \quad \text{and} \quad e^{-x}\,dx = dv$$

we get

$$(n-1)x^{n-2}\,dx = du \quad \text{and} \quad v = -e^{-x}$$

and hence

$$\int_0^\infty x^{n-1}e^{-x}\,dx = \lim_{A\to\infty} - \frac{x^{n-1}}{e^x}\bigg|_0^A + \int_0^\infty (n-1)x^{n-2}e^{-x}\,dx$$

Now, as $A \to \infty$ the first term in the right hand by l'Hospital's rule approaches zero and hence

$$\Gamma(n) = \int_0^\infty x^{n-1}e^{-x}\,dx = (n-1)\int_0^\infty x^{n-2}e^{-x}\,dx = (n-1)\Gamma(n-1)$$

Using successive integrations by parts, we obtain

$$\Gamma(n) = (n-1)(n-2)(n-3)\cdots(n-n) = (n-1)!$$

which was to be shown. Let us note that $\Gamma(1) = 0! = \int_0^\infty e^{-x}\,dx = 1$, and this is another reason for defining $0! = 1$. Let us also remark that n can assume any positive value. For example,

$$\Gamma(1.57) = (1.57 - 1)\Gamma(1.57 - 1) = (.57)\Gamma(.57)$$

and

$$\Gamma(2.5) = (1.5)(.5)\Gamma(.5) = .75\,\Gamma(.5)$$

It can be shown that $\Gamma(.5) = \sqrt{\pi}$ [See Prob. 4.3.2] and hence $\Gamma(2.5) = .75\sqrt{\pi}$.

To show that Eq. (4.3.1) satisfies the conditions of a p.f. we proceed as follows. First, $f(x) \geq 0$ because $x > 0$, $a > 0$ by assumption and

$\Gamma(n) > 0$ for $n > 0$. Second, it must be shown that

$$\int_0^\infty x^{n-1} e^{-x/a} dx = \Gamma(n) a^n$$

Making the substitution (transformation) $x = ay$, we obtain

$$\int_0^\infty x^{n-1} e^{-x/a} dx = \int_0^\infty (ay)^{n-1} e^{-y} a\, dy$$

$$= \int_0^\infty a^n y^{n-1} e^{-y} dy$$

$$= a^n \Gamma(n)$$

and therefore

$$\frac{1}{a^n \Gamma(n)} \int_0^\infty x^{n-1} e^{-x/a} dx = 1$$

which is the desired result.

In order to find the mean and the variance of the gamma p.f. we derive its moment-generating function.

$$M_x(t) = E[e^{tx}] = \int_0^\infty \frac{e^{tx}}{\Gamma(n) a^n} x^{n-1} e^{-x/a} dx$$

$$= \frac{1}{\Gamma(n) a^n} \int_0^\infty x^{n-1} e^{-x(1/a - t)} dx$$

Setting $k = \dfrac{1 - at}{a}$ and making the substitution $kx = y$, we get

$$M_x(t) = \frac{1}{\Gamma(n) a^n} \int_0^\infty \frac{y^{n-1}}{k^n} e^{-y} dy$$

$$= \frac{1}{(ak)^n} = (1 - at)^{-n} \tag{4.3.4}$$

Now, to obtain the mean of the gamma p.f. we evaluate the first derivative of $M_x(t)$ at $t = 0$; that is,

$$m_1 = \frac{d}{dt} M_x(0) = na(1 - at)^{-n-1}\Big|_{t=0} = na \tag{4.3.5}$$

To obtain the variance, one needs to calculate m_2 which equals

$$\frac{d^2}{dt^2} M_x(0) = n(n + 1) a^2$$

and therefore, $\sigma^2 = m_2 - (m_1)^2$ yields

$$\sigma^2 = a^2 n \tag{4.3.6}$$

Let us consider a specific gamma p.f.

Example 4.3.1 Sketch the gamma p.f. for the parameters $a = 1$ and $n = 2$. Find its mean and its variance.

Solution: Substituting $n = 2$ and $a = 1$ in Eq. (4.3.1) we obtain

$$\begin{aligned} f(x\colon 1, 2) &= x e^{-x} \qquad x > 0 \\ &= 0 \qquad \text{otherwise} \end{aligned}$$

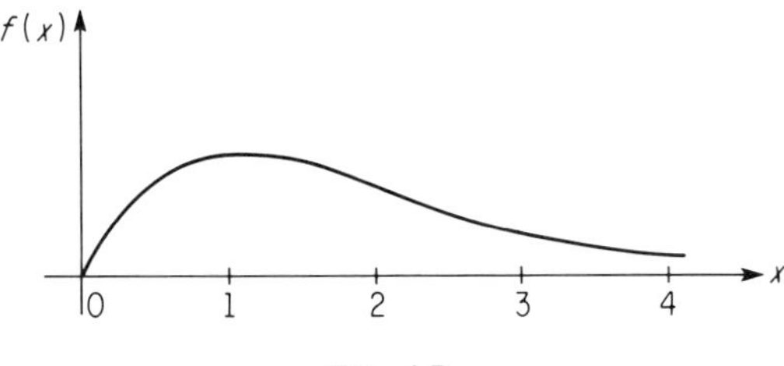

FIG. 4.7.

Using Eqs. (4.3.5) and (4.3.6) we get mean equal to 2 and variance equal to 2. The student is requested to refer and review Probs. 4.1.5 and 4.2.6 at this juncture.

A special case of the gamma p.f. plays an important role in many applications and is worth discussing separately. Consider the case where $n = 1$. If this substitution is made in Eq. (4.3.1) it reduces to

$$\begin{aligned} f(x{:}a,1) &= \frac{1}{\Gamma(1)\, a^1} e^{-x/a} = \frac{1}{a} e^{-x/a} \qquad x > 0, a > 0 \\ &= 0 \qquad \text{otherwise} \end{aligned} \tag{4.3.7}$$

and is called the *exponential p.f.* We shall refer to this special case as $f(x{:}a)$. The moment-generating function of the exponential p.f. is obtained from Eq. (4.3.4) and is

$$M_x(t) = (1 - at)^{-1} \tag{4.3.8}$$

Also, from Eqs. (4.3.5) and (4.3.6) we obtain the mean and the variance of this p.f. to be

$$\mu = a \tag{4.3.9}$$

and

$$\sigma^2 = a^2 \tag{4.3.10}$$

It is interesting to note that the mean and the standard deviation for this p.f. are equal to the single parameter a.

Example 4.3.2 The life of a certain make of light bulb obeys the exponential p.f. If the mean life is claimed to be 1,000 hours, what is the probability of the lamp lasting over 1,000 hours?

Solution: Equation (4.3.7) represents the required p.f. and from Eq.

(4.3.9) we obtain the value of the single parameter a. Therefore,

$$f(x\colon 1000) = \frac{1}{1000} e^{-x/1000} \qquad x > 0$$
$$= 0 \quad \text{otherwise}$$

is the desired p.f. Hence

$$\begin{aligned} P[x > 1000] &= \int_{1000}^{\infty} \frac{1}{1000} e^{-x/1000}\, dx \\ &= 1 - \int_{0}^{1000} \frac{1}{1000} e^{-x/1000}\, dx \\ &= 1 + [e^{-x/1000}] \Big|_{0}^{1000} \\ &= e^{-1} = .368 \end{aligned}$$

PROBLEMS

4.3.1 Using Eq. (4.3.3), evaluate (a) $\Gamma(4)$ (b) $\Gamma(6)$.

4.3.2 Show that $\Gamma\left(\frac{1}{2}\right) = \sqrt{\pi}$. [*Hint:* Use the transformation $x = y^2$ in Eq. (4.3.2). Square both sides of the resulting equation and transform the double integral into polar coordinates.]

4.3.3 Using Eq. (4.3.3) and the result of Prob. 4.3.2, evaluate

$$\text{(a) } \Gamma\left(\frac{3}{2}\right), \quad \text{(b) } \Gamma\left(\frac{7}{2}\right).$$

4.3.4 A random variable is distributed according to the gamma p.f. with $a = 2$ and $n = 3$. Find the probability that (a) $x > 1$. (b) $1 < x < 3$. (c) Find the *mode* of this p.f., that is, value of x where the p.f. attains its maximum.

4.3.5 A life insurance company assumes that the life expectancies for American males obey the exponential p.f., with mean 40 years. What proportion of American males are expected to be living between ages (a) 30 to 40? (b) 40 to 50? (c) over 60?

4.3.6 Repeat Prob. 4.3.5 for the life expectancies of American females, if their mean age is 50 years?

4.3.7 Brand A batteries are advertised to last at least 30 months or money back. The manufacturer knows that the life duration of these batteries obeys the exponential p.f. with mean 24 months. What proportion of the batteries does the manufacturer expect to replace?

4.3.8 The daily water consumption in a certain city obeys the gamma p.f. with parameters $a = 4$ and $n = 2$. Assuming the units are in millions of cubic feet, what is the probability that in a given day the water consumption will be between 7 and 15 million cubic feet?

4.4 TRANSFORMATION OF RANDOM VARIABLES

In many applications of statistical theory it is necessary to transform the given random variable in order to facilitate the solution or to be able to solve a problem. For instance, the transformation of the type $y = cx + k$ was used in Sec. 4.2 in conjunction with the moment-generating function. In subsequent chapters the derivation of certain important formulas will require other types of change in variable and hence a brief discussion of this topic will be presented here.

Let us introduce the notion of transformation with an example. Suppose the random variable x obeys the binomial probability function $f(x:n,p)$. What is the probability function of the random variable $y = x + 5$? Recalling that

$$f(x:n,p) = \binom{n}{x} p^x q^{n-x} \qquad x = 0, 1, \ldots, n$$

and substituting $x = y - 5$, we get

$$f(y - 5:n,p) = \binom{n}{y-5} p^{y-5} q^{n-y+5} \qquad y = 5, 6, 7, \ldots, n + 5$$

which represents the required probability function. Next suppose that the transformation is $y = x^2 + 2$ where the random variable x again obeys the binomial p.f. How could we describe the p.f. of this new transformed random variable? Clearly, $x = \pm\sqrt{y - 2}$. However, since $x \geq 0$ we need to consider only $\sqrt{y - 2} \geq 0$ and hence

$$f(x:n,p) = f(\sqrt{y-2}:n,p) = \binom{n}{\sqrt{y-2}} p^{\sqrt{y-2}} q^{n-\sqrt{y-2}}$$
$$y = 2, 3, 6, \ldots, n^2 + 2$$

is the p.f. associated with the transformed random variable. In general, the transformations of a random variable possessing a noncontinuous p.f. is straightforward, as shown in the above illustrations.

The approach to the problem of transformations is a little different if the random variable is continuous. The transformation $y = g(x)$ should possess an isomorphism, a one-to-one relationship, which implies that $g(x)$ is either a monotone increasing or a monotone decreasing function for all values of x in the defined range. Consequently, the inverse function $x = h(y)$ is unique and increases or decreases with x. That is, $a_1 \leq x \leq b_1$ and $g_1(a_1) \leq y \leq g_1(b_1)$ represent the same events for a monotone increasing function, and similarly $a_2 \leq x \leq b_2$ and $g_2(b_2) \leq y \leq g_2(a_2)$ represent the same events for a monotone decreasing function. These relationships are shown in Figs. 4.8*a* and 4.8*b*.

Consequently, for a monotone increasing function $g_1(x)$

$$\Pr[a_1 \leq x \leq b_1] = \Pr[g_1(a_1) \leq y \leq g_1(b_1)] \tag{4.4.1}$$

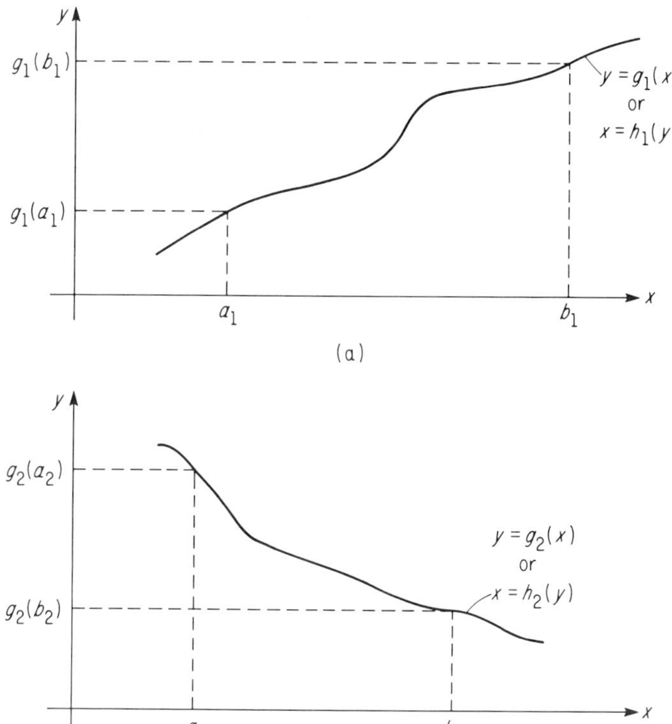

FIG. 4.8. (a) $g(x)$ a monotone increasing function of x. (b) $g(x)$ a monotone decreasing function of x.

which is to say

$$\int_{a_1}^{b_1} f_1(x)\, dx = \int_{g_1(a_1)}^{g_1(b_1)} j_1(y)\, dy \tag{4.4.2}$$

where $f_1(x)$ and $j_1(y)$ are the probability functions of the random variables x and y respectively. Similarly, for a monotone decreasing function $g_2(x)$,

$$\Pr[a_2 \leq x \leq b_2] = \Pr[g_2(b_2) \leq y \leq g_2(a_2)] \tag{4.4.3}$$

which is equivalent to

$$\int_{a_2}^{b_2} f_2(x)\, dx = \int_{g_2(b_2)}^{g_2(a_2)} j_2(y)\, dy \tag{4.4.4}$$

where $f_2(x)$ and $j_2(y)$ are the probability functions of x and y respectively.

We wish to find explicit expressions for $j_1(y)$ and $j_2(y)$. Since $y = g_1(x)$ and its inverse $x = h_1(y)$ are assumed to be monotone increasing functions, Eq. (4.4.2) can be expressed as

$$\int_{a_1}^{b_1} f_1(x)\,dx = \int_{g_1(a_1)}^{g_1(b_1)} f[h_1(y)] \frac{d}{dy} [h_1(y)]\,dy \tag{4.4.5}$$

Similarly, Eq. (4.4.4) reduces to

$$\int_{a_2}^{b_2} f_2(x)\,dx = \int_{g_2(b_2)}^{g_2(a_2)} - f[h_2(y)] \frac{d}{dy} [h_2(y)]\,dy \tag{4.4.6}$$

and therefore the integrands in the last two expressions represent the transformed probability functions $j_1(y)$ and $j_2(y)$. The following examples illustrate the concepts presented in this section.

Example 4.4.1 Given the triangular p.f.

$$f(x) = \frac{x}{2} \qquad 0 \le x \le 2$$
$$= 0 \qquad \text{otherwise}$$

Find the p.f. associated with the random variable $y = x^2 + 3$.

Solution: In this example $g(x) = x^2 + 3$, which is a monotone increasing function in the domain of defined x. Hence $x = \sqrt{y-3} = h_1(y)$ is the inverse function which also is an increasing function. Clearly, $g(0) = 3$ and $g(2) = 7$, and therefore

$$\int_0^2 \frac{x}{2}\,dx = \int_3^7 \left(\frac{\sqrt{y-3}}{2}\right) \frac{1}{2} \frac{1}{\sqrt{y-3}}\,dy$$

$$= \int_3^7 \frac{1}{4}\,dy = 1$$

That is, the given p.f. has been transformed to the uniform p.f.

$$f(y) = \frac{1}{4} \qquad 3 \le y \le 7$$
$$= 0 \qquad \text{otherwise}$$

Example 4.4.2 Given the exponential p.f.

$$f(x) = k\,e^{-kx} \qquad x > 0$$
$$= 0 \qquad \text{otherwise}$$

Find the probability function of the random variable $y = -2x$.

Solution: In this illustration $g(x) = -2x$ which is a monotone decreasing function. Its inverse, $x = -\dfrac{y}{2} = h(y)$ is a monotone decreasing function also. Therefore the required p.f. is

$$f(y) = \frac{k}{2}\,e^{ky/2} \qquad y < 0$$
$$= 0 \qquad \text{otherwise}$$

Clearly,

$$\int_{-\infty}^{0} f(y)\,dy = 1$$

The probability functions of the random variables x and y are sketched in Figs. 4.9a and 4.9b.

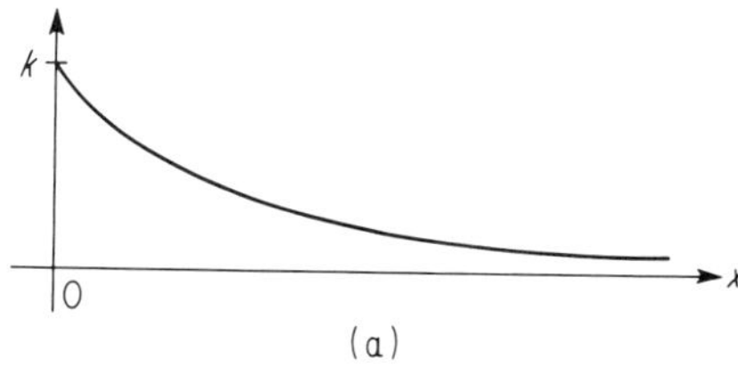

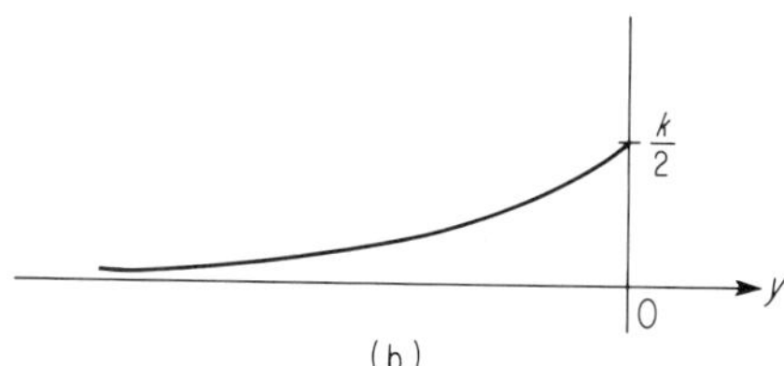

FIG. 4.9.

PROBLEMS

4.4.1 Given the triangular p.f. of Prob. 4.1.2
(a) Find the p.f. of the random variable $y = x^2$.
(b) Graph the p.f. of the random variable y.
(c) Find the mean and the variance of y's.
4.4.2 Given the uniform p.f. $f(x) = 1/\pi \quad 0 \le x \le \pi$
(a) Find the p.f. of the random variable $y = \cos x$.
(b) Graph the p.f. of the y's.
4.4.3 Given

$$f(x) = \frac{6x}{(1+x)^4} \qquad x > 0$$

represents a p.f. Find the p.f. of the random variable $y = 1/x$.
4.4.4 Given

$$f(x) = \frac{3}{4}\,x(2-x)^2 \qquad 0 \le x \le 2$$

represents a p.f. Find the p.f. of the random variable $y = 2x$.

4.4.5 Given

$$f(x) = \frac{1}{\sqrt{2\pi}} e^{-x^2/2} \qquad x \text{ real}$$

is a p.f. Find the p.f. of the random variable $y = x^2$.

4.4.6 Given

$$f(x) = \frac{1}{\sqrt{2\pi}} x^{-2} e^{-1/2x^2} \qquad x \text{ real}$$

(a) Graph $f(x)$.

(b) Find the p.f. of the random variable $y = 1/x$.

Compare the result of part (b) with $f(x)$ given in Prob. 4.4.5.

4.4.7 Given the exponential p.f. $f(x) = e^{-x} \quad x > 0$,

(a) Find the p.f. of the random variable $y = \sqrt[3]{x}$.

(b) Graph the p.f. of the y's.

(c) Find the *median* of the y's. That is, find the value of y which divides the area under p.f. curve into two equal parts.

4.5 THE NORMAL PROBABILITY FUNCTION

In this section we shall present a probability function which has occupied the attention of many mathematicians over a period of nearly two centuries. The reason for its being called the "normal" probability function is due to the fact that several attempts have been made to establish this probability function as the basis for all continuous probability functions. As early as 1733 Abraham De Moivre (or Demoivre) had given a derivation of it as the limiting form of the binomial probability function. Laplace is also attributed for deriving it and years later Gauss also was given credit for its derivation. However, in literature the normal p.f. is accepted to be synonymous with the Gaussian probability function. Even though the normal p.f. is not the basis for all continuous probability functions, it is nevertheless one of the most important probability functions in theoretical considerations. Here we shall give its definition and verify that it satisfies the conditions of a probability function.

Consider the gamma p.f. with parameter $a = 1$. That is,

$$f(x:1,n) = \frac{1}{\Gamma(n)} x^{n-1} e^{-x} \qquad x > 0 \quad n > 0$$
$$= 0 \quad \text{otherwise}$$

We wish to find the p.f. associated with the transformed random variable $z = \sqrt{2x}$. By the method discussed in Sec. 4.4, we note $z = g(x) = \sqrt{2x}$, $x = z^2/2 = h(z)$, $d/dz\,[h(z)] = z$ and therefore

$$f(z:1,n) = \frac{1}{\Gamma(n)} \left(\frac{z^2}{2}\right)^{n-1} e^{-[z^2/2]} z$$

$$= \frac{1}{\Gamma(n)} \frac{z^{2n-1}}{2^{n-1}} e^{-z^2/2} \qquad z > 0, n > 0 \tag{4.5.1}$$
$$= 0 \qquad \text{otherwise}$$

is the probability function of the random variable z. Next, consider the special case where $n = 1/2$. That is,

$$f\left(z : 1, \frac{1}{2}\right) = \frac{\sqrt{2}}{\Gamma\left(\frac{1}{2}\right)} e^{-z^2/2} \qquad z > 0 \tag{4.5.2}$$

Because this last expression represents a p.f. and is symmetric about the line $z = 0$, we can deduce that

$$\int_{-\infty}^{\infty} \frac{\sqrt{2}}{\Gamma\left(\frac{1}{2}\right)} e^{-z^2/2} dz = 2 \tag{4.5.3}$$

and therefore

$$f(z) = \frac{1}{\sqrt{2}\,\Gamma\left(\frac{1}{2}\right)} e^{-z^2/2} \qquad z \text{ real} \tag{4.5.4}$$

also represents a probability function. In Prob. 4.3.2 it was shown that $\Gamma(1/2) = \sqrt{\pi}$ and hence Eq. (4.5.4) becomes

$$f(z) = \frac{1}{\sqrt{2\pi}} e^{-z^2/2} \qquad z \text{ real} \tag{4.5.5}$$

which is called the *unit normal* or the *standard normal* probability function. The graph of this function is a bell-shaped curve and is shown in Fig. 4.10. It can be shown that the mean and the variance of the unit normal are 0 and 1 respectively. (Refer to Prob. 4.5.2.)

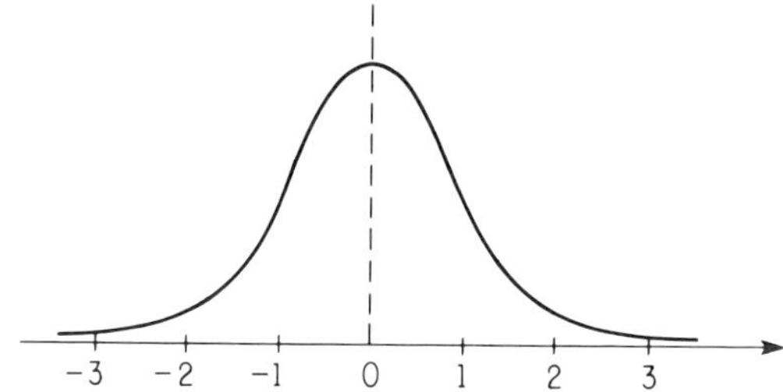

FIG. 4.10. The unit normal p.f.

Table IV(a) in the Appendix represents the tabulated values for

$$\int_0^z \frac{1}{\sqrt{2\pi}} e^{-x^2/2} dx$$

with increments of .01 of z. For example, for $z = 1.00$ we obtain .34134 and for $z = 2.00$ we read .47725.

The distribution function of the unit normal is by definition

$$F(z) = \int_{-\infty}^{z} \frac{1}{\sqrt{2\pi}} e^{-x^2/2}\,dx$$

$$= \int_{-\infty}^{0} \frac{1}{\sqrt{2\pi}} e^{-x^2/2}\,dx + \int_{0}^{z} \frac{1}{\sqrt{2\pi}} e^{-x^2/2}\,dx$$

$$= .50000 + \int_{0}^{z} \frac{1}{\sqrt{2\pi}} e^{-x^2/2}\,dx \tag{4.5.6}$$

and hence it can be obtained from Table IV(a). For instance, $F(1) = .84134$ and $F(2) = .97725$.

Table IV(b) in the Appendix represents the tabulated values for

$$\int_{z}^{\infty} \frac{1}{\sqrt{2\pi}} e^{-x^2/2}\,dx, \quad z > 0$$

which clearly is equivalent to

$$.50000 - \int_{0}^{z} \frac{1}{\sqrt{2\pi}} e^{-x^2/2}\,dx$$

Because the unit normal p.f. is symmetric about the line $z = 0$, the relationship

$$\int_{-\infty}^{-z} \frac{1}{\sqrt{2\pi}} e^{-x^2/2}\,dx = \int_{z}^{\infty} \frac{1}{\sqrt{2\pi}} e^{-x^2/2}\,dx \tag{4.5.7}$$

holds true. Therefore, Table IV(b) represents values of the distribution function for $z < 0$. Furthermore, for the unit normal p.f. Equation (4.1.8) becomes

$$P[a < z < b] = F(b) - F(a) \tag{4.5.8}$$

Example 4.5.1 If $f(x)$ is the unit normal p.f., use Tables IV(a) and IV(b) to evaluate (a) $P[x < 0.48]$ (b) $P[x > 2.31]$ (c) $P[1.1 < x < 1.4]$ (d) $P[-1.0 < x < 2.0]$

Solution: (a) In this situation the required probability corresponds to $F(.48)$ which according to Eq. (4.5.6) and Table IV(a) yields the value $.50000 + .18439 = .68439$.

(b) In this case we get the required value directly from Table IV(b). The value is .01045.

(c) Here we use Eq. (4.5.8), which yields $.91924 - .86433 = .05491$. This result could have been obtained from Table IV(a) without calculating $F(1.4)$ and $F(1.1)$. That is, the required value is $.41924 - .36433 = .05491$.

(d) In this case we need $F(2) - F(-1)$. We obtain $F(2)$ from Table IV(a) as $.50000 + .47725 = .97725$ and $F(-1)$ from Table IV(b) as .15866. Therefore, the required probability is

$$.97725 - .15866 = .81859$$

Suppose we wish to find the p.f. of the transformed random variable $z = \dfrac{x - k}{c}$, where c and k are arbitrary constants. It can be shown by the method of Sec. 4.4 that the required p.f. is

$$f(x:k,c^2) = \frac{1}{c\sqrt{2\pi}}\, exp - \frac{1}{2}\left(\frac{x - k}{c}\right)^2 \qquad x \text{ real} \qquad (4.5.9)$$

Furthermore, it can be shown that the mean and the variance of the p.f. given by Eq. (4.5.9) are k and c^2 respectively. Therefore, for any normally distributed random variable x, the transformation $z = \dfrac{x - \mu_x}{\sigma_x}$ reduces it to the unit normal. The next example illustrates this notion.

Example 4.5.2 The heights of college students are assumed to obey the normal p.f. with $\mu = 67$ inches and standard deviation $\sigma = 3$ inches. What percent of the students do have heights between 64 and 70 inches?

Solution: The transformation $z = \dfrac{x - \mu_x}{\sigma_x}$ reduces the given normal p.f. to the unit normal. Hence, $z_1 = \dfrac{x_1 - 67}{3}$ and $z_2 = \dfrac{x_2 - 67}{3}$ yield the required values of the unit normal, $z_1 = -1$ and $z_2 = 1$. Therefore,

$$P[-1 < z < 1] = F(1) - F(-1) = .68268$$

which implies that 68.268 percent of the students are expected to have heights between 64 and 70 inches.

In general for any normally distributed random variable x,

$$P[a < x < b] = P\left[\frac{a - \mu_x}{\sigma_x} < z < \frac{b - \mu_x}{\sigma_x}\right] \qquad (4.5.10)$$

which according to Eq. (4.5.8) reduces to

$$P[a < x < b] = F\left(\frac{b - \mu_x}{\sigma_x}\right) - F\left(\frac{a - \mu_x}{\sigma_x}\right) \qquad (4.5.11)$$

PROBLEMS

4.5.1 Show that the unit normal p.f. (a) is symmetric about the $z = 0$, (b) attains its maximum at $z = 0$ and (c) has inflection points at $z = \pm 1$.

4.5.2 Find the moment-generating function for the unit normal p.f. Using it verify that the mean and the variance are 0 and 1 respectively.

4.5.3 If the p.f. of a random variable x is assumed to be unit normal, find
(a) $P[0 < x < 1.5]$ (b) $P[x < 1.96]$
(c) $P[x > 1.28]$ (d) $P[-1.28 < x < 1.64]$
(e) $P[-.5 < x < .75]$

4.5.4 The IQ scores of elementary school pupils are assumed to be normally distributed with mean 100 and variance 225. What percent of grade school students are expected to have IQ points of (a) 115 or more? (b) 121 or less? (c) between 96 and 112? (d) less than 91 or more than 130?

4.5.5 The arrival time of employees to work is assumed to obey the normal p.f. with mean time at 7:30 a.m. and standard deviation of 2 minutes. What is the probability that an employee will arrive to work (a) on or before 7:30 a.m.? (b) between 7:29 and 7:33 a.m.? (c) after 7:32 a.m.?

4.5.6 In a large company the personnel department assumes that the period of employment for all their employees obeys the normal p.f. with mean 10 years and standard deviation of 3 years and 3 months. What percent of their employees have been working for (a) 12 years or less? (b) 13 years or more? (c) between 8 to 11 years?

4.5.7 In a residential section of a certain city the assessed values of homes are assumed to be normally distributed. If the mean and the variance are 5,000 and 1,000,000 respectively and the assessed values are 25 percent of the appraised values, what percent of the homes have appraised values of (a) 25,000 or more? (b) 15,000 or less? (c) between 17,000 and 21,000?

4.5.8 The Chamber of Commerce of city A informs an inquiring tourist that 15.87 percent of the motels in town charge $10 or more and the average price is $8. Assuming that motel rates in this town obey the normal p.f., find the variance of the motel rates.

4.5.9 The inquiring motorist in Prob. 4.5.8 receives information from another source that 15.87 percent of the motels in city B (next town on the road) charge $10.50 or more and 2.28 percent charge $6 or less. Assuming that the quality, room availability and services are comparable in cities A and B, in which city will the tourist have a better chance of finding a room if he can spend (a) $7.50–8.50? (b) $6.50–7.50?

4.6 THE NORMAL APPROXIMATION TO THE BINOMIAL

The use of a normal p.f. to approximate the probabilities associated with a binomial p.f. constitutes a powerful technique and reduces laborious computations to a minimum. For instance, one might obtain the expression for the probability of exactly 54 heads showing in the toss of 100 unbiased coins from the binomial $f(x{:}100,.5)$ to be $f(54{:}100,.5)$ which is equal to $C(100,54)(.5)^{100}$. However, the exact evaluation of this expression represents a difficult task.

The following example demonstrates the method of approximation we shall adopt in general for problems of this type. Suppose the binomial $f(x{:}4,.5)$ is evaluated for $x = 0, 1, 2, 3,$ and 4 and the corresponding

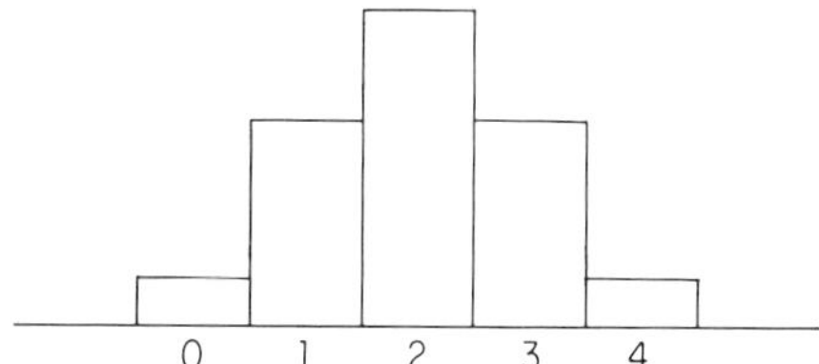

FIG. 4.11. Histogram for the binomial $f(x: 4, .5)$.

histogram drawn. The result will be $f(0) = 1/16$, $f(1) = 4/16$, $f(2) = 6/16$, $f(3) = 4/16$ and $f(4) = 1/16$. Figure 4.11 represents the associated histogram.

Recalling that the mean and the variance of any binomial are np and $np(1 - p)$ respectively, [refer to Eqs. (3.2.4) and (3.2.18)], we get $\mu = 2$ and $\sigma^2 = 1$ for this case.

Next let us consider the normal p.f. with $\mu = 2$ and $\sigma^2 = 1$. From Eq. (4.5.9) we get $f(x) = \dfrac{1}{\sqrt{2\pi}} e^{-(x-2)^2/2}$ to represent the requested normal p.f. Figure 4.12 is the graph of this normal probability function.

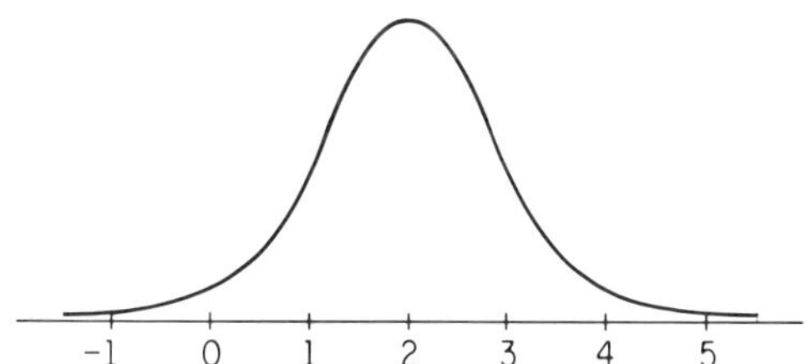

FIG 4.12. Normal p.f. with $\mu = 2$ and $\sigma^2 = 1$.

Now let us superimpose Figs. 4.11 and 4.12 such that the respective means coincide. See Fig. 4.13. Let us recall that both probability functions have equal areas, means and standard deviations. Furthermore, $P[a < x < b]$ associated with the normal p.f. is approximately equal to $P[a < x < b]$ associated with the binomial. Any error is due to the difference in the amount of areas included under the normal probability function which is not part of the binomial (areas shaded with vertical lines in Fig. 4.13) and the amount of areas excluded by the normal probability function which is part of the binomial (areas shaded with horizontal lines in Fig. 4.13).

For instance, $P[2.5 < x < 3.5]$ for the binomial p.f. is equal to

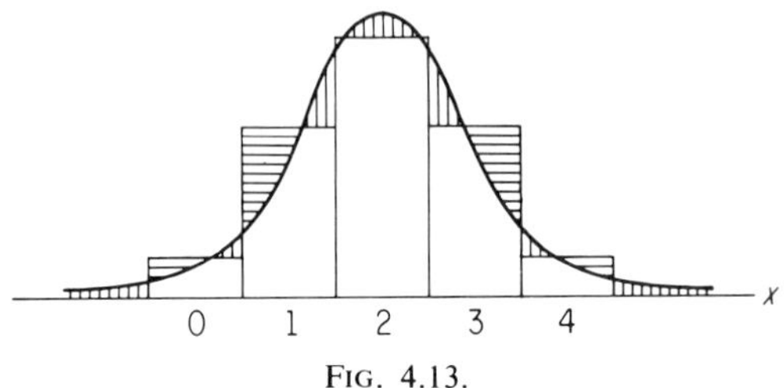

FIG. 4.13.

$f(3{:}4,.5) = 4/16 = 0.25000$, while $P[2.5 < x < 3.5]$ for the corresponding normal p.f. according to Eq. (4.5.11) is

$$F\left(\frac{3.5 - 2}{1}\right) - F\left(\frac{2.5 - 2}{1}\right) = F(1.5) - F(.5) = .24173$$

This is an approximation for the exact value of .25000.

Let us remark that the binomial p.f. is discrete where as the normal p.f. is continuous. The use of $P[2.5 < x < 3.5]$ instead of $P[x = 3]$ is necessary to make the appropriate approximation. The interval $3 \pm .5$ associated with the continuous p.f. corresponds to the discrete value of 3 and .5 is referred to as the *correction for continuity*. In using the normal approximation to the binomial, the correction for continuity must be accounted for.

Now, if n is large and the histograms of the binomial $f(x{:}n,p)$ and the graph of the normal p.f. with mean equal to np and variance equal to $np(1 - p)$ are superimposed, the error indicated above will be very small. That is, the included and the excluded portions in the two graphs will be very close for all subintervals and hence the binomial will approach the normal probability function. The following theorem, which we state without proof, is quite useful and we shall refer to it in subsequent chapters.

Theorem 4.6.1 The random variable $z = \dfrac{x - np}{\sqrt{np(1 - p)}}$ has a distribution function (d.f.) which approaches the d.f. of the unit normal as n gets large without bound.

Example 4.6.1 Using Theorem 4.6.1, evaluate the binomial expression $f(54{:}100,.5)$.

Solution: $np = 100(.5) = 50$ and $np(1 - p) = 100(.5)^2 = 25$. Hence

$$\begin{aligned} P[53.5 < x < 54.5] &= P\left[\frac{53.5 - 50}{5} < z < \frac{54.5 - 50}{5}\right] \\ &= P[.7 < z < .9] = F(.9) - F(.7) \\ &= .05790 \end{aligned}$$

PROBLEMS

4.6.1 An honest coin is tossed 64 times. *Write* only the exact expression for (a) the probability of obtaining 38 heads, (b) the probability of obtaining 35 or more heads, (c) the probability of obtaining between 25 and 35 heads.

4.6.2 Using the normal approximation to the binomial evaluate the probabilities of Prob. 4.6.1.

4.6.3 Refer to Prob. 2.4.7. Using the method of this section evaluate parts (a), (c), and (e).

4.6.4 Over the last several years a TV shop has found that 10 percent of TV picture tubes burn out prior to the expiration of the guaranty period. Use the method of Sec. 4.6 to find (a) the probability of replacing at most 4 if 100 picture tubes have been sold, (b) the probability of replacing at least 5 but not more than 15.

4.6.5 Repeat Prob. 2.4.12 by using the method of this section.

4.7 THE LOG-NORMAL PROBABILITY FUNCTION

A probability function with many practical applications, particularly in engineering, is the log-normal p.f. defined by the expression

$$f(x:\alpha,\beta) = \frac{1}{(x-c)\beta\sqrt{2\pi}} e^{-(1/2\beta^2)[\ln(x-c)-\alpha]^2} \qquad x > c$$

$$= 0 \qquad \text{otherwise} \tag{4.7.1}$$

where c is an arbitrary constant.

As suggested by its name, this p.f. is related to the normal p.f. presented in Sec. 4.5. Referring to Eq. (4.5.9), one recognizes the analogy between the two—namely, the logarithm of the random variable $(x - c)$ obeys the normal probability function. It can be readily shown that Eq. (4.7.1) satisfies Eqs. (4.1.1) and (4.1.2), which we show next.

Since $x > c$, clearly $f(x) \geq 0$. To show that $\int_c^\infty f(x)\,dx = 1$, we make the transformation $y = \ln(x - c)$, which yields

$$\int_{-\infty}^{\infty} \frac{1}{\beta\sqrt{2\pi}} e^{-(1/2)[(y-\alpha)/\beta]^2}\,dy \tag{4.7.2}$$

This expression is the integral of the normal p.f. with mean α and variance β^2 which according to Eq. (4.5.9) represents a probability function.

We obtain the mean of the log-normal p.f. by finding $E[x]$.

$$E[x] = \int_c^\infty \frac{x}{(x-c)\beta\sqrt{2\pi}} e^{-(1/2\beta^2)[\ln(x-c)-\alpha]^2}\,dx \tag{4.7.3}$$

Again, setting $y = \ln(x - c)$ and expressing it equivalently as

$$e^y = e^{\ln(x-c)} = x - c \tag{4.7.4}$$

we get $x = c + e^y$ and therefore Eq. (4.7.3) reduces to

$$E[x] = \int_{-\infty}^{\infty} \frac{(c + e^y)}{\beta\sqrt{2\pi}} e^{-(1/2)[(y-\alpha)/\beta]^2}\, dy \tag{4.7.5}$$

$$= c\int_{-\infty}^{\infty} \frac{1}{\beta\sqrt{2\pi}} e^{-(1/2)[(y-\alpha)/\beta]^2}\, dy + \int_{-\infty}^{\infty} \frac{e^y}{\beta\sqrt{2\pi}} e^{-(1/2)[(y-\alpha)/\beta]^2}\, dy \tag{4.7.6}$$

Now the first integral of Eq. (4.7.6) has the normal p.f. with mean α and standard deviation β for its integrand and hence has a value of $c(1) = c$. The exponent of the integrand in the second integral may be written as

$$\left[y - \frac{1}{2}\left(\frac{y-\alpha}{\beta}\right)^2\right] = -\frac{1}{2}\left[\frac{y - (\alpha + \beta^2)}{\beta}\right]^2 + \alpha + \frac{\beta^2}{2} \tag{4.7.7}$$

and hence

$$e^{[y-(1/2\beta^2)(y-\alpha)^2]} = e^{(\alpha+\beta^2/2)}\left\{e^{-(1/2\beta^2)[y-(\alpha+\beta^2)]^2}\right\} \tag{4.7.8}$$

Therefore the value of the second integral is $e^{(\alpha+\beta^2/2)}$, which yields for Eq. 4.7.6

$$E[x] = c + e^{\alpha+\beta^2/2} \tag{4.7.9}$$

For instance, the log-normal p.f. with parameters $\alpha = 1$ and $\beta = 0.4$ has a mean of $c + e^{1.08}$. Furthermore, if we assume $c = 0$, $\mu = e^{1.08} = 2.9447$.

To find the variance of the log-normal p.f. we use the relationship $\sigma^2 = E[x^2] - (E[x])^2$. Making the same transformation as in Eq. (4.7.4) and pursuing a similar approach as in the derivation of Eq. (4.7.9), it can be shown that

$$E[x^2] = e^{2(\alpha+\beta^2)} + 2ce^{(\alpha+\beta^2/2)} + c^2 \tag{4.7.10}$$

The expression $E[x^2] - (E[x])^2$ for the log-normal when simplified reduces to

$$\sigma^2 = [e^{(2\alpha+\beta^2)}][e^{\beta^2} - 1] \tag{4.7.11}$$

Next let us consider the distribution function of the log-normal probability function. Just as the probability function of the log-normal was related to that of the normal, the distribution function of log-normal has similarities with that of the normal. The distribution function of the log-normal is

$$F(x) = \int_c^x \frac{1}{(t-c)\beta\sqrt{2\pi}} e^{-(1/2\beta^2)[\ln(t-c)-\alpha]^2}\, dt \tag{4.7.12}$$

The transformation $z = \ln(t - c)$ reduces this last equation to

$$F(\ln x) = \int_{-\infty}^{\ln x} \frac{1}{\beta\sqrt{2\pi}} e^{-(1/2)[(z-\alpha)/\beta]^2}\, dz \tag{4.7.13}$$

Therefore, analogous to Eq. (4.5.11), we get

$$P[\ln a < z < \ln b] = F\left(\frac{\ln b - \alpha}{\beta}\right) - F\left(\frac{\ln a - \alpha}{\beta}\right)$$

$$= F(z_1) - F(z_2) \tag{4.7.14}$$

Example 4.7.1 The current gains in microamperes of a certain type of transitors is assumed to obey the log-normal p.f. with parameters $c = 0, \alpha = 1$, and $\beta = .2$. (a) Find the variance of this probability function. (b) Find the probability that the current gain for a transistor will be between 3.00 and 3.32 microamperes.

Solution: Using Eq. (4.7.11) we obtain for part (a) $\sigma^2 = (e^{.04} - 1)(e^{2+.04}) = .314$.

For part (b), we use Eq. (4.7.14) and substituting $b = 3.32$, $a = 3.00$, $\alpha = 1, \beta = .2$ and $c = 0$, we get for the required probability

$$F\left(\frac{\ln 3.32 - 1}{.2}\right) - F\left(\frac{\ln 3.00 - 1}{.2}\right) = F(1) - F(.5)$$

$$= .84134 - .69146 = .14988$$

PROBLEMS

4.7.1 Verify Eq. (4.7.10) and thus show that Eq. (4.7.11) holds true.

4.7.2 Find the parameters α and β of the log-normal p.f. whose mean and variance are 2 and .2 respectively. Assume $c = 1$.

4.7.3 The distribution of monthly electricity consumption by the residents of a certain community is assumed to be log-normal, with mean 925 and standard deviation 25 kilowatt hours. What percent of the homes consume between 700 to 1,000 kilowatt hours of electricity per month?

5

Treatment of Samples from a Population

5.0 INTRODUCTION

In Chapter 2 some important probability functions of the discrete type were discussed and in Chapter 3 a few of their measurable characteristics such as the mean and the variance defined. The same notions were extended to probability functions of the continuous type in Chapter 4. What one should look for next might become evident if we refer once again to the analogy of suit models for men and probability functions. We are at that stage where suit patterns or sizes are available, production has started and suits are sent out to stores for sale. Now the suit buyer has arrived and wishes to make a purchase. Suppose that this customer is from another culture where men's clothing is not described in the same fashion as in this country and he therefore does not know what size "suit" he should look for. The salesman immediately comes to his rescue and attempts to determine the proper size (model) needed by measuring his actual physical characteristics. If for example, his chest measurement is 44 inches and he is a tall person, the suit model "44 long" might be suggested. Analogously, in our considerations of probability functions we are at the juncture where theoretical probability models are available and a random phenomenon is observed, but we do not know which probability function describes the random phenomenon. We confront now the problem of finding comparable characteristics between the observed data and the corresponding probability function, and when we have found such characteristics we will use them to choose a suitable probability function.

In actual practice, however, the order of this procedure is reversed. That is, usually it is assumed that a given probability function with specific parameters describes the random phenomenon in question and then some of the measurable characteristics of the data are examined or tested to verify whether or not the assumptions made are true. In this chapter we shall discuss methods of selecting data and define some measurable characteristics such as the range, the mean and the standard deviation, which will constitute the basis for testing the assumptions made regarding a probability function.

5.1 SAMPLING FROM A FINITE POPULATION

In statistical considerations the term *population* is used in a different sense than the common connotation of the word. Here the words population and universe are used synonymously; that is, a population is the totality of measurements of objects possessing some measurable criterion. For example, if in a production process 50 circular containers are manufactured and the measurable criterion is the diameter of the containers, the 50 measurements of diameters represent a population. Again, if in a given coeducational college there are 600 students and the measurable characteristic is their heights, the 600 measurements of heights represent another population. In the statistical usage of the term, a population is a set of measurements or counts. As another example, if on an acre of land there are 250 apple trees and the measurable criterion is the number of apples on each tree, the counts from all the trees form a population.

It should be evident from the definition that a population can be either finite, countably infinite, or infinite. Since the number of apples on each tree in the above example is finite it represents a finite population. The number of tosses necessary in flipping a coin until a head shows represents a countably infinite population, while the measurements of heights of college students represents an infinite population. Infinite population in this context implies that the measurements are located in a continuum. In all subsequent discussions the term population, finite, countably infinite or infinite, will imply reference to some measurable characteristics. In this section finite and countably infinite populations will be considered and in Sec. 7.1 the concept of samples from infinite populations will be presented.

It is seldom desirable or feasible to examine a population in order to describe one of its measurable criteria. For instance, in the above example of 600 students, merely to arrive at the conclusion that half of the population has measurements of less than 68.75 inches or that 40 percent of the population is located in the interval 66 to 70 inches, one would not consider measuring all 600 heights. Similarly, one would not test (measure) all available missiles in order to conclude that 98.9 percent of them function properly. Again, it would be very impractical to count all the apples on the 250 trees mentioned above. The fundamental notion underlying statistical theory is to obtain information regarding a population by examining a subset of measurements or counts from that population. The process of selecting a subset of a given population is called *sampling.* Sampling is a powerful tool and is used in everyday statistical practice. The reader may be familiar with the sampling procedures of opinion polling organizations. On the basis of a relatively few interviews with the voting public they predict the outcome of an election—that is, they seek an answer to the question, "What proportion of the voters are for or against a candidate or a proposition?" prior to the actual voting.

A *sample* is defined as any subset of a population and the number of elements in it is referred to as the *sample size.* For instance, eight trees from the population of 250 represent a sample of size 8 and four containers from the population of 50 containers are a sample size of 4. All samples from a population are not equally useful or meaningful, however. Referring to the example of 600 students once again, it would be illogical to guess or predict with reasonable accuracy any characteristic of the population if the sample is comprised of only 25 coeds or 25 engineers. A sample including only coeds or being predominantly composed of men students does not represent a typical case. The information used as a basis for prediction should come from samples that are representative of the population. Since in this population both male and female students are included, a representative sample should contain men and women students.

In Sec. 1.4 it was shown that there were $\binom{N}{k}$ different ways of choosing k objects from the N available. Now, if there are N elements in a population and it is required to take a sample of size k from it, there would be a total of $\binom{N}{k}$ different samples of size k. To illustrate, suppose that $N = 5$ and $k = 2$. There are $\binom{5}{2} = 10$ different samples of size 2 possible in the population of size 5. If the elements are denoted by $A, B, C, D,$ and E, the 10 samples of size 2 would be $AB, AC, AD, AE, BC, BD, BE, CD, CE,$ and DE.

Whenever each of the $\binom{N}{k}$ samples of size k drawn from a population of size N has the same probability, the sample is said to be *random.* Equivalently, a *random sample* is one in which each sample of size k has a probability equal to $1/\binom{N}{k}$. This probability can be expressed as the product of k terms; that is,

$$\frac{1}{\binom{N}{k}} = \frac{k!(N-k)!}{N!} = \left(\frac{k}{N}\right)\left(\frac{k-1}{N-1}\right)\left(\frac{k-2}{N-2}\right)\cdots\left(\frac{1}{N-k+1}\right) \tag{5.1.1}$$

Furthermore, while defining combinations in Sec. 1.4, it was indicated that $\binom{N}{k}$ represents selection regardless of order. The definition of random sample implies that all possible selections of k items from N, regardless of order, have equal probabilities. We illustrate this in the following example.

Example 5.1.1 The instructor of a class with 25 students plans to invite 2 of his pupils to a dinner party. Suggest a method of selecting a pair of students randomly.

Solution: If each student is assigned a number between 1 and 25 inclusive, then (1,2), (1,3), . . . , (1,25), (2,3), (2,4), (2,5), . . . , (2,25), . . . , (24,25) represent the 300 possible 2-tuples corresponding to all unordered pairs. Since (i,j) and (j,i) represent the same pair of students only one of the two is listed. If these 300 pairs of numbers are written on small and identical cardboard chips, placed in an urn, then mixed well and one drawn blindfolded; the numbers on the drawn chip satisfy the randomness criterion.

The notion of randomness was tacitly implied in many examples and problems of the previous chapters. For instance, in Example 2.4.5 the condition "4 items are selected at random" was stipulated to precisely imply that all $\binom{N}{4}$ possible samples of size 4 from a day's output of N (a population) had the same probability of being chosen. Again, in Example 2.6.2 the stipulation ". . . will test two items at random" was stated to imply that $\binom{24}{2} = 276$ samples of size two had equal probabilities.

If in a sampling procedure the order in which the individual elements are selected is relevant, the process of selection is referred to as *ordered sampling*. Let us recall that there are $P(N,k) = \binom{N}{k}k!$ different-ordered selections of size k possible from a population of size N. Hence an ordered sample is defined as one for which each sample of size k has a probability equal to $1/P(N,k)$. This probability can be expressed as

$$\frac{1}{P(N,k)} = \frac{(N-k)!}{N!} = \frac{1}{N(N-1)(N-2)\cdots(N-k+1)} \tag{5.1.2}$$

The following example illustrates the concept of ordered sampling.

Example 5.1.2 Repeat Example 5.1.1 if the selection is carried out by ordered sampling method.

Solution: Write on 25 identical cardboard chips the numbers 1 to 25 inclusive, place them in an urn, mix well the content of the urn, and draw one chip blindfolded. Without replacing the drawn chip, repeat once again the process of mixing and draw a second chip. The two numbers selected according to this scheme represent ordered samples of size two. This method takes into account the order of choice—that is, (i,j) and (j,i) are considered different pairs.

The special case in which $k = 1$ is noteworthy for both the random and the ordered sampling. If $k = 1$ both the expressions (5.1.1) and (5.1.2) reduce to $1/N$. Therefore, in this particular case the two types of sampling obey the uniform probability function. (See Prob. 3.1.8) For this reason a random sample is sometimes defined as one in which each element has a probability equal to $1/N$.

Whenever a sample has to be taken from a large population, the process of either random or ordered sampling presents computational as well as methodical difficulties. Suppose for example, it is required to obtain a random sample of 2 from a population of 1,000,000. One would not think of writing out the $\binom{1,000,000}{2}$ possible samples of size 2 on cardboard chips, place them in an urn, thoroughly mix the contents, and draw one chip to obtain a random sample. (See Example 5.1.1.)

Similarly, for an ordered sampling one would not draw 2 chips without replacement from an urn containing 1,000,000 cardboard chips with the numbers 1, 2, 3, . . . , 1,000,000 inscribed on them. (See Example 5.2.2.) A more convenient method commonly employed for sampling from large populations is that of *random numbers*. The notion underlying this approach is straightforward. It merely assumes that each digit in the number associated with the sample is random. For instance, 743,106 and 215,911 represent one of the $\binom{1,000,000}{2}$ different samples of size 2 in a population of 1,000,000. The method of random numbers requires that the digits, 7, 4, 3, 1, 0, and 6 in the first and 2, 1, 5, 9, 1, and 1 in the second number are selected *individually* at random, which is to say that the digits 0, 1, 2, . . . , 9 obey the uniform probability function. Sampling with this assumption simplifies greatly the process of selection and can be achieved by a simple experiment, namely ten identical objects (cardboard chips, balls, hexagons, etc.) are placed in an urn with one of the digits 0, 1, 2, . . . , 9 inscribed on each. The content of the urn is thoroughly mixed, one object is drawn blindfolded and the inscribed digit on it observed and recorded. Suppose the number is 7. If the sample needed is of size one from a population of size 10, the 7th element is taken as the sample of size one.

If a sample of size one is required from a population of size 100, this procedure is repeated once again after the previously drawn object is replaced. This time say the observed digit is 1. Hence, the 71st element is the required sample from this population. When a sample of size one is needed from a population of size 475, one would proceed similarly. The only difference in this situation is that the first digit drawn (the hundreds digit) should be less than or equal to 4. If the first drawn digit is 5 or greater, it is discarded and the process continued until a digit less than 5 shows. For instance, the digit 3 might be the first number drawn in this process. After this object is replaced and mixed thoroughly, an object is drawn for the second time, which might show the digit 9 on it. When this process is carried out for the third time, the obtained digit might be 9 also. Therefore, the 399th element is the required sample of size one from this population with 475 elements. When a sample of size 5 is required from the same population, the process of selecting 3 digit number described is repeated 5 times and a possible sample might be 194, 086, 407, 148, and 265. Hence the 86th, 148th, 194th, 265th, and 407th elements represent a random sample of size 5. Clearly, this procedure can be applied to selecting samples of any size from a population of size N.

Selection of samples by this procedure has an obvious disadvantage; namely, it is time consuming. The same objective can be achieved if an experiment of this type is carried out once and the result tabulated for reference. If 450 such drawings were made and the observed digits recorded, a table of random numbers might have been as in Table 5.1.1.

TABLE 5.1.1

450 Random Numbers

44135	74218	66384	03320	51967	77343	72234	25292	65782	71484
11655	76171	82247	43652	97027	83203	21777	70996	41186	03515
59218	13964	77681	15234	56254	15039	96171	75292	71593	26171
41182	98339	21965	20996	30004	39453	64581	66503	50403	19824
73416	01562	91596	92382	26511	47460	26398	92578	52511	96289
15171	99707	05042	35839	52226	44042	29790	28320	50917	96875
57235	71777	78195	43457	34113	40332	23380	37109	04920	53222
65090	57617	62697	38769	48970	82519	88393	43261	21394	16503
59241	08886	57822	99804	47144	04296	88805	66406	53516	35742

Because these digits are drawn from a uniform probability function, one anticipates finding equal frequencies of each digit and unrelatedness among them in the long run. In this particular case, each digit should appear approximately 45 times and no pattern should be evident. Table 5.1.1 corresponds and is analogous to an urn containing 45 of each digit (chip) and the contents having been thoroughly mixed. Now one simply refers to some column and row entry in this Table and reads the digit found in that location. If a sample of size one is required from a population of size 10, and if, say, the 13th column and the 3rd row entry were chosen, then the 6th element of the population is taken as the sample.

If a sample of size two is needed from a population of 25, such as mentioned in Example 5.1.1, the same technique is pursued. Suppose this time we refer to the 6th column and the 4th row location. The digit in that location is 9 which is discarded because we need in the tens digit position numbers less than 3. The choice for the next digit may be made from either reading down or reading across the row of reference. Reading down the column of reference we find the digit 0, which is acceptable. Next we need to choose a digit corresponding to the unit's position. Again, a new reference location may be selected, or reading of the digits continued from the last location. Continuing to read down from the last reference location, we find 9, and hence the 9th element is selected for one of the two required. Repeating this procedure once again and choosing a new reference location, say the 25th column and the 5th row, and reading across, we note that the 14th element would be selected as the second in the sample of size two. Clearly this approach can be applied for sampling from any population and for samples of all sizes.

In actual practice, random numbers of considerable magnitude are not obtained by the experimental method described above. List of random numbers such as Table 5.1.1 can be generated more conveniently by means of modern digital computers, and there are many methods available to generate random digits. Lists of this type are referred to as *pseudorandom numbers.* For a short list of random numbers the following technique is sometimes used. An arbitrary integer of even digit length is multiplied by itself, which yields a result of either $(2k - 1)$ or $2k$ digits, where k is the number of digits in the constant. If the result has $(2k - 1)$ digits, a zero is added to the leftmost position to make it an even digit number and the middle k digits of the number thus formed is considered to represent a sequence of k random digits. For example, when a six-digit number such as 241383 is multiplied by itself, the result is 58265752689. This number has $2(6) - 1 = 11$ digits which is odd, and therefore a zero is added to the leftmost position. The middle 6 digits of this result, namely 265752, are considered to represent a sequence of 6 random numbers or digits. Next, this six-digit number is squared, if the result has an odd number of digits, a zero is added to the leftmost position and the middle 6 digits taken as another sequence of 6 random digits. This process is continued until sufficient numbers are generated. Table 5.1.1 was generated by this approach and only a partial list is included there. This method has a disadvantage in that the sequence of k digits begin to repeat or become all zeros within a short cycle. For instance, when the six digit number 625000 is squared the result is 390625000000, whose middle 6 digits are 625000 and hence the sequence will repeat itself from here on. Or when 100000 is squared, the middle 6 digits of the result yield all zeros.

Another technique more commonly used to generate pseudorandom numbers is the following. An arbitrary constant, say C, composed of an even-digit number is multiplied with a different even-digit number of equal length. As in the previous method here also the result will yield either $(2k - 1)$ or $2k$ digits. If the result has $(2k - 1)$ digits again a zero is added to the leftmost position (in the computer register the leftmost position contains a zero already) and the middle k digits are assumed to represent a sequence of random digits. Next, this number of k-digit length is multiplied with the same constant C and the process continued in the same manner as described previously. For instance, if a four-digit number such as 3125 is taken as the constant C and 1111 as the other four-digit number, then $(3125)(1111) = 03471875$ yields 4718 for a sequence of four random digits. Next 3125 is multiplied by 4718 and the middle four digits of the answer is taken as another sequence of four random digits and the process continued. The random number lists in the Appendix were computed by such a technique.

PROBLEMS

5.1.1 The statisticians of a public polling organization use the following methods for sampling public opinions. Discuss the advantages and disadvantages of each method.

(a) Names are selected at random from the local telephone directories.

(b) Names are taken from motor vehicle registration lists at random.

(c) Names are taken at random from the mailing list of Times magazine.

(d) Names are selected from last year's income tax returns.

(e) Interviewing persons on the street.

(f) Sending questionnaires to those who reply to a paid political announcement made over the radio.

5.1.2 From the list of all colleges in a given state, 10 colleges are selected at random. From each of these colleges two professors are selected at random to form a committee. Does the selection of 20 professors in this manner satisfy the criterion of randomness? Why?

5.1.3 How would you select at random 10 students from a senior class of 100, without the use of a random-number table?

5.1.4 Using the random-number table in the Appendix, select 20 two-digit numbers.

5.1.5 Using the random-number table in the Appendix, select at random 10 numbers from a population of size 2,575.

5.1.6 In a manufacturing process, every 7th item produced is examined for defect. Is this method of selection random? Explain.

5.1.7 Suppose you were on the President's Advisory Board. Suggest methods that will ensure randomness in drafting the 19-year-olds.

5.1.8 A questionnaire is to be sent to 5 percent of the alumni of a certain university. Suggest methods of selecting names to ensure randomness.

5.2 ORGANIZATION OF DATA

Suppose for instance, a random sample of size 25 from a student body of 1,000 yields height measurements as shown in Table 5.2.1.

TABLE 5.2.1

HEIGHT MEASUREMENTS IN INCHES OF 25 STUDENTS

67, 68, 64, 67, 64, 65, 69, 70, 74, 71,
72, 74, 68, 72, 74, 76, 72, 62, 64, 68,
67, 71, 70, 68, 65

Certain information available in this sample becomes more evident if the data is ordered according to some scheme. If these 25 measurements are arranged in ascending order of magnitude, the result will be as shown in Table 5.2.2.

TABLE 5.2.2

HEIGHT MEASUREMENTS IN INCHES OF 25 STUDENTS

62, 64, 64, 64, 65, 65, 67, 67, 67, 68,
68, 68, 68, 69, 70, 70, 71, 71, 72, 72,
72, 74, 74, 74, 76

It becomes immediately apparent for example that the largest and smallest values of the variable in this data are 76 and 62 inches respectively, which information is useful in locating the interval where the sample values are found.

When the ordered data is presented in a frequency table, further information may become evident. The frequency table for the data in Table 5.2.2 is given in Table 5.2.3, from which one observes, among other things, that 68 inches is the most frequently occuring measurement.

TABLE 5.2.3

FREQUENCY TABLE FOR HEIGHT MEASUREMENTS, IN INCHES, OF 25 STUDENTS

Frequency:	1	0	3	2	0	3	4	1	2	2	3	0	3	0	1
Measurement:	62	63	64	65	66	67	68	69	70	71	72	73	74	75	76

The information in Table 5.2.3 can be immediately used to construct a line graph. Figure 5.1 represents the line graph associated with Table 5.2.3.

The interval 62 to 76 inches represents one of many observable and measurable characteristics of this sample. The population from which these 25 measurements were selected has a corresponding characteristic: namely, some interval, say $\alpha \leq x \leq \beta$, (α and β are unknown), such that all possible height measurements of students are found in this interval. Now x represents a random variable whose values represent the population in question, and there exists a probability function $f(x)$ which describes it. Because of this association, the expression "a sample from a

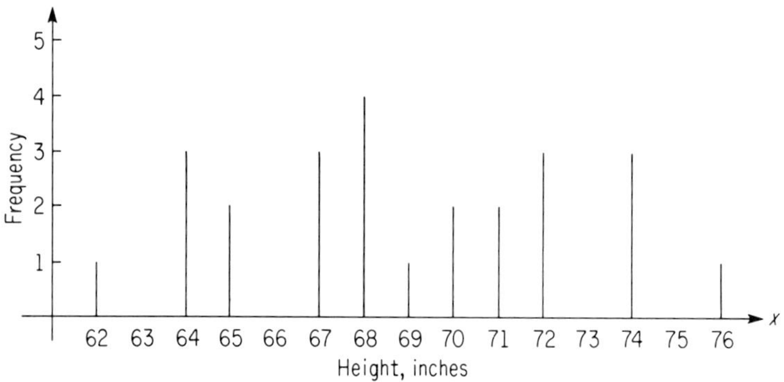

FIG. 5.1. Line graph of 25 height measurements.

population" and "a sample from a probability function" are often used synonymously. More precisely, however, a sample is from a population which obeys a given probability function. For example, a sample of size two from 24 articles of which 50 percent are defective implies a sample of size two drawn from a population which obeys the binomial probability function with parameters $p = 0.5$ and $n = 24$. Furthermore, since the measurable characteristics of the probability functions were defined as parameters, the same term, *parameter*, will be used to indicate the measurable or observable characteristics of the populations. In order to distinguish between the observable characteristics of a population and a sample, we define a *statistic* as an observable or measurable characteristic of a sample. Each parameter of a population has a corresponding sample statistic, and vice versa. Conventionally Greek letters are used to represent parameters and Roman letters to denote statistics. For example, σ^2 and s^2 are commonly used to designate the population and sample variances respectively.

PROBLEMS

5.2.1 The following are the temperature readings in degrees centigrade during 15 consecutive days in a certain city.

27, 18, 17, 20, 16, 17, 22, 22, 21, 23, 19, 20, 18, 22, 24

(a) Arrange this data in ascending order.
(b) Make a line graph representing this data.

5.2.2 Given the following weight distribution of 49 high school students in pounds:

121, 122, 124, 124, 125, 125, 125, 127, 127,
128, 128, 128, 128, 129, 129, 130, 130, 130,
131, 131, 131, 131, 131, 131, 131, 132, 132,

134, 134, 134, 134, 134, 134, 134, 134, 135,
135, 137, 137, 137, 137, 137, 138, 138, 138,
138, 140, 140, 142

Make a line graph for this data.

5.2.3 The following data represents the monthly expenditures of a small business firm. Figures are in dollars.

149.00,	16.00,	12.50,	158.50,	80.00,	15.70,	55.57,
219.00,	63.73,	44.35,	5.53,	79.83,	10.00,	81.42,
4.25,	117.75,	7.95,	201.20,	11.24,	12.25,	25.00,
375.00,	110.00,	119.50,	321.00,	90.95,	38.47,	64.50,
33.00,	113.45					

Arrange this data in descending order.

(a) What is the value of the 10th entry in the ordered list? of the 20th?

(b) What is the value of one half the sum of the 15th and 16th entries equal to?

(c) What is the value of the difference between the largest and the smallest entries?

5.2.4 Given the following data:

1.00	4.00	8.50	3.00	2.25	2.00	7.75	6.75
2.00	5.25	2.25	5.00	1.25	5.00	5.25	8.75
4.75	8.25	4.00	7.00	8.75	2.50	3.50	1.75
4.25	4.00	0.25	2.25	6.00	3.00	9.00	4.00
4.00	4.00	9.00	7.75	6.00	6.50	5.00	5.50
7.50	6.00	3.75	4.75	7.25	6.00	1.25	0.25
7.00	2.50	8.75	5.75	4.25	4.75	3.00	5.50
6.50	7.00	3.00	4.75	4.25	3.25	5.00	2.00

(a) Arrange these 64 observations in ascending order.

(b) Put the result in a frequency table.

(c) What is the value of one half the sum of the 10th and the 55th observations?

5.3 COMPUTATION OF SOME STATISTICS

(a) The Range. When the sample measurements are arranged in ascending order and the random variable denoted by some letter, the individual measurements can be identified with further ease. Referring once again to the height measurements of the 25 students shown in Table 5.2.2, we notice the x_1 and x_{25} represent the smallest (62 inches) and the largest (76 inches) observations respectively. A statistic associated with the largest and the smallest observations is the *range*. The range is usually denoted by the letter R and is defined as the largest minus the smallest observations plus one unit of measurement (u.m.). In symbols,

$$R = x_n - x_1 + 1 \text{ u.m.} \tag{5.3.1}$$

where x_n is the largest and x_1 is the smallest measurements.

The unit of measurement should be included in the definition of the range, in order to account for the length within one observation. That is, if all observations are equal in a sample or a population, $x_n = x_1$ and hence the range is correctly represented by 1 unit of measurement. The range for the data in Table 5.2.2 is $76 - 62 + 1 = 15$ inches.

If the measurements of a sample or a population are arranged in descending order, x_n and x_1 are merely interchanged. The choice of ordering—either ascending or descending order—is often a matter of convenience. Unless otherwise stated, for consistency, ascending order will be assumed in all subsequent discussions. It should be clear that the range is a statistic describing the degree of variation among the individual measurements. The larger the range, the greater the variability among the observations.

(b) The Mean

1. The Arithmetic Mean. The mean of the sample is a statistic which indicates the central tendency or central values of the measurements and is defined as

$$\bar{x} = \frac{\sum_{i=1}^{n} x_i}{n} \tag{5.3.2}$$

The mean is a very important statistic and much discussion will be given as to its usefulness in the following chapters. Conventionally, the sample mean is denoted by a bar over the letter signifying the random variable in question. If some random variable is denoted by w and a sample of size 50 is taken from its population,

$$\bar{w} = \frac{\sum_{i=1}^{50} w_i}{50}$$

represents the mean for this sample. The mean defined by Eq. (5.3.2) is also called the *arithmetic mean* to distinguish from the geometric and harmonic means which also will be defined in this Section. Unless otherwise stated, the mean will be used to imply the arithmetic mean. The mean for the height measurements of the 25 students given in Table 5.2.2 is $(62 + 64 + \cdots + 76)/25 = 68.64$ inches.

2. The Geometric Mean. The geometric mean of a sample of size n is defined as the nth root of the product of the sample measurements. In symbols,

$$\bar{x}_g = \left(\prod_{i=1}^{n} x_i\right)^{1/n} \tag{5.3.3}$$

where

$$\prod_{i=1}^{n} x_i = x_1 \cdot x_2 \cdots x_n$$

For example, if 3 measurements of a random variable yield the values 4, 20, and 100, the geometric mean for this data is

$$\bar{x}_g = \sqrt[3]{4 \cdot 20 \cdot 100} = 20$$

It is obvious from the definition that if any of the measurements are zero, the geometric mean becomes zero. Also, if some values are negative, the geometric mean might be an imaginary number. Therefore, the geometric mean is not used except for positive values.

The computation of the geometric mean is greatly simplified if logarithms are used. Recalling that $\log \prod_{i=1}^{n} x_i = \sum_{i=1}^{n} \log x_i$, one gets

$$\log \bar{x}_g = \frac{1}{n} \log \prod_{i=1}^{n} x_i = \frac{1}{n} \sum_{i=1}^{n} \log x_i \tag{5.3.4}$$

In Sec. 3.2 the mean was interpreted to represent the line upon which the center of gravity of a histogram is located. Such a concrete interpretation is not available for the geometric mean except to deduce from Eq. (5.3.4) that the logarithm of a geometric mean is equal to the arithmetic mean of the logarithms of the measurements. The geometric mean is a statistic that can be best applied to measurements which represent pure ratios. Suppose for example, we wish to compare the male-female ratios of two colleges and the data in the table is available:

	College A	College B
Male	6,000	2,000
Female	4,000	1,000

Consider the ratios of male students in these two schools, A/B yields 3 and B/A yields 1/3. Similarly, the ratios of female students yield 4 and 1/4, respectively. In the first case, the geometric mean is $\sqrt{3 \times 4} = \sqrt{12}$, while the second geometric mean is $\sqrt{1/3 \times 1/4} = \sqrt{1/12}$. Clearly, the reciprocals of these two means are equal and hence the order of the ratios are not relevant. This reciprocal relationship does not hold for the mean. That is,

$$\frac{3+4}{2} \neq \frac{1}{1/2(1/3 + 1/4)}$$

For another illustration let us consider a compound interest rate problem. Suppose an investor has two choices to invest his money for 3 years. Plan A will yield 5 percent the first, 6 percent the second, and 7 percent the third year. Plan B will yield 6 percent each year. If the interest is compounded annually, which plan should he prefer? Clearly, the mean for both plans is 6 percent. However, notice that $\sqrt[3]{(5)(6)(7)} = 5.92$ while

$\sqrt[3]{(6)(6)(6)} = 6$. Therefore, Plan B is the better. The reader is requested to verify this with an investment of $1,000.

3. The Harmonic Mean. The harmonic mean of a sample of size n is defined as the reciprocal of the mean for reciprocals of each measurement and is denoted by $\bar{x}_h$. That is,

$$\bar{x}_h = \frac{1}{\frac{1}{n}\sum_{i=1}^{n}\frac{1}{x_i}} = \frac{n}{\sum_{i=1}^{n}\frac{1}{x_i}} \tag{5.3.5}$$

The harmonic mean for the values 2, 3, 4, and 12 is equal to

$$\frac{4}{1/2 + 1/3 + 1/4 + 1/12} = \frac{4}{\frac{6 + 4 + 3 + 1}{12}} = \frac{48}{14} = \frac{24}{7}$$

The harmonic mean is best applicable to data that are mixed ratios, such as speeds (miles per hour), prices (dollars per item), etc. Suppose for example a person drives his car to a city 500 miles away at 50 miles per hour and flies his airplane on the return trip at a speed of 200 miles per hour. What is the average speed for the round trip? Since speed is defined as total distance traveled divided by total time, the correct answer is $(500 + 500)/(10 + 2.5) = 1{,}000/12.5 = 80$ miles per hour, and *not* the mean $(50 + 200)/2 = 125$ mph. However, the harmonic mean for this data is $2/(1/50 + 1/200) = 400/5 = 80$, which represents the correct answer.

(c) The Variance and Standard Deviation. The variance of a sample is denoted by s_x^2 and is defined as

$$s_x^2 = \frac{\sum_{i=1}^{n}(x_i - \bar{x})^2}{n - 1} \tag{5.3.6}$$

The reason for using $n - 1$ instead of n in the denominator of Eq. (5.3.6) is a technical one and will be discussed in Sec. 7.5.

The square root of the variance is defined as the sample standard deviation and the symbol s_x is associated with it. In symbols,

$$s_x = \sqrt{\frac{\Sigma(x_i - \bar{x})^2}{n - 1}} \tag{5.3.7}$$

The subscript letter x in s_x is merely to identify the random variable. If the random variable is designated by y, its standard deviation is denoted by s_y.

Example 5.3.1 A sample of 5 students from a class had the following ages: 18, 19, 20, 21, and 22 years. Calculate the standard deviation.

Solution: Since $\bar{x} = 20$, the variance yields

$$s^2 = \frac{(18 - 20)^2 + (19 - 20)^2 + \cdots + (22 - 20)^2}{4}$$

$$= \frac{4 + 1 + 0 + 1 + 4}{4} = \frac{5}{2}$$

Hence

$$s = \sqrt{5/2} = 1.58$$

The expression for the variance given by Eq. (5.3.6) can be put into an equivalent form which is more convenient to use when calculators are available. Now

$$\sum_{i=1}^{n} (x_i - \bar{x})^2 = \Sigma x_i^2 - 2\bar{x}\,\Sigma x_i + \Sigma \bar{x}^2$$

$$= \Sigma x_i^2 - 2\bar{x}n\frac{\Sigma x_i}{n} + n\bar{x}^2$$

$$= \Sigma x_i^2 - n\bar{x}^2 \qquad (5.3.8)$$

$$= \Sigma x_i^2 - \frac{(\Sigma x_i)^2}{n}$$

$$= \frac{n\,\Sigma x_i^2 - (\Sigma x_i)^2}{n} \qquad (5.3.9)$$

and hence

$$s_x^2 = \frac{1}{n(n - 1)}[n\,\Sigma x_i^2 - (\Sigma x_i)^2] \qquad (5.3.10)$$

The expression for the variance given in Eqs. (5.3.6) and (5.3.10) are referred to as the *definitional* and *computational forms*, respectively. Equation (5.3.9), the numerator of Eq. (5.3.6), is called *sums of squares* and commonly designated by SS. That is,

$$s_x^2 = \frac{SS}{n - 1}$$

(d) The Median and the Fractiles. Another statistic of practical importance is the *median*. This, like the mean, is a statistic describing the central tendency of the measurements and is defined as the middle observation of an ordered sample or population. If the sample size n is odd, the median is the $\left(\frac{n + 1}{2}\right)$th measurement and no calculation is necessary. For instance, when a sample of size 5 yields the measurements 49.6, 47.4, 49.9, 52.5, and 50.1 pounds, the ordered data would be 47.4, 49.6, 49.9, 50.1, and 52.5 pounds. The median for these observations is the $\left(\frac{5 + 1}{2}\right)$th $=$ 3rd measurement, namely, 49.9 pounds.

If, however, the sample size n is even, the median is defined as the mean of the $\left(\frac{n}{2}\right)$th and $\left(\frac{n}{2}+1\right)$th measurements. Suppose that a sixth measurement in the above example yields the value of 54.7 pounds, then the median is the mean of the $\frac{6}{2} = 3$rd and $(3 + 1) = 4$th observations; namely, $\frac{1}{2}[49.9 + 50.1] = 50.0$ lb. The median is usually denoted by $\tilde{x}$. In symbols, then

$$\begin{aligned} \tilde{x} &= x_{(n+1)/2} \quad \text{if } n \text{ is odd} \\ &= \frac{1}{2}(x_{n/2} + x_{n/2+1}) \quad \text{if } n \text{ is even} \end{aligned} \tag{5.3.11}$$

Suppose next that a sample (or a population) is divided into j equal parts after the measurements are arranged in *ascending* order. The i/jth *fractile* of observations is defined as the value of the variable below which the i/jth fractional part of the measurements are located. The symbol $F_{i/j}$ is used for the i/jth fractile. For example, $F_{1/4}$, the 1/4th fractile, indicates the value of the variable below which 1/4th of the sample measurements are located and $F_{4/5}$ represents the value below which 4/5th of the measurements are found. For a given set of n observations, $F_{i/j}$ is calculated as the $(i/j)[n + 1]$th ordered observation. For instance, $F_{1/2}$ for the data of Table 5.2.2 is $\frac{1}{2}(25 + 1)$th $= 13$th ordered measurement which is 68 inches. Again, $F_{3/13}$ for the same data is $\frac{3}{13}(25 + 1)$th $= 6$th ordered observation which is 65 inches. Whenever $(i/j)(n + 1)$ has a fractional part, linear interpolation is used to find the appropriate value. In the computation of $F_{3/4}$ for the data of Table 5.2.2 one gets $(3/4)(25 + 1) = 78/4 = 19.5$, which indicates to find half the distance between the 19th and the 20th observations. Since the 19th and the 20th observations are 71 and 72 inches respectively, 71.5 inches represents $F_{3/4}$ of this data.

The most commonly used fractiles are those for which j assumes the values 4, 10, and 100. The fractiles $F_{1/4}$, $F_{2/4}$, and $F_{3/4}$ are usually referred to as the 1st, 2nd, and 3rd *quartiles*, respectively. Similarly, the fractiles $F_{1/10}, F_{2/10}, \ldots, F_{9/10}$ are called the 1st, 2nd, . . . , 9th *deciles* and the fractiles $F_{1/100}, F_{2/100}, \ldots, F_{99/100}$ are called the 1st, 2nd, . . . , 99th *percentiles*. Sometimes the quartiles, the deciles, and the percentiles are designated by Q_i, D_i, and P_i, respectively. That is, $Q_i = F_{i/4}$, $D_i = F_{i/10}$, and $P_i = F_{i/100}$. It should be clear that $F_{1/2} = Q_2 = P_{50} = D_5$ represent the same quantity; namely, the median.

(e) The Mode. Another statistic which usually describes the central values of measurements is the *mode*. The mode is defined as the most frequently occurring value in a set of measurements. For example, for the data represented in Table 5.2.2, 68 inches is the mode. In case of tie, that is, situations where two values have equal frequencies, the data are said to be *bimodal.* The data 12, 13, 13, 13, 14, 14, 15, 16, 17, 17, 17, and 18 are bimodal because the measurements 13 and 17 occur with equal frequencies. If there are more than two values which have equal frequencies, the data are said to be *multimodal.* There are practical situations where the mode is more meaningful than either the mean or the median. For instance, if the following data represent the sizes of shoes sold in a certain store, $7\frac{1}{2}$, 8, 8, 8, 8, $8\frac{1}{2}$, 9, and 11, the salesman might be more interested in the mode, which is 8, rather than the mean, which is $8\frac{1}{2}$.

PROBLEMS

5.3.1 For the data given in Prob. 5.2.1 find (a) the range, (b) the mode, (c) the mean, (d) the median, (e) the variance, and (f) $F_{1/4}$ and $F_{3/4}$.

5.3.2 Answer the same questions as in Prob. 5.3.1 for the data given in Prob. 5.2.2.

5.3.3 If x_1 and x_2 are two measurements with positive but unequal values, show that $\overline{x}_h < \overline{x}_g < \overline{x}$.

5.3.4 If for a set of n observations $x_1 = x_2 = \cdots = x_n$, verify that $\overline{x}_h = \overline{x}_g = \overline{x}$.

5.3.5 Verify that for a set of n measurements which have all positive values, $\overline{x}_h \le \overline{x}_g \le \overline{x}$.

5.3.6 The following values represent the weekly wages of 20 college students in dollars:

40.30,	77.60,	99.00,	18.60,	61.20,
29.70,	67.40,	80.60,	76.30,	27.40,
12.80,	15.60,	22.80,	39.20,	35.90,
54.10,	27.60,	44.40,	82.70,	19.80

Calculate for this data: (a) the range, (b) the mean, (c) $F_{1/2}$, $F_{3/4}$, and $F_{9/10}$, and (d) the standard deviation.

5.3.7 Using logarithms calculate the geometric mean for the data of Prob. 5.3.6.

5.3.8 Find the mean and the standard deviation for the data of Prob. 5.2.3.

5.3.9 Show that for a set of data if the mean equals the median the data is not necessarily symmetric. (*Hint:* Try to construct a counterexample to the assumption that it is symmetric.)

5.3.10 Show that for a set of symmetric data the mean equals the median.

5.4 CALCULATIONS WITH LINEARLY TRANSFORMED DATA

(a) The Mean. Consider the transformation $u_i = \dfrac{x_i - k}{c}$, where both c and k are arbitrary constants. Applying the definition of the mean to the u's, we obtain

$$\bar{u} = \frac{\Sigma u_i}{n} = \frac{\Sigma\left(\frac{x_i - k}{c}\right)}{n}$$

$$= \frac{1}{cn}\Sigma(x_i - k)$$

$$= \frac{1}{cn}(\Sigma x_i - nk)$$

$$= \frac{\bar{x}}{c} - \frac{k}{c} = \frac{\bar{x} - k}{c} \tag{5.4.1}$$

or equivalently,

$$c\bar{u} + k = \bar{x} \tag{5.4.2}$$

The last two relationships are convenient and facilitate the computation of a mean greatly.

Example 5.4.1 Evaluate the mean for the data of Table 5.2.2 using the transformation $u_i = x_i - 68$.

Solution: $x_i - 68$ yields the following values of u_i's:

$$-6, -4, -4, -4, -3, -3, \ldots, 6, 8 \quad \text{and}$$

$$\bar{u} = \frac{\Sigma u_i}{25} = \frac{16}{25} = .64$$

Therefore, $\bar{x} = .64 + 68 = 68.64$. In this example $k = 68$ and $c = 1$.

Example 5.4.2 Using a linear transformation, evaluate the mean for the following set of data x_i:

1225, 1225, 1230, 1245, 1250, 1260,
1265, 1270, 1280, 1290

Solution: Let

$$u_i = \frac{x_i - 1225}{5}$$

$$u_i\text{: } 0, 0, 1, 4, 5, 7, 8, 9, 11, 13$$

$$\bar{u} = 5.8$$

Hence

$$\bar{x} = 5(5.8) + 1225$$

$$= 1254.0$$

(b) The Variance. As in the computation of the mean, a linear transformation facilitates the calculation of the variance also. If

$$u_i = \frac{x_i - k}{c}$$

$$s_u^2 = \frac{\Sigma(u_i - \bar{u})^2}{n - 1}$$

$$= \frac{\Sigma\left(\frac{x_i - k}{c} - \frac{\overline{x} - k}{c}\right)^2}{n - 1}$$

$$= \frac{1}{n - 1} \frac{\Sigma (x_i - \overline{x})^2}{c^2} = \frac{1}{c^2} s_x^2 \tag{5.4.3}$$

and therefore

$$c^2 s_u^2 = s_x^2 \tag{5.4.4}$$

Taking the square root of the expression in Eq. (5.4.4), we obtain

$$cs_u = s_x \tag{5.4.5}$$

Considering the special case where $c = 1$, one notes that $s_u = s_x$. That is, the variance and hence the standard deviation, is not altered by a translation in the x-axis.

Example 5.4.3 Calculate the standard deviation for the following data, by using the computational form and a linear transformation.

$$x_i\text{: } 60, 65, 70, 75, 80, 85$$

Solution: Let

$$u_i = \frac{x_i - 70}{5}$$

then

$$u_i\text{: } -2, -1, 0, 1, 2, 3$$

Now $\Sigma u_i = 3$ and $\Sigma u_i^2 = 19$. Using Eq. (5.3.10), one obtains

$$s_u^2 = \frac{1}{6 \times 5} [6(19) - 3^2] = 3.5$$

Therefore

$$s_x = 5\sqrt{3.5} = 9.35$$

PROBLEMS

5.4.1 Using the transformation $u_i = x_i - 131$, calculate the mean and the standard deviation for the data in Prob. 5.2.2.

5.4.2 Using a convenient transformation, find the mean and the standard deviation of the data in Prob. 5.3.6.

5.4.3 Use the transformation $u_i = \dfrac{x_i - 4.00}{.25}$ to find the mean and the variance of the data given in Prob. 5.2.4.

5.5 GROUPING OF DATA

When the number of measurements in a given data is large, ordering and constructing a frequency table could be very laborious if each value of the observations is to be recorded. Consider the situation, for instance, where 1,000 measurements are made and the range for the data is 250 units of measurements. To construct a frequency table for this data one needs to enter on the frequency table 250 numbers and their corresponding frequencies. This is unnecessary and impractical. A method often used to partly overcome this difficulty is *grouping*. To illustrate the notion of grouping with a concrete example, let us refer to Table 5.2.3. One notes there that several values of the variable have zero frequencies. If, however, each successive measurement were combined or *grouped* together in units of two, three, four, etc., most groups thus formed will contain some frequency, and no important information will be destroyed. Table 5.5.1 represents a grouped-frequency table for the data of Table 5.2.3 in which successive three units are grouped together.

TABLE 5.5.1

GROUPED-FREQUENCY TABLE FOR HEIGHT MEASUREMENTS OF 25 STUDENTS

x_i	Frequency
62–64	4
65–67	5
68–70	7
71–73	5
74–76	4

In Table 5.5.1 the various intervals, 62–64, 65–67, etc. are called *class intervals* which in this case represent equal subintervals of the range. The magnitude of these subintervals is called the *class interval size*. In this particular example there are five class intervals of size three. The number of class intervals and hence the class interval size basically depend on the range and the size of the sample. If the data are large and warrant grouping, a common rule of thumb is to construct a frequency table with k class intervals, where k is an integer between 5 and 15. A technical formula known as Sturges' rule suggests that k be taken approximately equal to $1 + 3.3 \log (n)$, where n is the sample size and log is the common logarithm. For example, if there were 100 observations in a set of data, Sturges' rule suggests that k be approximately equal to $1 + 3.3 \log (100) = 7.6$. Once the value of k is decided upon, the expression R/k, or range divided by k, gives an approximate value to consider for the choice of the class-interval size. The range for the data of Table 5.2.2 is 15, and suppose we arbitrarily choose the value of 8 for k, the class interval size should be approximately equal to 15/8. Therefore, 2 is the desired class-interval size. The frequency table with class-interval sizes of 2 for the data in Table 5.2.2 is presented in Table 5.5.2.

TABLE 5.5.2

HEIGHT MEASUREMENTS (INCHES) OF 25 STUDENTS

x_i	Frequency
62–63	1
64–65	5
66–67	3
68–69	5
70–71	4
72–73	3
74–75	3
76–77	1

Whenever data are grouped, certain information is lost. For instance, the class interval 64–65 in Table 5.5.2 shows a frequency of 5. Without referring to the original data one could not find the exact values for these five observations. That is, were these observations, 64, 64, 65, 65, and 65 inches, or were they 64, 64, 64, 64, and 65 inches or some other combination of possible values? Two assumptions are commonly made in grouping data: (1) either all frequencies in a particular class interval are situated at the *midmark* of that class interval, or (2) the frequencies in a class interval are equidistant from one another.

According to the first assumption, the five observations in the above class interval in question are located at 64.5, the midmark of the class interval 64–65. The value of 64.5 may seem questionable at first since the unit of measurement was an inch and there were no fractional values in the original data. In order to interpret fractional values in grouped data, another assumption is necessary. In grouping it is further assumed that the variable in question is continuous, although the measurements themselves might be discrete. That is, each discrete value of the measurement is assumed to represent the midmark of an interval. For example, the value of 62 inches represents a measurement between 61.5 and 62.5 and the value of 63 inches, a measurement between 62.5 and 63.5. It should be emphasized that the values 61.5, 62.5, 63.5, etc. cannot occur during the actual measurements because the unit of measurement is an inch. If the values 61.5, 62.5, 63.5, etc., could have been measured, then the unit of measurement for these observations is 0.1 inch, not 1 inch.

When data are grouped, a class interval such as 62–63 is assumed to represent the interval 61.5–63.5, a class interval 64–65 the interval 63.5–65.5, etc. The values of class interval which are represented in multiples of the unit of measurement are called *class limits*. The smaller number in such a class interval is called the *lower class limit* and the larger, the *upper class limit*. In Table 5.5.2 the values 62, 64, 66, . . . , 76 represent lower class limits and 63, 65, 67, . . . , 77 upper class limits. However, if one half a unit of measurement is subtracted from the lower class limit and one half a unit of measurement is added to the upper class limit, one obtains the *lower* and *upper class boundaries* respectively of a given class interval.

The lower class boundaries for Table 5.5.2 are 61.5, 63.5, . . . , 75.5 and the upper class boundaries are 63.5, 65.5, . . . , 77.5. The upper class boundaries of a class interval is equal to the lower class boundary of the next higher class interval, and hence there are no gaps in the continuum. Let us also notice that the upper class boundary minus the lower class boundary of a given class interval gives the class-interval size. For instance, 63.5–61.5 = 2 inches, which is the class-interval size for the data in Table 5.5.2.

Next let us consider the second assumption stated above. If the frequencies are to be equidistant from one another in a given class interval, one simply divides the class-interval size by the frequency in question, and distributes them evenly. Referring once again to the class interval 64–65 in Table 5.5.2, we note that the frequencies should be located at 2/5 = .4 unit apart. Therefore

$$\begin{aligned} 63.5 + .4 &= 63.9 \text{ inches} \\ 63.9 + .4 &= 64.3 \text{ inches} \\ 64.3 + .4 &= 64.7 \text{ inches} \\ 64.7 + .4 &= 65.1 \text{ inches} \\ 65.1 + .4 &= 65.5 \text{ inches} \end{aligned}$$

represent the values where the five frequencies are assumed to be found. Also, as seen from this example, it is also assumed that the last frequency is located at the upper class boundary. In the computation of the mean and the variance of the grouped data, it will be assumed that the frequencies are located at the midmarks, while in the computation of the fractiles it will be assumed that the frequencies are located equidistant from one another.

PROBLEMS

5.5.1 Using equal class intervals and making the smallest class interval 121–123, group the data of Prob. 5.2.2. Represent the result of this grouping in a histogram.

5.5.2 Using equal class intervals and making the smallest class interval .00–39.99, group the data of Prob. 5.2.3. Represent the result in a histogram.

5.5.3 Repeat Prob. 5.5.2, making the smallest class interval .00–24.99. Compare your results and comment on the differences between the histograms.

5.5.4 Using equal class intervals and beginning with .00–.99, group the data of Prob. 5.2.4.

5.6 COMPUTATION OF SOME STATISTICS FOR GROUPED DATA

(a) The Mean. Suppose one is required to calculate the mean for a set of grouped measurements. As a concrete example, consider the following data which represents the examination scores of 50 students.

TABLE 5.6.1

EXAMINATION SCORES OF 50 STUDENTS

Class Limits	Class Boundaries	Frequency	x_i
60–64	59.5–64.5	1	62
65–69	64.5–69.5	2	67
70–74	69.5–74.5	4	72
75–79	74.5–79.5	13	77
80–84	79.5–84.5	14	82
85–89	84.5–89.5	6	87
90–94	89.5–94.5	7	92
95–99	94.5–99.5	3	97
		50	

Now, in order to calculate the mean for this data, we assume that the frequencies in each interval are located at the respective midmarks and use the definition for the mean of ungrouped data. The class midmarks of Table 5.6.1 are 62, 67, 72, . . . , 97 respectively, which are designated by x_i. Denoting the corresponding frequencies by f_i, then, $\sum_{i=1}^{k} f_i x_i$ yields the total sum of the 50 scores, where k represents the number of class intervals. Therefore, $\sum_{i=1}^{8} f_i x_i = 1(62) + 2(67) + \cdots + 3(97) = 4090$ for the data of Table 5.6.1. Dividing the total sum by the total number of scores, gives the mean. That is, $4090/50 = 81.8$ is the mean of the 50 examination scores.

In general the mean for a grouped data is given by the relationship

$$\bar{x} = \frac{\sum_{i=1}^{k} f_i x_i}{\sum_{i=1}^{k} f_i} \tag{5.6.1}$$

where k is the number of class intervals, x_i the respective class midmark and $\sum_{i=1}^{k} f_i$ the number of measurements in the data.

In Sec. 5.4, the computation of the mean with linear transformation was presented. We shall apply the same principles in the calculation of the mean for grouped data.

Suppose $u_i = \dfrac{x_i - t}{c}$, where x_i's represent the class midmarks, (t and c are arbitrary constants). The mean of the u_i's according to Eq. (5.6.1) yields

$$\bar{u} = \frac{\sum_{i=1}^{k} f_i u_i}{\sum_{i=1}^{k} f_i} = \frac{\sum_{i=1}^{k} f_i\left(\frac{x_i - t}{c}\right)}{\sum_{i=1}^{k} f_i}$$

$$= \frac{\sum_{i=1}^{k} (f_i x_i - f_i t)}{c\sum_{i=1}^{k} f_i}$$

$$= \frac{\bar{x}}{c} - \frac{t}{c} = \frac{\bar{x} - t}{c} \qquad (5.6.2)$$

Equation (5.6.2) is identical with Eq. (5.4.1) which when solved for $\bar{x}$ yields

$$\bar{x} = c\bar{u} + t \qquad (5.6.3)$$

Suppose we apply the linear transformation $u_i = \dfrac{x_i - 77}{5}$ to the data of Table 5.6.1. The result is given in Table 5.6.2.

TABLE 5.6.2

EXAMINATION SCORES OF 50 STUDENTS

Class Limits	Frequency	x_i	$u_i = \frac{x_i - 77}{5}$
60–64	1	62	−3
65–69	2	67	−2
70–74	4	72	−1
75–79	13	77	0
80–84	14	82	1
85–89	6	87	2
90–94	7	92	3
95–99	3	97	4

Therefore,

$$\bar{u} = \frac{\sum_{i=1}^{8} f_i u_i}{\sum_{i=1}^{8} f_i} = \frac{-3 - 4 - 4 + 0 + 14 + 12 + 21 + 12}{50}$$

$$= \frac{48}{50} = .96$$

Hence by Eq. (5.6.3),

$$\bar{x} = 5(.96) + 77 = 81.8$$

which agrees with the previous calculation.

Let us remark that the arbitrary constants t and c could have been given any value whatsoever, but carefully chosen values would simplify the computation tremendously. In this example, t was equal to 77 and c equal to 5, which were a class midmark and the class interval size, respectively. In general, if t and c were chosen in this fashion, the linear transformation yields integer values which are more convenient to use in computations. Furthermore, if c is the class interval size, any midmark value used for t would yield integer values for u_i's. However, if t is selected from the central values of the midmarks, the u_i's would be centered near zero, which further simplifies the computations.

Example 5.6.1 Given the accompanying frequency table, calculate the mean for the data using the linear transformation $u_i = \dfrac{x_i - 59.5}{10}$.

TABLE 5.6.3

	f_i	$u_i = \dfrac{x_i - 59.5}{10}$	$f_i u_i$
25–34	1	−3	−3
35–44	9	−2	−18
45–54	12	−1	−12
55–64	11	0	0
65–74	5	1	5
75–84	2	2	4
	40		$\Sigma f_i u_i = -24$

Solution:

$$\bar{u} = \frac{-24}{40} = -.6$$

$$\bar{x} = 10(-.6) + 59.5 = -6 + 59.5 = 53.5$$

(b) Variance and Standard Deviation for Grouped Data. As in the computation of the mean for grouped data, in the computation of variance, it is assumed that the frequencies are located at the respective midmarks. With this assumption Eq. (5.3.6) becomes

$$s_x^2 = \frac{\sum_{i=1}^{k} f_i(x_i - \bar{x})^2}{\left(\sum_{i=1}^{k} f_i\right) - 1} \tag{5.6.4}$$

Denoting $\sum_{i=1}^{k} f_i = n$, Eq. (5.6.4) can be expressed as

$$s_x^2 = \frac{\sum_{i=1}^{k} f_i(x_i - \bar{x})^2}{n - 1} \tag{5.6.5}$$

Expanding the numerator of Eq. (5.6.5) and simplifying yields

$$s_x^2 = \frac{n\sum_{i=1}^{k} f_i x_i^2 - \left(\sum_{i=1}^{k} f_i x_i\right)^2}{n(n - 1)} \tag{5.6.6}$$

Expressions (5.6.5) and (5.6.6) are called the *definitional and computational forms of the variance* for the grouped data, respectively.

The computation of the variance for grouped data can be further simplified with a linear transformation. Making the transformation $x_i = cu_i + t$, and recalling that $\bar{x} = c\bar{u} + t$, expression (5.6.5) yields

$$\begin{aligned} s_x^2 &= \frac{\sum_{i=1}^{k} f_i\,[cu_i + t - (c\bar{u} + t)]^2}{n - 1} \\ &= \frac{c^2\sum_{i=1}^{k} f_i(u_i - \bar{u})^2}{n - 1} \\ &= c^2 s_u^2 \end{aligned} \tag{5.6.7}$$

which implies

$$s_x = cs_u \tag{5.6.8}$$

Substituting the computational form of s_u^2 in Eq. (5.6.7) one gets

$$s_x^2 = c^2\left[\frac{n\sum_{i=1}^{k} f_i u_i^2 - \left(\sum_{i=1}^{k} f_i u_i\right)^2}{n(n - 1)}\right] \tag{5.6.9}$$

or equivalently

$$s_x = c\sqrt{\frac{n\sum_{i=1}^{k} f_i u_i^2 - \left(\sum_{i=1}^{k} f_i u_i\right)^2}{n(n - 1)}} \tag{5.6.10}$$

Example 5.6.2 Using Eq. (5.6.10), evaluate the standard deviation for the data of Example 5.6.1.

Solution: One needs to calculate additionally the term $\sum_{i=1}^{k} f_i u_i^2$. Referring to Example 5.6.1, we obtain

$$\sum_{i=1}^{k} f_i u_i^2 = 1(-3)^2 + 9(-2)^2 + 12(-1)^2 + 11(0) + 5(1)^2 + 2(2)^2 = 70$$

Hence

$$s_x = 10 \sqrt{\frac{40(70) - (-24)^2}{40(39)}}$$

$$= 10 \sqrt{\frac{2224}{40 \times 39}} = 11.94$$

(c) **Fractiles.** $F_{i/j}$ for grouped data is similarly defined as for the ungrouped case. $F_{i/j}$ is the value of the variable below which i/j th fractional part of the measurements are located. Referring to Table 5.6.1, for example, we note that $F_{1/50} = 64.5$ and $F_{7/50} = 74.5$, since 1/50 of the data have values below 64.5 and 7/50 of the measurements values below 74.5. For the same data $F_{2/5} = 79.5$ and $F_{4/5} = 89.5$. In computing the fractiles for grouped data, it is assumed that the frequencies are equidistant within the class interval. $F_{1/4}$, $F_{1/2}$, and $F_{3/4}$ are defined as the first quartile, the median, and the third quartile, respectively. Here too, the first quartile is denoted by Q_1, the median by $\tilde{x}$, and the third quartile by Q_3. The median for the data under consideration is the value corresponding to the 25th measurement. Now, the 20th measurement is 79.5 and the 5 extra measurements needed should be taken from the next class interval, which has a frequency of 14. Therefore

$$\tilde{x} = 79.5 + \frac{5}{14}(5) = 81.29$$

Similarly, $F_{7/10}$ for the same data represents the value of the variable corresponding to the 35th measurement. The 34th measurement is 84.5 and the one extra measurement needed should come from the next class interval, which has a frequency of 6. Hence

$$F_{7/10} = 84.5 + \frac{1}{6}(5) = 85.33$$

PROBLEMS

5.6.1 Referring to Prob. 5.5.1, find (a) the mean, (b) the variance, (c) the median, (d) $F_{1/4}$, and (e) $F_{3/4}$.

5.6.2 Referring to Prob. 5.5.2, find (a) the mean, (b) the standard deviation, (c) $F_{6/10}$, and (d) the median.

5.6.3 Find the same quantities requested in Prob. 5.6.2 for the data of Prob. 5.5.4.

5.7 COMPUTATION OF STATISTICS FOR COMPOSITE DATA

Sometimes it is desirable and advantageous to combine two samples of sizes n_1 and n_2 taken from the same population or from different populations to form a single sample of size $n_1 + n_2$. The difficulty however, is that the original data might not be available to compute the required statistic. The information to which one might have access are statistics pertaining to the original samples. For instance, suppose that $n_1 = 15$, $n_2 = 10$, $\overline{x}_1 = 100$ and $\overline{x}_2 = 125$ for two samples. What is the mean of the composite data of the 25 observations. Furthermore, if $s_1^2 = 25$ and $s_2^2 = 50$, what is the value for the variance of the composite data of the 25 observations? We shall derive general expressions to answer questions of these types.

(a) The Mean of Composite Data. From the definition of the mean we have

$$\overline{x}_1 = \frac{\sum_{i=1}^{n_1} x_i}{n_1}$$

and

$$\overline{x}_2 = \frac{\sum_{i=1}^{n_2} x_i}{n_2}$$

Therefore

$$n_1\overline{x}_1 = \sum_{i=1}^{n_1} x_i$$

and

$$n_2\overline{x}_2 = \sum_{i=1}^{n_2} x_i$$

which when added yield

$$n_1\overline{x}_1 + n_2\overline{x}_2 = \sum_{i=1}^{n_1} x_i + \sum_{i=1}^{n_2} x_i \qquad (5.7.1)$$

Now the right-hand side of Eq. (5.7.1) represents the total sum for the composite data. Hence dividing it by $n_1 + n_2$ yields the value for the composite mean, $\bar{x}_c$. That is,

$$\bar{x}_c = \frac{n_1\bar{x}_1 + n_2\bar{x}_2}{n_1 + n_2} = \frac{\sum_{i=1}^{n_1} x_i + \sum_{i=1}^{n_2} x_i}{n_1 + n_2} \tag{5.7.2}$$

The student is asked to generalize Eq. (5.7.2) for the case of k samples of size $n_1, n_2, \ldots, n_k$. (See Prob. 5.7.5.)

The following example illustrates the concepts being discussed.

Example 5.7.1 Three samples of size 10, 15, and 20 have means 100, 110, and 95 respectively. Find the composite mean for the 45 observations.

Solution: Applying Eq. (5.7.2) twice one gets

$$\bar{x}_c = \frac{10(100) + 15(110) + 20(95)}{10 + 15 + 20}$$

$$= 101.11$$

(b) The Variance of Composite Data. The variance of a set of observations is defined by

$$s^2 = \frac{\sum_{i=1}^{n} (x_i - \bar{x})^2}{n - 1}$$

Applying this definition to a composite data of n_1 and n_2 observations yields

$$s_c^2 = \frac{\sum_{i=1}^{n_1} (x_i - \bar{x}_c)^2 + \sum_{i=1}^{n_2} (x_i - \bar{x}_c)^2}{n_1 + n_2 - 1} \tag{5.7.3}$$

which reduces to

$$s_c^2 = \frac{\sum_{i=1}^{n_1} x_i^2 + \sum_{i=1}^{n_2} x_i^2 - (n_1 + n_2)\bar{x}_c^2}{n_1 + n_2 - 1} \tag{5.7.4}$$

[Refer to Eq. (5.3.8)]. Likewise we get

$$(n_1 - 1)\, s_1^2 = \sum_{i=1}^{n_1} x_i^2 - n_1\bar{x}_1^2 \tag{5.7.5}$$

or equivalently,

$$\sum_{i=1}^{n_1} x_i^2 = (n_1 - 1)\, s_1^2 + n_1\bar{x}_1^2 \tag{5.7.6}$$

Similarly,

$$\sum_{i=1}^{n_2} x_i^2 = (n_2 - 1)\, s_2^2 + n_2\overline{x}_2^2 \tag{5.7.7}$$

Substituting Eqs. (5.7.6) and (5.7.7) in Eq. (5.7.4) we obtain

$$s_c^2 = \frac{(n_1 - 1)\, s_1^2 + (n_2 - 1)\, s_2^2 + n_1\overline{x}_1^2 + n_2\overline{x}_2^2 - (n_1 + n_2)\overline{x}_c^2}{n_1 + n_2 - 1} \tag{5.7.8}$$

This expression can be generalized for the case of k samples of size $n_1, n_2, \ldots, n_k$ which yields

$$s_c^2 = \frac{\sum_{i=1}^{k} (n_i - 1)s_i^2 + \sum_{i=1}^{k} n_i\overline{x}_i^2 - \sum_{i=1}^{k} n_i\overline{x}_c^2}{\left[\sum_{i=1}^{k} n_i\right] - 1} \tag{5.7.9}$$

Example 5.7.2 Suppose in Example 5.7.1 the respective variances are 30, 40, and 50. Find the composite variance for this data.

Solution: Applying Eq. (5.7.9) to this data we get

$$s_c^2 = \frac{9(30) + 14(40) + 19(50) + 10(100)^2 + 15(110)^2 + 20(95)^2 - 45(101.11)^2}{10 + 15 + 20 - 1}$$

$$= 84.876$$

PROBLEMS

5.7.1 Given two set of data:

$$n_1 = 20 \qquad n_2 = 50$$
$$\overline{x}_1 = 300 \qquad \overline{x}_2 = 280$$
$$s_1 = 12 \qquad s_2 = 15$$

Calculate the mean and the variance of the composite data.

5.7.2 Given a set of n observations with mean $\overline{x}$ and standard deviation s. Derive an expression for the variance of the new set, if a single observation x_k is (a) added to the original set, (b) deleted from the original set.

5.7.3 Generalize Prob. 5.7.2, part (b), for the case where k observations are deleted from the original set. $k < n$.

5.7.4 For a set of 46 observations the mean and the variance are computed to be 75 and 100 respectively. What will be the mean and the variance of the new set if the values 71, 59, 68, and 62 are (a) added to the set? (b) Deleted from the original observations?

5.7.5 Generalize Eq. (5.7.2) for the case of k samples of size $n_1, n_2, \ldots, n_k$.

5.8 RAPID ESTIMATION

In the introduction of this chapter it was mentioned that the primary purpose of statistics obtained from samples is for estimating and making inferences concerning the corresponding characteristics of the population. The notion of estimation and making inferences will be discussed in detail in Chapters 7 and 8 respectively. However, a brief introduction to these topics at this point, with an intuitive approach, we believe will be of practical value.

In Secs. 5.3 and 5.6 methods of computation for the sample mean, median, fractiles, standard deviation, etc., were presented. If the underlying population is known to be symmetric, then we know that the population mean and median are equal and consequently it is reasonable to infer that random samples from a symmetric population will yield approximately equal values for the mean and the median. Since the computation of the median [Eq. (5.3.11)], is relatively simple, one could use the value of the sample median for a rapid estimation of the population mean. In general, a rapid estimation for the mean of a symmetric population, based upon fractiles, is the expression

$$M = \frac{1}{2}[F_{i/j} + F_{(j-i)/j}] \tag{5.8.1}$$

Specific cases of Eq. (5.8.1) are:

1. $M = \frac{1}{2}[F_{1/(n+1)} + F_{n/(n+1)}]$ (5.8.2)

which represents the average of the smallest and the largest observations.

2. $M = \frac{1}{2}[F_{5/100} + F_{95/100}] = \frac{1}{2}[P_5 + P_{95}]$ (5.8.3)

which represents the mean of the 5th and 95th percentile, and

3. $M = \frac{1}{2}[P_{25} + P_{75}] = \frac{1}{2}[Q_1 + Q_3]$ (5.8.4)

which is the mean of the first and the third quartiles. Because interval $Q_1 < x < Q_3$ represents the middle portion of the range, Eq. (5.8.4) is commonly called the *mid-quartile*. It should be clear that $M = \frac{1}{2}[P_{50} + P_{50}] = P_{50} =$ median.

Although the above relations were assumed to be applicable to symmetric distributions, Eq. (5.8.4) could be applied to any moderately skewed distribution to yield a rapid estimate of the population mean.

In Sec. 3.2 Chebyshev's theorem was stated. This theorem could be looked upon as a rapid estimation for the population standard deviation. That is, the converse of the theorem implies that for *any* distribution,

$\left(1 - \frac{1}{k^2}\right)$ fractional part of the total distribution is located within k standard deviations of the mean. We can improve upon this conservative assertion if further information is available regarding the probability function. For instance, if the underlying distribution is unimodal and symmetric, it can be shown that at least $\left[1 - \frac{1}{(1.5k)^2}\right]$ fractional part of the distribution is located within k standard deviations of the mean. If the distribution of the random variable is assumed to obey a specific probability function, one could determine the standard deviation precisely. Suppose the distribution of the random variable in question is normal. Then, $P_{84.134} - P_{50}$ is equal to the standard deviation of the normal probability function. Again, $P_{97.5} - P_{2.5}$ is approximately 4 times the standard deviation and hence

$$S = \frac{1}{4}[P_{97.5} - P_{2.5}] \tag{5.8.5}$$

represents a rapid estimate for the population standard deviation. Sometimes the mean of P_{100} and P_{95} is substituted for $P_{97.5}$ and the mean of P_5 and P_0 for $P_{2.5}$ which reduces Eq. (5.8.5) to

$$S = \frac{1}{8}[(P_{100} + P_{95}) - (P_5 + P_0)] \tag{5.8.6}$$

where P_0 and P_{100} represent the smallest and the largest observations in an ungrouped data or the smallest lower class boundary and the largest upper class boundary in a grouped data, respectively.

Among other possible estimates for the standard deviation of a normally distributed population, based upon fractiles, are

$$S = \frac{1}{3}[F_{15/16} - F_{1/16}] \tag{5.8.7}$$

and

$$S = \frac{3}{4}[Q_3 - Q_1] \tag{5.8.8}$$

The last expression is particularly useful because Q_1 and Q_3 could also be used for rapid estimation of the population mean as shown above. Some authors refer to $Q_3 - Q_1$ as *interquartile range*.

Example 5.8.1 Assuming that the data in Example 5.6.1 is from a normally distributed population, use a rapid estimation to find an approximate value for the population mean and standard deviation.

Solution: From Example 5.6.1 we calculate

$$P_{25} = 44.5 + 10\left(\frac{0}{12}\right) = 44.5$$

$$P_{75} = 54.5 + 10\left(\frac{8}{11}\right) = 61.77$$

Therefore using Eqs. (5.8.4) and (5.8.8) we get

$$M = \frac{1}{2}[44.5 + 61.77] = 53.13$$

$$S = \frac{3}{4}[61.77 - 44.5] = 12.95$$

for rapid estimates of the population mean and the standard deviation. The actual sample mean and standard deviation were calculated to be 53.5 and 11.94.

PROBLEMS

5.8.1 Give a rapid estimate of the mean and the standard deviation for the data given in Prob. 5.2.1 using (a) $F_{1/16}$ and $F_{15/16}$, (b) $F_{1/4}$ and $F_{3/4}$. (c) Compare your results with Prob. 5.3.1, parts (c), (d), and (e).

5.8.2 Give a rapid estimate of the mean and the standard deviation for the data given in Prob. 5.2.2 using Eqs. (5.8.3) and (5.8.6).

5.8.3 Use Eq. (5.8.5) to obtain a rapid estimate for the standard deviation of the data in Prob. 5.3.6 and compare your result with part (d) of that problem.

6

Multivariate or Joint Probability Function

6.0 INTRODUCTION

In Chapters 2 and 4 when presenting the probability functions of discrete and continuous types respectively, the outcomes of an experiment were described by giving values to a single random variable. For instance, in the experiment of the breakdown voltage of transistors only the random variable v, the breakdown voltage, was the criterion of discussion. Suppose in that experiment the temperature of the transistor at the instant of breakdown is also considered. That is, the experimenter might consider a composite experiment with 2 random variables; namely, breakdown voltage and temperature. If the time at breakdown is also considered, there will be three variables under investigation. Probability functions describing two variables are referred to as *bivariate probability functions.* If there are more than two variables, the probability function is referred to as *multivariate probability function.* Both bivariate and multivariate probability functions are also referred to as *joint probability functions.* In this chapter we shall discuss joint probability functions and show some of their applications. The concepts of joint probability functions are necessary in the derivation of probability functions for sample statistics, which will be considered in Chapter 7.

6.1 JOINT PROBABILITY FUNCTION

If x_1 and x_2 are any two random variables and a function $f(x_1,x_2)$ exists, the conditions

$$f(x_1,x_2) \geq 0 \tag{6.1.1}$$

$$\sum_{x_1} \sum_{x_2} f(x_1,x_2) = 1 \tag{6.1.2}$$

define a bivariate probability function of the discrete type. Similarly, if an integrable function exists, the conditions

$$f(x_1,x_2) \geq 0 \tag{6.1.3}$$

$$\int_{-\infty}^{\infty} \int_{-\infty}^{\infty} f(x_1,x_2)\,dx_1\,dx_2 = 1 \tag{6.1.4}$$

define a bivariate probability function of the continuous type. Expressions (6.1.1) and (6.1.2) are analogous to Eqs. (2.1.1) and (2.1.2) respectively. The analogy between Eqs. (6.1.3) and (4.1.1) also between Eqs. (6.1.4) and (4.1.2) should be evident.

In Chapter 4 the total probability of a univariate probability function was compared to the area under a curve. The total probability in the continuous bivariate probability function can be compared to the volume under a surface.

Examples 6.1.1 and 6.1.2 illustrate bivariate probability functions of the continuous and the discrete type respectively.

Example 6.1.1 Given

$$\begin{aligned} f(x,y) &= e^{-x-y} \qquad x > 0, y > 0 \\ &= 0 \quad \text{otherwise} \end{aligned}$$

(a) Graph this function.
(b) Verify that $f(x,y)$ is a joint probability function.

Solution:

(a)

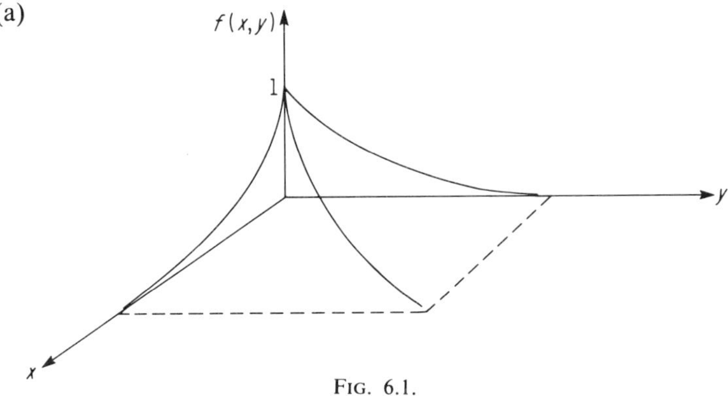

FIG. 6.1.

(b) Clearly, $e^{-x-y} \geq 0$, if $x > 0$ and $y > 0$. Now

$$\int_0^{\infty} \int_0^{\infty} e^{-x} e^{-y}\,dx\,dy = \int_0^{\infty} e^{-x}\,dx \int_0^{\infty} e^{-y}\,dy = 1$$

Therefore, $f(x,y)$ thus defined is a bivariate probability function.

Example 6.1.2 Given

$$\begin{aligned} f(x,y) &= \frac{1}{12} \qquad x = 1,2;\; y = 1,2,\ldots,6 \\ &= 0 \quad \text{otherwise} \end{aligned}$$

(a) Graph this function.
(b) Verify that it is a bivariate probability function.
Solution:
(a)

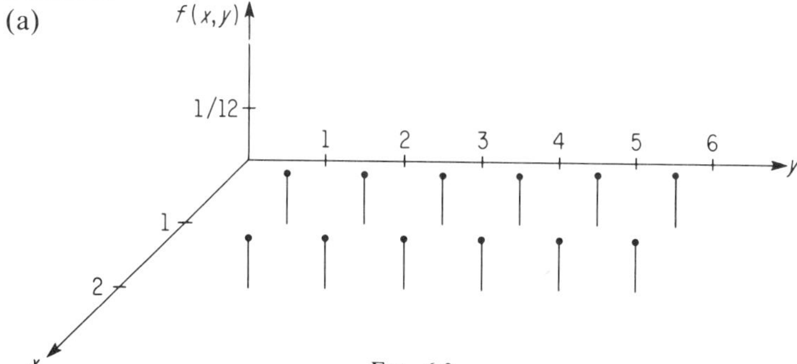

FIG. 6.2.

(b) $f(x,y) = \frac{1}{12} > 0$ and

$$\sum_{x=1}^{2} \sum_{y=1}^{6} f(x,y) = 2 \sum_{y=1}^{6} \frac{1}{12} = 2(6)\,\frac{1}{12} = 1$$

Referring to Fig. 1.4, we note that this bivariate probability function corresponds to the composite experiment of tossing a fair coin and a die simultaneously.

Following a similar discussion as in Sec. 4.1, the expression

$$P[a < x_1 < b; c < x_2 < d] = \int_a^b \int_c^d f(x_1,x_2)\,dx_1\,dx_2 \qquad (6.1.5)$$

represents the probability of x_1 falling in the interval between a and b, and x_2 in the interval between c and d.

Example 6.1.3 Given the bivariate probability function of Example 6.1.1, find $P[0 < x < 1, 1 < y < 2]$.
Solution:

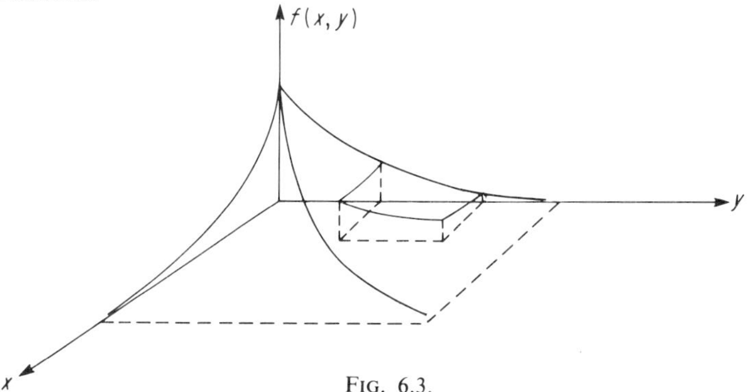

FIG. 6.3.

$$P[0 < x < 1, 1 < y < 2] = \int_0^1 \int_1^2 e^{-x-y}\,dx\,dy$$

$$= \int_0^1 e^{-x}\,dx \int_1^2 e^{-y}\,dy$$

$$= [1 - e^{-1}][e^{-1} - e^{-2}] = .147$$

The notion of a joint probability function can readily be extended to three or more variables. That is, if $x_1, x_2, \ldots, x_k$ are any k random variables and an integrable function $f(x_1, x_2, \ldots, x_k)$ exists, the two conditions

$$f(x_1, x_2, \ \ldots \ , x_k) \geq 0 \tag{6.1.6}$$

$$\int_{-\infty}^{\infty} \int_{-\infty}^{\infty} \cdots \int_{-\infty}^{\infty} f(x_1, x_2, \ \ldots \ , x_k)\,dx_1\,dx_2 \cdots dx_k = 1 \tag{6.1.7}$$

define a joint probability function of the continuous type. The discrete joint probability functions are defined similarly, the only difference being in Eq. (6.1.7), integration is replaced by summation over all values of each variable. The following is an example of a joint probability function of three random variables.

Example 6.1.4 Given $f(x,y,z) = kxyz \qquad 0 \leq x,y,z \leq 1$
$= 0 \quad$ otherwise

For what value of k is $f(x,y,z)$ a joint probability function?

Solution:

$$k \int_0^1 \int_0^1 \int_0^1 xyz\,dx\,dz = \frac{k}{8}$$

Therefore, $k = 8$ makes the given function a joint probability function.

The probability that x_1 falls between a_1 and b_1, x_2 between a_2 and b_2, x_3 between a_3 and b_3, etc., is given by the expression

$$P[a_1 < x_1 < b_1, a_2 < x_2 < b_2, \ \ldots \ , a_k < x_k < b_k]$$

$$= \int_{a_1}^{b_1} \int_{a_2}^{b_2} \cdots \int_{a_k}^{b_k} f(x_1, x_2, \ldots, x_k)\,dx_1\,dx_2 \cdots dx_k \tag{6.1.8}$$

Example 6.1.5 Find $P\left[0 < x < \frac{1}{2}, \frac{1}{4} < y < 1, z < 1\right]$ for the joint probability function of Example 6.1.4.

Solution:

$$8 \int_0^{1/2} \int_{1/4}^1 \int_0^1 xyz\,dx\,dy\,dz = 8 \int_0^{0.5} x\,dx \int_{0.25}^1 y\,dy \int_0^1 z\,dz = \frac{15}{64}$$

PROBLEMS

6.1.1 Given $f(x,y) = \frac{1}{4}xy \quad 0 < x < 1, 0 < y < 4$
$= 0 \quad$ otherwise
Show that $f(x, y)$ is a joint probability function and sketch the surface.

6.1.2 Given $f(x,y) = 2 - x - y \quad 0 < x < 1, 0 < y < 1$
$= 0 \quad$ otherwise
Show that $f(x,y)$ is a joint probability function.

6.1.3 Given $f(x,y) = \dfrac{k}{(1 + x + y)^3} \quad x > 0,\ y > 0.$ For what value of k is $f(x,y)$ a joint probability function?

6.1.4 Given $f(x,y,z) = ke^{-2x-3y-5z} \quad x,y,z > 0$
$= 0 \quad$ otherwise
Find the value of k for which $f(x,y,z)$ will represent a probability function.

6.1.5 Given $f(x,y) = k(1 - x^2 - y^2) \quad 0 \le x^2 + y^2 \le 1$
$= 0 \quad$ otherwise
Find the value of k for which $f(x,y)$ will be joint probability function and sketch the surface.

6.2 EXPECTATION OF JOINT PROBABILITY FUNCTION

The notion of expectation for a joint probability function is similar to the expectation discussed in conjunction with the univariate probability function. Here however, one needs to include all the variables under consideration instead of just a single variable. The expectation for the bivariate case is defined as

$$E[g(x_1,x_2)] = \sum_{x_1} \sum_{x_2} g(x_1,x_2) f(x_1,x_2) \tag{6.2.1}$$

if $f(x_1,x_2)$ is discrete and

$$E[g(x_1,x_2)] = \int_{-\infty}^{\infty} \int_{-\infty}^{\infty} g(x_1,x_2) f(x_1,x_2)\, dx_1\, dx_2 \tag{6.2.2}$$

if $f(x_1,x_2)$ is continuous.

The following example illustrates this definition.

Example 6.2.1 Find $E[xw]$ if

$$f(x,w) = \frac{4}{3}xw \qquad 0 \le x \le 1,\ 1 \le w \le 2$$
$$= 0 \qquad \text{otherwise}$$

Solution:

$$E[xw] = \int_0^1 \int_1^2 (xw)\frac{4}{3}(xw)\, dxdw = \int_0^1 \int_1^2 \frac{4}{3} x^2 w^2\, dxdw = 28/27$$

28/27 represents the average value of the product of the two random variables x and w.

The expectation for any probability function is defined as

$$E[g(x_1,x_2,\ldots,x_k)] = \sum_{x_1}\sum_{x_2}\cdots\sum_{xk} g(x_1,x_2,\ldots,x_k)f(x_1,x_2,\ldots,x_k) \tag{6.2.3}$$

if the joint probability function is discrete and

$$E[g(x_1,x_2,\ldots,x_k)] = \int_{-\infty}^{\infty}\int_{-\infty}^{\infty}\cdots\int_{-\infty}^{\infty} g(x_1,x_2,\ldots,x_k) \times f(x_1,x_2,\ldots,x_k)\,dx_1\,dx_2\cdots dx_k \tag{6.2.4}$$

if the joint probability function is continuous.

The moments of a joint probability function are also defined by a generalization of Eqs. (4.2.2), (4.2.3), and (4.2.4). For the bivariate case the moments about points a_1 and a_2 are defined as

$$m_{(k_1,a_1)(k_2,a_2)} = E[(x_1 - a_1)^{k_1}(x_2 - a_2)^{k_2}] \tag{6.2.5}$$

and the moments about the origin

$$m_{k_1,k_2} = E[x_1^{k_1})(x_2^{k_2})] \tag{6.2.6}$$

The moments about the respective means are defined as

$$u_{k_1,k_2} = E[(x_1 - E[x_1])^{k_1}(x_2 - E[x_2])^{k_2}] \tag{6.2.7}$$

In the expressions (6.2.5), (6.2.6), and (6.2.7) the quantity $k_1 + k_2$ is referred to as the *order of the moments.* The moments of order 0, 1, and 2 have already been encountered. For instance, $m_{1,0} = E[x_1]$, which represents the mean of the random variable x_1 and similarly $m_{0,1} = E[x_2]$ which is the mean of the random variable x_2. $m_{1,1} = E[x_1x_2]$ is called the *product moment.* In Example 6.2.1, a product moment was calculated. Again, $\mu_{2,0} = E[(x_1 - E[x_1])^2]$ which is the variance of the random variable x_1 and likewise $\mu_{0,2}$ represents the variance of the random variable x_2. The only other second-order moment about the mean that remains is $\mu_{1,1}$, that is, $E[(x_1 - E[x_1])(x_2 - E[x_2])]$. $\mu_{1,1}$ is called the *covariance* of the random variables x_1 and x_2. The symbol $\operatorname{cov}(x_1,x_2)$ is also used to designate $\mu_{1,1}$. The covariance plays an important role in regression analysis which will be discussed in Chapter 9.

PROBLEMS

6.2.1 Referring to the joint probability function of Prob. 6.1.1, find (a) $E[x]$, (b) $E[y]$, (c) $E[xy]$, and (d) the covariance.

6.2.2 Find $E[x]$ and $E[y]$ for the joint probability function of Prob. 6.1.2.

6.2.3 Find $E[xy]$ for the joint probability function of Prob. 6.1.5.

6.3 JOINT DISTRIBUTION FUNCTION

Associated with each joint probability function there is a joint distribution function. For the bivariate continuous case the joint distribution function is defined by

$$F(x_1,x_2) = \int_{-\infty}^{x_1} \int_{-\infty}^{x_2} f(t_1,t_2)\,dt_1\,dt_2 \tag{6.3.1}$$

The joint distribution function represented by the last equation possesses similar properties as the univariate distribution functions, namely

$$F(-\infty, x_2) = 0 \tag{6.3.2}$$

$$F(x_1, -\infty) = 0 \tag{6.3.3}$$

$$F(\infty, \infty) = 1 \tag{6.3.4}$$

$$P[a < x_1 < b, c < x_2 < d] = F(b,d) - F(a,d) - F(b,c) + F(a,c) \tag{6.3.5}$$

The student is requested to verify the above properties.

Example 6.3.1 (a) Find the joint distribution function of

$$f(x,y) = e^{-(x+y)} \qquad x > 0, y > 0$$
$$= 0 \qquad \text{otherwise}$$

(b) Using the result of part (a), evaluate $F(1,1)$ and sketch the corresponding volume.

Solution: In Example 6.1.1, it was shown that $f(x,y)$ represents a joint probability function. Hence

$$F(x,y) = \int_0^x \int_0^y e^{-(t+s)}\,dt\,ds$$

$$= \int_0^x e^{-t}\,dt \int_0^y e^{-s}\,ds$$

$$= [1 - e^{-x}][1 - e^{-y}]$$

is the joint distribution function, and hence

$$F(1,1) = [1 - e^{-1}][1 - e^{-1}] = [1 - e^{-1}]^2 = .399$$

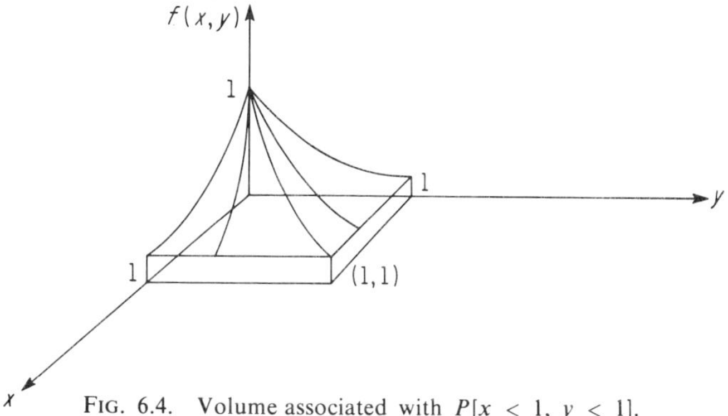

FIG. 6.4. Volume associated with $P[x < 1, y < 1]$.

Let us remark that the joint distribution function when partially differentiated with respect to x_1 and x_2 yields $f(x_1,x_2)$. That is,

$$\frac{\partial^2}{\partial x_1 \partial x_2} F(x_1,x_2) = f(x_1,x_2) \tag{6.3.6}$$

Equation (6.3.1) states that the joint distribution function can be derived if a joint probability function is given. Conversely, Eq. (6.3.6) states that the joint probability function can be derived from the joint distribution function.

PROBLEMS

6.3.1 Find the joint distribution function for the joint p.f. of Prob. 6.1.4 and evaluate $P[x < 1, 1 < y < 2, z < 1]$.

6.3.2 Find the joint distribution function for $f(x,y)$ given in Prob. 6.1.1.

6.3.3 Find the joint distribution function for the joint p.f. of Prob. 6.1.2.

6.4 MARGINAL AND CONDITIONAL PROBABILITY FUNCTIONS

Often an experimenter may perform a composite experiment, record all data, but examine only one variable at a time. For instance, an instructor might teach two subject matters, English and history, to the same group of students, test his students in both disciplines, but only examine the result of the English examination. Again, both height and weight of Navy personnel may be recorded, but at a given situation only the weight might be the criterion of consideration. If in the joint probability function $f(x_1,x_2)$ the random variable x_1 is considered irrespective of the second variable x_2, the resulting probability function is called the *marginal probability function of* x_1. If x_1 and x_2 are two discrete random variables, the marginal probability function of x_1 and x_2 are defined

$$f_1(x_1) = \sum_{x_2} f(x_1,x_2) \tag{6.4.1}$$

and

$$f_2(x_2) = \sum_{x_1} f(x_1,x_2) \tag{6.4.2}$$

respectively. In the case of continuous random variables,

$$f_1(x_1) = \int_{-\infty}^{\infty} f(x_1,x_2)\,dx_2 \tag{6.4.3}$$

$$f_2(x_2) = \int_{-\infty}^{\infty} f(x_1,x_2)\,dx_1 \tag{6.4.4}$$

Example 6.4.1 Find the marginal probability functions for the joint probability function given in Example 6.1.1.

Solution:

$$f_1(x) = \int_0^\infty f(x,y)\,dy$$

$$= \int_0^\infty e^{-x-y}\,dy = e^{-x}$$

$$f_2(y) = \int_0^\infty e^{-x-y}\,dx = e^{-y}$$

In general, given any joint probability function $f(x_1,x_2, \ldots, x_k)$, the marginal probability function of any subset of the variables is defined as the integral of the joint probability function with respect to all other variables between the limits of $-\infty$ and ∞. For example,

$$f_{2,4,5}(x_2,x_4,x_5) = \int_{-\infty}^{\infty}\int_{-\infty}^{\infty}\cdots\int_{-\infty}^{\infty} f(x_1,x_2, \ldots, x_k)\, dx_1\,dx_3\,dx_6 \cdots dx_k \qquad (6.4.5)$$

Example 6.4.2. Find $f_{12}(x,y)$, $f_{23}(y,z)$ and $f_{13}(x,z)$ for the joint probability function of Example 6.1.4.

Solution:

$$f_{12}(x,y) = \int_0^1 8xyz\,dz = \frac{8}{2}\,xyz^2\Big|_0^1 = 4xy$$

$$f_{23}(y,z) = \int_0^1 8xyz\,dx = 4yz$$

$$f_{13}(x,z) = \int_0^1 8xyz\,dy = 4xz$$

If the joint probability function is of the discrete type, the marginal probability function is defined as in Eq. (6.4.5) except integration is replaced by summation.

Related to the notion of marginal probability function is the *conditional probability function.* For the bivariate case, the conditional probability function of x_1 given x_2 is defined as

$$f(x_1 \mid x_2) = \frac{f(x_1,x_2)}{f_2(x_2)} \qquad (6.4.6)$$

where $f_2(x_2)$ is the marginal p.f. of x_2. Likewise,

$$f(x_2 \mid x_1) = \frac{f(x_1,x_2)}{f_1(x_1)} \qquad (6.4.7)$$

is the conditional p.f. of x_2 given x_1. In Eqs. (6.4.6) and (6.4.7) it is assumed that $f_1(x_1)$ and $f_2(x_2)$ are greater than zero.

The conditional probability function for the multivariate cases are similarly defined. For instance,

$$f(x_1,x_2 \mid x_3) = \frac{f(x_1,x_2,x_3)}{f_3(x_3)}$$

and

$$f(x_1,x_2 \mid x_3,x_4) = \frac{f(x_1,x_2,x_3,x_4)}{f_{34}(x_3,x_4)}$$

where $f_3(x_3)$ is the marginal probability function of x_3 and $f_{34}(x_3,x_4)$ represents the marginal probability function of x_3 and x_4.

Example 6.4.3 Find $f(x \mid y)$ and $f(y \mid x)$ for the probability function of Example 6.1.1.

Solution: From Example 6.4.1,

$$f_1(x) = e^{-x} \text{ and } f_2(y) = e^{-y}$$

Therefore,

$$f(x \mid y) = \frac{e^{-x-y}}{e^{-y}} = e^{-x} \text{ and } f(y \mid x) = e^{-y}$$

Example 6.4.4 Find $f(x \mid yz)$ and $f(xy \mid z)$ for the probability function of Example 6.4.2.

Solution:

$$f(x \mid yz) = \frac{f(x,y,z)}{f_{23}(y,z)} = \frac{8xyz}{4yz} = 2x$$

$$f(xy \mid z) = \frac{f(x,y,z)}{f_3(z)} = \frac{8xyz}{2z} = 4xy$$

PROBLEMS

In Probs. 6.4.1 to 6.4.4 find the marginal and the conditional probability functions for the specified joint probability functions.

6.4.1 The joint p.f. of Prob. 6.1.1.

6.4.2 The joint p.f. of Prob. 6.1.2.

6.4.3 The joint p.f. of Prob. 6.1.3.

6.4.4 The joint p.f. of Prob. 6.1.4.

6.4.5 Two random variables have the *circular bivariate normal* probability function represented by

$$f(x,y) = \frac{1}{2\pi\sigma^2} e^{-1/2\sigma^2[(x-\mu_1)^2+(y-\mu_2)^2]}$$

defined for all real x and y.

(a) Assuming $\mu_1 = \mu_2 = 0$ and $\sigma = 1$, sketch $f(x,y)$.

(b) Under the assumptions of part (a), find the probability that $f(x,y)$ is contained in the region between $x^2 + y^2 = 1$ and $x^2 + y^2 = 4$.

6.4.6 Hitting the target by an ICBM is assumed to obey the circular bivariate normal p.f. with $\sigma = 2$ miles. [See Prob. 6.4.5 above.] Find the probability that a missile will hit the target within a 1.5 mile radius.

6.4.7 Given

$$f(x,y) = \frac{1}{4} \qquad -1 < x < 1, -1 < y < 1$$
$$= 0 \quad \text{otherwise}$$

to be a joint probability function.

Find the probability associated with the region

(a) between the circles $x^2 + y^2 = 1/9$ and $x^2 + y^2 = 1/4$.
(b) $x > 1/2$ and $y > 1/2$.
(c) $x + y > 1$.

6.5 STOCHASTIC INDEPENDENCE

In Sec. 1.5 a random experiment was said to be stochastically independent if for all events in the partitioned sample space $P(E_i)P(E_j) = P(E_i E_j)$, $i \neq j$. Now we present the notion of stochastic independence for joint probability functions. A joint probability function $f(x_1,x_2, \ldots, x_k)$ is said to be stochastically independent if

$$f(x_1,x_2, \ldots, x_k) = f_1(x_1) f_2(x_2) \cdots f_k(x_k) \tag{6.5.1}$$

This definition simply states that the joint probability function can be expressed as the product of the marginal probability functions. For example, $f(x,y) = e^{-x-y}$ is a stochastically independent bivariate probability function because $e^{-x-y} = e^{-x}e^{-y}$, that is, $f(x,y) = f_1(x) f_2(y)$ where $f_1(x) = e^{-x}$ and $f_2(y) = e^{-y}$. Stochastically independent joint probability functions play a vital role in many applications. If the assumption of stochastic independence can be made for a given joint probability function, the computation of moments reduces to the product of univariate expectations. Referring to Example 6.2.1, one observes that

$$f(x,w) = \frac{4}{3} xw \qquad 0 \leq x \leq 1; 1 \leq w \leq 2$$
$$= 0 \qquad \text{otherwise}$$

is a stochastically independent joint probability function and hence

$$E[xw] = \int_0^1 \int_1^2 (xw) \frac{4}{3} xw \, dx dw = \int_0^1 2x^2 \, dx \int_1^2 \frac{2}{3} w^2 \, dw = E[x]E[w]$$

In general, if $f(x_1,x_2, \ldots, x_k)$ is a stochastically independent joint probability function and $g_1(x_1)$, $g_2(x_2)$, $\ldots$, $g_k(x_k)$ depend only on x_1, $x_2, \ldots, x_k$; respectively, then

$$E[g_1(x_1)g_2(x_2) \cdots g_k(x_k)] = E[g_1(x_1)]E[g_2(x_2)] \cdots E[g_k(x_k)] \tag{6.5.2}$$

To verify Eq. (6.5.2) we proceed as follows: Let $f(x_1,x_2, \ldots, x_k)$ be a continuous joint probability function. Then

$$E[g_1(x_1)g_2(x_2)\cdots g_k(x_k)] = \int_{-\infty}^{\infty}\cdots\int_{-\infty}^{\infty}[g_1(x_1)\cdots g_k(x_k)] \times [f(x_1,\ldots,x_k)]\,dx_1\cdots dx_k$$

$$= \int_{-\infty}^{\infty} g_1(x_1)f_1(x_1)\,dx_1 \cdots \int_{-\infty}^{\infty} g_k(x_k)f_k(x_k)\,dx_k$$

$$= E[g_1(x_1)]\cdots E[g_k(x_k)]$$

As a special case of Eq. (6.5.2), let us remark that the moment-generating function of k stochastically independent random variables is the product of the separate moment-generating functions—that is,

$$M_{x_1+x_2+\cdots+x_k}(t) = M_{x_1}(t)\cdot M_{x_2}(t)\cdots M_{x_k}(t) \tag{6.5.3}$$

which represents a very useful relationship.

PROBLEMS

6.5.1 Show that

$$f(x,y,z) = ke^{-x^2-y-4z} \qquad y,z > 0$$
$$= 0 \qquad \text{otherwise}$$

is stochastically independent.

6.5.2 Show that

$$f(x_1,x_2,\ldots,x_n) = ke^{-\left[\sum_{i=1}^{n}(a_ix_i-b_i)^2\right]}$$

where k, a_i's and b_i's are constants, is stochastically independent joint probability function.

6.5.3 State whether

$$f(x,y) = 2 - x - y \qquad 0 < x < 1,\ \ 0 < y < 1$$
$$= 0 \qquad \text{otherwise}$$

is a stochastically independent joint probability function.

6.5.4 Show that cov $(x,x) = \sigma_x^2$.

6.5.5 Show that if x and y are stochastically independent, cov $(x,y) = 0$.

6.5.6 Show that if a, b, c and d are constants, then cov $(ax + c, by + d) = ab$ cov (x,y).

7

Sampling Distributions—Probability Functions of Sample Statistics

7.0 INTRODUCTION

In Chapter 5 it was pointed out that the fundamental notion underlying statistical theory was to obtain information regarding a population by examining a subset of that population. Also, methods of selecting random samples from a population and the computation of certain statistics such as the mean, the median, the variance, etc., were discussed. However, no inductive inference from a statistic regarding the corresponding population parameter was considered. For instance, the random sample of 25 height measurements from a population of 1,000 yielded a mean of 68.64 inches. (Data of Table 5.5.2.) What assertion can one make regarding the mean of the population, based on the value of this statistic? Is the population mean 68.64 or 68.70 or 69.80 or 67.50 inches? In any inductive inference there is uncertainty present which can be best expressed by a probability statement. The major difficulty, however, lies in that the p.f. describing the statistic in question is in general unknown. In this chapter we shall present some probability functions describing the distribution of two statistics, the *mean* and the *variance*. The probability functions related to the mean are the unit normal and the Student's t and those related to the variance are the chi-square and the F probability functions.

7.1 SAMPLING DISTRIBUTION OF THE MEAN WHEN σ^2 IS KNOWN

(a) The Mean of the Sampling Distribution of the Mean. The theory associated with the probability functions of the sample means might be easier to discuss if we present a simple example first. Suppose a population has four elements x_1,x_2, x_3, and x_4 from which all samples of size 2 are taken. There will be $\binom{4}{2} = 6$ such samples if order is not taken into consideration. These samples are (x_1, x_2), (x_1, x_3), (x_1, x_4), (x_2, x_3), (x_2, x_4), and (x_3, x_4). Consider the means of these 6 samples and designate by $\bar{x}_1, \bar{x}_2, \bar{x}_3, \bar{x}_4, \bar{x}_5$, and $\bar{x}_6$. That is, $\bar{x}_1 = \dfrac{x_1 + x_2}{2}$ and so on. How are these sample means related to the population mean? Clearly, the mean of the population denoted by μ_x is equal to

$$\mu_x = \frac{1}{4}[x_1 + x_2 + x_3 + x_4]$$

Now the mean of the 6 sample means denoted by $\mu_{\bar{x}}$ is equal to

$$\begin{aligned}\mu_{\bar{x}} &= \frac{1}{6}(\bar{x}_1 + \bar{x}_2 + \bar{x}_3 + \bar{x}_4 + \bar{x}_5 + \bar{x}_6) \\ &= \frac{1}{6}\left[\frac{1}{2}(x_1 + x_2) + \frac{1}{2}(x_1 + x_3) + \cdots + \frac{1}{2}(x_3 + x_4)\right] \\ &= \frac{1}{12}[3(x_1 + x_2 + x_3 + x_4)] = \frac{1}{4}(x_1 + x_2 + x_3 + x_4)\end{aligned}$$

which for this example shows that the mean of the population is equal to the mean of the sample means. This relationship is true in general. That is, the mean of all possible samples of size n from a finite population is equal to the population mean. We show this next.

There are $\binom{N}{n} = K$ possible samples of size n from a population with N elements. Denote these sample means by $\bar{x}_1, \bar{x}_2, \ldots, \bar{x}_K$. That is

$$\bar{x}_1 = \frac{\sum_{i=1}^{n} x_{i1}}{n}, \quad \bar{x}_2 = \frac{\sum_{i=1}^{n} x_{i2}}{n}, \quad \ldots, \quad \bar{x}_K = \frac{\sum_{i=1}^{n} x_{iK}}{n}$$

The mean of the K sample means is equal to

$$\mu_{\bar{x}} = \frac{\sum_{j=1}^{K} \bar{x}_j}{K} = \frac{\sum_{i=1}^{n} x_{i1} + \sum_{i=1}^{n} x_{i2} + \cdots + \sum_{i=1}^{n} x_{iK}}{nK} \tag{7.1.1}$$

The numerator of Eq. (7.1.1) when expanded contains $\binom{N-1}{n-1}$ of each element of the population and hence can be expressed as

$$\mu_{\bar{x}} = \frac{\binom{N-1}{n-1}\sum_{i=1}^{N} x_i}{n\binom{N}{n}} \tag{7.1.2}$$

Recalling that $n\binom{N}{n} = N\binom{N-1}{n-1}$ and substituting into Eq. (7.1.2), we obtain

$$\mu_{\bar{x}} = \frac{\sum_{i=1}^{N} x_i}{N} = \mu_x \tag{7.1.3}$$

(b) The Variance of the Sampling Distribution of the Mean. Next we wish to show the relationship between the population variance and the variance of the distribution of the sample means. Let us introduce this notion by an example also. Referring to the aforementioned example, we note that

$$\sigma_x^2 = \frac{\sum_{i=1}^{4} (x_i - \mu_x)^2}{4} = \frac{1}{4}\sum_{i=1}^{4} x_i^2 - \mu_x^2 \tag{7.1.4}$$

The reason for dividing by 4 instead of by (4 − 1) is because the variance of a population is being calculated. Similarly, the variance of the $\bar{x}$'s is

$$\sigma_{\bar{x}}^2 = \frac{\sum_{i=1}^{6} (\bar{x}_i - \mu_x)^2}{6} = \frac{1}{6}\sum_{i=1}^{6} \bar{x}_i^2 - \mu_x^2 \tag{7.1.5}$$

Recalling that

$$\bar{x}_1 = \frac{x_1 + x_2}{2}, \bar{x}_2 = \frac{x_1 + x_3}{2}, \ldots, \bar{x}_6 = \frac{x_3 + x_4}{2}$$

and substituting these values in Eq. (7.1.5) we get

$$\sigma_{\bar{x}}^2 = \frac{1}{6}\left[\left(\frac{x_1 + x_2}{2}\right)^2 + \left(\frac{x_1 + x_3}{2}\right)^2 + \cdots + \left(\frac{x_3 + x_4}{2}\right)^2\right] - \mu_x^2$$

$$= \frac{1}{24}[3(x_1^2 + x_2^2 + x_3^2 + x_4^2)$$

$$+ 2(x_1x_2 + x_1x_3 + x_1x_4 + x_2x_3 + x_2x_4 + x_3x_4)] - \mu_x^2$$

$$= \frac{1}{24}\left[\sum_{i=1}^{4} x_i^2 + 2(x_1x_2 + x_1x_3 + \cdots + x_3x_4) + 2\sum_{i=1}^{4} x_i^2\right] - \mu_x^2$$

$$= \frac{1}{24}\left[\left(\sum_{i=1}^{4} x_i\right)^2 + 2\sum_{i=1}^{4} x_i^2\right] - \mu_x^2$$

$$= \frac{2}{3}\left(\frac{\sum_{i=1}^{4} x_i}{4}\right)^2 + \frac{1}{12}\sum_{i=1}^{4} x_i^2 - \mu_x^2$$

$$= \frac{2}{3}\mu_x^2 + \frac{1}{12}\sum_{i=1}^{4} x_i^2 - \mu_x^2$$

$$= \frac{1}{3}\left[\frac{1}{4}\sum_{i=1}^{4} x_i^2 - \mu_x^2\right] = \frac{\sigma_x^2}{3} \tag{7.1.6}$$

In general, following similar steps as in the above example, it can be shown that the variance of the sample means is related to the population variance by the following formula:

$$\sigma_{\bar{x}}^2 = \frac{N-n}{N-1}\frac{\sigma_x^2}{n} \tag{7.1.7}$$

where N is the population size and n the sample size. As $N \longrightarrow \infty$, Eq. (7.1.7) becomes

$$\sigma_{\bar{x}}^2 = \frac{\sigma_x^2}{n} \tag{7.1.8}$$

That is, the variance of the sample means from an infinite population is equal to the population variance divided by the sample size. A direct proof of Eq. (7.1.8) requires the definition of random samples from an infinite population. However, prior to stating this definition, let us illustrate the concept to be defined by an example. Consider the exponential p.f.:

$$\begin{aligned} f(x) &= e^{-x} & x > 0 \\ &= 0 & \text{otherwise} \end{aligned}$$

Since the positive real numbers do not form a denumerable set, this $f(x)$ is the p.f. of an infinite population. Let (x_{11}, x_{12}, x_{13}) represent a sample of size three from $f(x)$. Let (x_{21}, x_{22}, x_{23}) represent another sample of size three from $f(x)$ and repeating this process k times we obtain (x_{k1}, x_{k2}, x_{k3}) as our kth sample of size three.

Now suppose we consider $(x_{11}, x_{21}, \ldots, x_{k1})$ as a sample of size k and designate its p.f. by $f_1(x_1)$. Clearly $f(x) = f_1(x_1)$. Similarly, $(x_{12}, x_{22}, \ldots, x_{k2})$ and $(x_{13}, x_{23}, \ldots, x_{k3})$ constitute samples of size k whose probability functions $f_2(x_2)$ and $f_3(x_3)$ respectively are equal to $f(x)$. That is,

$$\begin{aligned} f(x) = f_1(x_1) = f_2(x_2) = f_3(x_3) &= e^{-x} & x > 0 \\ &= 0 & \text{otherwise} \end{aligned}$$

Furthermore, assuming stochastic independence, the joint p.f. of the random variables x_1, x_2, x_3 becomes

$$\begin{aligned} f(x_1, x_2, x_3) &= e^{-x_1-x_2-x_3} & x_i > 0, \quad i = 1,2,3 \\ &= 0 & \text{otherwise} \end{aligned}$$

The definition of a random sample of size n from an infinite population is based upon the assumption of stochastic independence. If the joint p.f. of n random variables $x_1, x_2, \ldots, x_n$, say $f(x_1, x_2, \ldots, x_n)$ is equal to $f(x_1)\cdot f(x_2)\cdots f(x_n)$ then $x_1, x_2, \ldots, x_n$ is said to be a random sample of size n from $f(x)$.

At this point let us verify that the mean of all possible samples of size n from an infinite population is equal to the population mean—that is, $E[\bar{x}] = \mu_x$:

$$\begin{aligned}
\mu_{\bar{x}} = E[\bar{x}] &= E\left[\frac{1}{n}\sum_{i=1}^{n} x_i\right] = \frac{1}{n} E\left[\sum_{i=1}^{n} x_i\right] \\
&= \frac{1}{n}\int_{-\infty}^{\infty}\int_{-\infty}^{\infty}\cdots\int_{-\infty}^{\infty}(x_1 + x_2 + \cdots + x_n) \\
&\quad \times f(x_1, x_2, \ldots, x_n)\,dx_1\,dx_2\cdots dx_n \\
&= \frac{1}{n}\left[\int_{-\infty}^{\infty} x_1 f(x_1)\,dx_1 + \int_{-\infty}^{\infty} x_2 f(x_2)\,dx_2 + \cdots \right. \\
&\quad \left. + \int_{-\infty}^{\infty} x_n f(x_n)\,dx_n\right] \\
&= \frac{1}{n}[n\mu_x] = \mu_x
\end{aligned} \tag{7.1.9}$$

which agrees with expression (7.1.3). The result of the following theorem is needed in the proof of Eq. (7.1.8).

Theorem 7.1.1 Let $x_1, x_2, \ldots, x_k$ represent k stochastically independent random variables with respective means $\mu_1, \mu_2, \ldots, \mu_k$ and variances $\sigma_1^2, \sigma_2^2, \ldots, \sigma_k^2$. Then for any arbitrary constants $c_1, c_2, \ldots, c_k$,

$$y = \sum_{i=1}^{k} c_i x_i$$

is a random variable with mean

$$\mu_y = \sum_{i=1}^{k} c_i \mu_i$$

and variance

$$\sigma_y^2 = \sum_{i=1}^{k} c_i^2 \sigma_i^2$$

Proof: By the definition of the mean we have

$$\mu_y = E[y] = E\left[\sum_{i=1}^{k} c_i x_i\right]$$

$$= \sum_{i=1}^{k} c_i E[x_i] \quad \text{(See Prob. 4.2.1c)}$$

$$= \sum_{i=1}^{k} c_i \mu_i \tag{7.1.10}$$

Similarly, for the composite variance we have

$$\sigma_y^2 = E[(y - \mu_y)^2]$$

$$= E\left[\left(\sum_{i=1}^{k} c_i x_i - \sum_{i=1}^{k} c_i \mu_i\right)^2\right]$$

$$= E\left[\left(\sum_{i=1}^{k} c_i (x_i - \mu_i)\right)^2\right]$$

$$= E\left[\sum_{i=1}^{k} c_i^2 (x_i - \mu_i)^2\right] \quad \text{(because of stochastic independence)}$$

$$= \sum_{i=1}^{k} c_i^2 E(x_i - \mu_i)^2$$

$$= \sum_{i=1}^{k} c_i^2 \sigma_i^2 \tag{7.1.11}$$

Example 7.1.1 A random variable x has a mean of 17.5 and a variance of 14.4. Another random variable y has a mean 10.0 and variance of 19.6. If x and y are stochastically independent, find the mean and the variance of the random variable $w = 3x - 2y$.

Solution: Applying Eq. (7.1.10) to obtain the composite mean, we get

$$\mu_w = 3\mu_x - 2\mu_y$$
$$3(17.5) - 2(10.0) = 32.5$$

Similarly, applying Eq. (7.1.11) we obtain

$$\sigma_w^2 = 3^2(14.4) + (-2)^2(19.6)$$
$$= 129.6 + 78.4 = 208.0$$

Now we show that $\sigma_{\bar{x}}^2 = \dfrac{\sigma_x^2}{n}$:

$$\sigma_{\bar{x}}^2 = E[(\bar{x} - \mu)^2]$$

$$= E\left[\left(\frac{1}{n}\sum_{i=1}^{n} x_i - \mu\right)^2\right]$$

$$= \frac{1}{n^2} E\left[\left(\sum_{i=1}^{n} x_i - n\mu\right)^2\right] \quad \text{(by Theorem 7.1.1)}$$

$$= \frac{1}{n^2} E\left[\sum_{i=1}^{n} (x_i - \mu)^2\right] \quad \text{(because of stochastic independence)}$$

$$= \frac{1}{n^2}\sum_{i=1}^{n} E[(x_i - \mu)^2]$$

$$= \frac{1}{n^2}\sum_{i=1}^{n} \sigma_x^2 = \frac{1}{n^2} n\,\sigma_x^2 = \frac{\sigma_x^2}{n}$$

The next theorem is given as a résumé of the preceeding discussions.

Theorem 7.1.2 If from a population with mean μ_x and variance σ_x^2 a random sample of size n is taken, the sample mean $\bar{x}$ is a random variable whose probability function has a mean μ_x. If the population is finite, the variance of the distribution of $\bar{x}$'s is $\left(\frac{N-n}{N-1}\right)\left(\frac{\sigma_x^2}{n}\right)$ and if the population is infinite, the variance is $\frac{\sigma_x^2}{n}$.

Example 7.1.2 If samples of size 4 are taken from the unit normal p.f., find the mean and the variance of the probability function of the sample means.

Solution: The mean and the variance of the unit normal p.f. are 0 and 1 respectively. By theorem 7.1.2 we get $\mu_{\bar{x}} = 0$ and $\sigma_{\bar{x}}^2 = 1/4$.

(c) Central Limit Theorem. One of the most useful and important theorems in statistical theory and applications is the following theorem, commonly referred to as the *central limit theorem.* We state this theorem without proof.

Theorem 7.1.3 Let $\bar{x}$ be the mean of a random sample of size n, taken from a population with mean μ and finite variance σ^2. The probability function of the random variable $\frac{(\bar{x} - \mu)}{\sigma/\sqrt{n}}$ approaches the unit normal, as n increases without bound. More precisely,

$$\lim_{n \to \infty} P\left[\frac{\bar{x} - \mu}{\sigma/\sqrt{n}} < z\right] = \frac{1}{\sqrt{2\pi}} \int_{-\infty}^{z} e^{-t^2/2}\, dt = F(z)$$

The most significant characteristic of the central limit theorem is that it applies to any population whether discrete or continuous, symmetric or skewed. Let us consider the following example to illustrate this theorem.

Example 7.1.3 Given the uniform probability function

$$f(x) = \frac{1}{6} \qquad x = 1,2,3,4,5,6$$

$$= 0 \qquad \text{otherwise}$$

Consider all possible samples of size 2 (order relevant) from this population and draw the line graph for $f(x)$ and the distribution of the sample means.

Solution: There are $6^2 = 36$ possible samples of size 2. Frequencies of these sample means are as follows:

$\bar{x}$	f_i
1.0	1
1.5	2
2.0	3
2.5	4
3.0	5
3.5	6
4.0	5
4.5	4
5.0	3
5.5	2
6.0	1

The line graph of $f(x)$ is given in Fig. 7.1 and the line graph for $f(\bar{x})$ is presented in Fig. 7.2.

Comparing Figs. 7.1 and 7.2 one notes the probability function of the sample means approaching the normal probability function. Also note that the means of $f(x)$ and $f(\bar{x})$ are equal to 3.5. Furthermore, the variance of $f(\bar{x})$ is smaller than the variance of $f(x)$, which the reader is requested to verify.

Let us remark for emphasis that the population moments must be finite and σ^2 known in order to apply the central limit theorem. This restrictive condition may seem alarming at first, but since most probability functions do have finite variances the central limit theorem still remains as one of the most useful tools in statistical theory.

Example 7.1.4 A random sample of size 25 from a large population yields $\bar{x} = 68.64$ inches. Assuming $\sigma_{\bar{x}}^2 = 9$, find the probability that μ_x does exceed 70 inches.

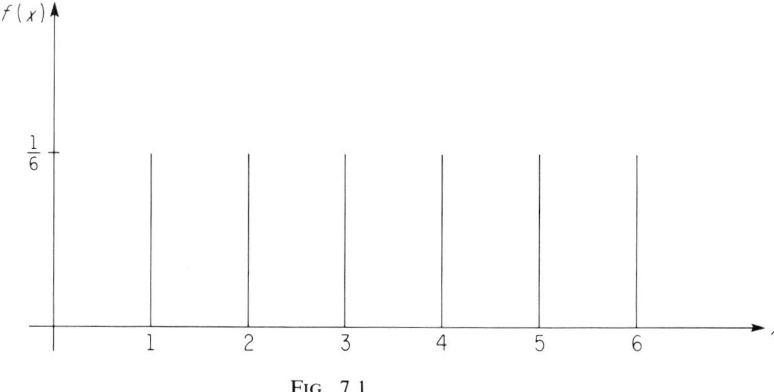

FIG. 7.1.

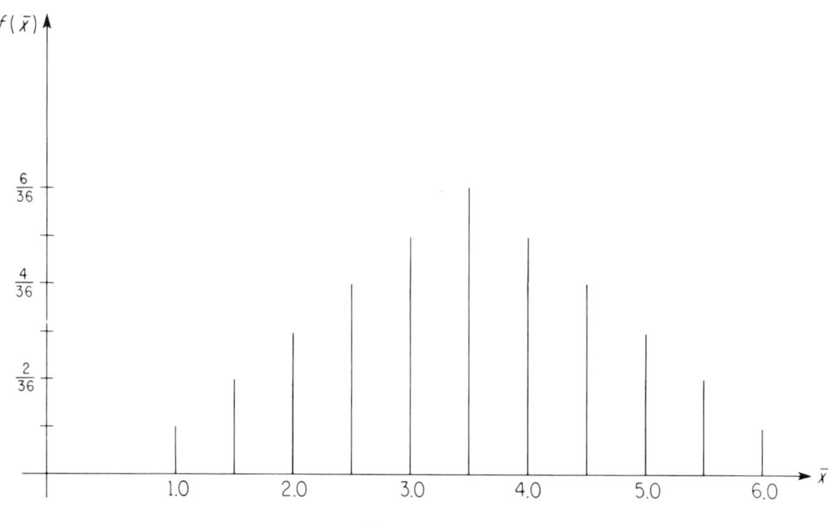

FIG. 7.2.

Solution: Since $E[\overline{x}] = \mu_x$, we let $\mu_x = 68.64$ and apply the central limit theorem. That is,

$$z = \frac{70 - 68.64}{\dfrac{3}{\sqrt{25}}} = 2.266$$

From Table IV(b) in the Appendix we get $P[z > 2.266] = 0.01170$

(d) Law of Large Numbers. If we apply Chebyshev's theorem to the sampling distribution of the mean, a stronger probability statement could be made about the concentration of probability around its mean than merely stating that Eq. (7.1.8) holds.

According to Chebyshev's Theorem [Eq. (3.2.30)] for any $k > 0$,

$$P[\,|\,\overline{x} - \mu\,| > k\,\sigma_{\overline{x}}] < \frac{1}{k^2}$$

or equivalently

$$P\left[\,|\,\overline{x} - \mu\,| > k\,\frac{\sigma_x}{\sqrt{n}}\right] < \frac{1}{k^2}$$

Letting $\epsilon = k\,\dfrac{\sigma_x}{\sqrt{n}}$ and noting that $\epsilon > 0$ we get

$$P[\,|\,\overline{x} - \mu\,| > \epsilon] < \frac{\sigma^2}{\epsilon^2 n} \tag{7.1.12}$$

In words, Eq. (7.1.12) states that the probability of the sample mean $\overline{x}$ does not differ from the population mean by more than ϵ, and could be made as small as we please, provided the sample size n is chosen sufficiently large.

The following example illustrates the notion under consideration.

Example 7.1.5 Using the result of Eq. (7.1.12) determine the minimum value of the sample size n, taken from the Poisson p.f. $f(x\colon\ 16)$ such that

$$\text{(a)}\quad P[\,|\,\overline{x} - \mu_x\,| > 1] < .80$$
$$\text{(b)}\quad P[\,|\,\overline{x} - \mu_x\,| > .5] < .90$$

Solution: (a) Recalling that the mean and the variance of the Poisson p.f. $f(x\colon\ 16)$ are both equal to 16, and noting that $\epsilon = 1$, we get

$$\frac{\sigma^2}{\epsilon^2 n} = \frac{16}{n} = .8$$

and hence $n = 20$. That is, a sample size of at least 20 is necessary to satisfy the stipulation of the problem.

(b) In this case we note that $\epsilon = .5$ and $\dfrac{\sigma^2}{\epsilon^2 n} = .90$ which yields $\dfrac{16}{(.5)^2 n}$ $= .90$ or

$$n = \frac{16}{(.25)(.9)} = \frac{64}{.9} = 71.11$$

Since n cannot be a fraction, n should be the next larger integer; namely 72.

Next let us apply the Chebyshev's theorem to the binomial p.f. which yields

$$P[\,|\,x - np\,| > k\sqrt{npq}] < \frac{1}{k^2}$$

Dividing the inequality inside the brackets by n we get

$$P\left[\left|\frac{x}{n} - p\right| > k\sqrt{\frac{pq}{n}}\right] < \frac{1}{k^2}$$

where $q = 1 - p$. Letting $\epsilon = k\sqrt{\frac{pq}{n}}$ and recalling that $k > 0$, Chebyshev's theorem becomes

$$P\left[\left|\frac{x}{n} - p\right| > \epsilon\right] < \frac{pq}{\epsilon^2 n}$$

and hence

$$\lim_{n \to \infty} P\left[\left|\frac{x}{n} - p\right| > \epsilon\right] = 0 \qquad (7.1.13)$$

This last equation is called *the (weak) law of large numbers*, which states that if trials are of the binomial type and n is large, the sample proportion of successes, x/n, is an approximation of the true proportion p, and that this approximation improves as n increases without bound. For example, if we wish to find the true proportion of registered Republicans in a community without having access to the registration list, we can take samples of size, say 100 or 200, and estimate the true proportion of Republicans. Intuitively, as the sample size increases one gets better approximation of the true proportion. We shall consider inferences of this type further in Sec. 7.4.

PROBLEMS

7.1.1 Verify Eq. (7.1.3) for a population with 5 elements from which samples of size 3 are taken.

7.1.2 Verify Eq. (7.1.7) under the same conditions of Prob. 7.1.1

7.1.3 The random digits have a mean value of 4.5. Using Table I in the Appendix, take 25 samples of 2 digits, calculate the means for the 2 digits and put your result in a tabular form. Will the mean of these 25 means be equal to 4.5? Why?

7.1.4 If it is assumed that the heights of Air Force personnel are distributed according to the normal probability function with variance 2.25, what is the probability that among 10 randomly selected officers the sample mean does not differ from the population mean by more than 1 inch?

7.1.5 Graph the sampling distribution of the means of all possible samples of size 16 from a normal p.f. with mean 15 and variance 64.

7.1.6 Within what interval would you expect to find the middle 80 percent of the sample means in Prob. 7.1.5?

7.1.7 Two stochastically independent random variables x and y are distributed according to the normal p.f. with equal means and standard deviations of 6 and 8 respectively. (a) Find the sampling distributions of the random variable

$w = x - y$. (b) Find $P[w > 5]$ and $P[-10 < w < 10]$. [*Hint:* $M_w(t) = M_x(t) \cdot M_y(t)$.]

7.1.8 The standard deviation of a population comprised of 101 elements is 16. Find the variance of the sample means if random samples of size (a) 4, (b) 8, and (c) 16 are taken from the given population.

7.2 SAMPLING DISTRIBUTION OF THE MEAN WHEN σ^2 IS UNKNOWN

In order to apply the central limit theorem it was indicated in the previous section that the population variance must be known. However, the assumption of the knowledge of the population variance is unrealistic in most situations. Therefore, a probability function independent of the population variance is required for the probability function describing the sample mean. W. S. Gosset discovered the use of the t distribution at the beginning of the twentieth century (1907–1908). This probability function is commonly referred to as the Student's t distribution because Gosset wrote under the pseudonym "Student." In the following section the Student's p.f. will be derived. Prior to doing that, we will discuss the beta and the Cauchy p.f.'s which will be needed in the derivation of the t distribution.

(a) The Beta p.f. The beta probability function has many practical applications and is related to the gamma probability function discussed in Sec. 4.3. Consider Eq. (4.3.2)—that is,

$$\Gamma(n) = \int_0^\infty x^{n-1} e^{-x}\, dx$$

When the transformation $x = u^2/2$ is substituted in this expression it becomes

$$\Gamma(n) = \int_0^\infty \left(\frac{u^2}{2}\right)^{n-1} e^{-u^2/2}\, u\, du \tag{7.2.1}$$

Similarly for another parameter m,

$$\Gamma(m) = \int_0^\infty \left(\frac{v^2}{2}\right)^{m-1} e^{-v^2/2}\, v\, dv \tag{7.2.2}$$

The product of Eqs. (7.2.1) and (7.2.2) can be expressed as

$$\Gamma(m)\Gamma(n) = \int_0^\infty \int_0^\infty \left(\frac{u^2}{2}\right)^{n-1} \left(\frac{v^2}{2}\right)^{m-1} e^{-u^2/2 - v^2/2}\, uv\, du\, dv \tag{7.2.3}$$

which can be readily integrated in polar coordinates. If we let $u = r\cos\theta$ and $v = r\sin\theta$, it becomes

$$\Gamma(m)\Gamma(n) = \int_0^{\pi/2} \int_0^\infty \left[\frac{(r\cos\theta)^2}{2}\right]^{n-1} \left[\frac{(r\sin\theta)^2}{2}\right]^{m-1} \cdot e^{-r^2/2} \sin\theta\cos\theta\, r^3\, dr\, d\theta \tag{7.2.4}$$

which is equivalent to

$$\Gamma(m)\Gamma(n) = 2\int_0^{\pi/2} (\cos\theta)^{2n-1}(\sin\theta)^{2m-1}\,d\theta \int_0^{\infty} \left(\frac{r^2}{2}\right)^{m+n-1} e^{-r^2/2}\,r\,dr \tag{7.2.5}$$

The second integral in Eq. (7.2.5) is by definition $\Gamma(m + n)$ and therefore

$$\frac{\Gamma(m)\Gamma(n)}{\Gamma(m+n)} = 2\int_0^{\pi/2} (\cos\theta)^{2n-1}(\sin\theta)^{2m-1}\,d\theta \tag{7.2.6}$$

The transformation $x = \cos^2\theta$ reduces Eq. (7.2.6) to

$$\frac{\Gamma(m)\Gamma(n)}{\Gamma(m+n)} = \int_0^1 x^{n-1}(1-x)^{m-1}\,dx \tag{7.2.7}$$

Equations (7.2.6) and (7.2.7) are equivalent forms and define the beta function. The symbol $B(n,m)$ is used for beta function with parameters n and m. That is,

$$B(n,m) = \frac{\Gamma(n)\Gamma(m)}{\Gamma(n+m)} = \int_0^1 x^{n-1}(1-x)^{m-1}\,dx \tag{7.2.8}$$

Dividing Eq. (7.2.8) by $B(n,m)$ yields

$$\frac{1}{B(n,m)}\int_0^1 x^{n-1}(1-x)^{m-1}\,dx = 1$$

or

$$\frac{1}{B(n,m)}\int_0^{\pi/2} 2(\cos\theta)^{2n-1}(\sin\theta)^{2m-1}\,d\theta = 1$$

Therefore

$$\begin{aligned} f(x\colon n,m) &= \frac{1}{B(n,m)}\,x^{n-1}(1-x)^{m-1} \qquad n > 0, m > 0, 0 \le x \le 1 \\ &= 0 \qquad \text{otherwise} \end{aligned} \tag{7.2.9}$$

or,

$$\begin{aligned} f(\theta\colon n,m) &= \frac{2}{B(n,m)}(\cos\theta)^{2n-1}(\sin\theta)^{2m-1} \qquad n > 0, m > 0, 0 \le \theta \le \frac{\pi}{2} \\ &= 0 \qquad \text{otherwise} \end{aligned} \tag{7.2.10}$$

define the beta probability function.

Example 7.2.1 Evaluate the beta probability function for $n = 2$ and $m = 2$.

Solution: For $n = 2$ and $m = 2$, Eq. (7.2.9) becomes

$$\begin{aligned} f(x) &= \frac{1}{B(2,2)}\,x^{2-1}(1-x)^{2-1} \qquad 0 \le x \le 1 \\ &= 0 \quad \text{otherwise} \end{aligned}$$

Now

$$B(2,2) = \frac{\Gamma(2)\Gamma(2)}{\Gamma(4)}$$

From Eq. (4.3.3) we know $\Gamma(\alpha) = (\alpha - 1)!$ and therefore $\Gamma(2) = 1$ and $\Gamma(4) = 3!$. Hence

$$\begin{aligned} f(x) &= 6x(1 - x) \qquad 0 \le x \le 1 \\ &= 0 \qquad \text{otherwise} \end{aligned}$$

is the beta probability function for $n = m = 2$. The student is requested to review Example 4.1.1 at this juncture.

In order to find the mean and the variance of the beta p.f., let us first derive a general expression for the kth moment about the origin. By definition,

$$\begin{aligned} m_k &= \frac{1}{B(n,m)} \int_0^1 x^k \cdot x^{n-1}(1 - x)^{m-1}\,dx \\ &= \frac{1}{B(n,m)} \int_0^1 x^{n+k-1}(1 - x)^{m-1}\,dx \\ &= \frac{1}{B(n,m)} \cdot B(n + k,m) \\ &= \frac{\Gamma(m + n)}{\Gamma(m)\Gamma(n)} \cdot \frac{\Gamma(n + k)\Gamma(m)}{\Gamma(n + k + m)} \\ &= \frac{\Gamma(n + k)\Gamma(m + n)}{\Gamma(n + k + m)\Gamma(n)} \end{aligned} \tag{7.2.11}$$

Therefore, for the special case $k = 1$, Eq. (7.2.11) yields

$$m_1 = \mu = \frac{\Gamma(n + 1)\Gamma(n + m)}{\Gamma(n + m + 1)\Gamma(n)} = \frac{n!\,(n + m - 1)!}{(n + m)!\,(n - 1)!} = \frac{n}{n + m} \tag{7.2.12}$$

Similarly, for $k = 2$ one gets

$$m_2 = \frac{(n + 1)n}{(n + m + 1)(n + m)} \tag{7.2.13}$$

Therefore the expression for the variance becomes

$$\sigma^2 = \frac{(n + 1)n}{(n + m + 1)(n + m)} - \left(\frac{n}{n + m}\right)^2 \tag{7.2.14}$$

Example 7.2.2 Using the results of Eq. (7.2.12) and (7.2.14), find the mean and the variance for beta p.f. with parameters $n = m = 2$.

Solution: Equation (7.2.12) states $\mu = \frac{2}{2 + 2} = \frac{1}{2}$, and Eq. (7.2.14) yields

$$\sigma^2 = \frac{(3)(2)}{(5)(4)} - \left(\frac{2}{4}\right)^2 = \frac{1}{20}$$

The student is requested to review Examples 4.2.1 and 4.2.2 at this point.

Some special cases of the beta p.f. are worth noting. If $n = m = 1$, the probability function reduces to $f(x) = 1$ which is the uniform probability function. If $n = 1$ and $m = 2$ the probability function becomes a triangular distribution, which is skewed to the right. If $n = 2$ and $m = 1$, the probability function yields another triangular distribution which is skewed to the left. In general, the beta p.f. is symmetrical about the line $x = 1/2$ if $n = m$, skewed to the right (positively skewed) if $n < m$ and skewed to the left (negatively skewed) if $n > m$.

The distribution function of the beta p.f. is called the *incomplete beta function* and is commonly denoted by $I_x(n,m)$. That is,

$$F(x\colon n,m) = I_x(n,m) = \frac{1}{B(n,m)} \int_0^x u^{n-1}(1 - u)^{m-1}\,du \qquad (7.2.15)$$

The incomplete beta function and the cumulative binomial function are related as follows:

$$\sum_{x=k}^{N} \binom{N}{x} p^x q^{N-x} = I_p(k, N - k + 1) \qquad (7.2.16)$$

For instance, suppose we wish to evaluate

$$\int_0^{0.5} \frac{1}{B(4,4)}\, x^3(1 - x)^3\,dx$$

to obtain $F(.5{:}4,4)$ of the beta probability function. According to Eq. (7.2.16) this integral is equal to

$$\sum_{x=4}^{7} \binom{7}{x} (.5)^x (.5)^{7-x}$$

since $k = 4$, $N - k + 1 = 4$ yield $N = 7$. However, in Table II the tabulated value for this expression is .5000, which agrees with the above remark that the beta p.f. is symmetric about the line $x = .5$, if $n = m$.

Again, in order to evaluate $F(.4{:}5,7)$ for the beta p.f. we need to evaluate the integral

$$\int_0^{.4} \frac{\Gamma(12)}{\Gamma(5)\Gamma(7)}\, x^4(1 - x)^6\,dx$$

In this situation $k = 5$ and $N - k + 1 = 7$ which yield $N = 11$. Hence

$$I_{.4}(5,11) = \sum_{x=5}^{11} \binom{11}{x} (.4)^x (.6)^{11-x}$$

From Table II in the Appendix we obtain .4672 to be the value of the required integration.

Since Table II is tabulated with increments .05 of p, linear interpolation could be used to evaluate the distribution function of the beta probability function with other values of p. For instance,

$$\int_0^{.37} \frac{\Gamma(12)}{\Gamma(5)\Gamma(7)} x^4(1 - x)^6\,dx = .3859$$

since $I_{.4}(5,11) = .4672$ and $I_{.35}(5,11) = .3317$.

(b) The Cauchy Probability Function. Let

$$f(x) = \frac{K}{(1 + x^2)^{k/2}} \qquad k = 1, 2, \ldots;\ x \text{ real}$$

In order to find the constant K which makes $f(x)$ a probability function, consider the integral

$$K \int_{-\infty}^{\infty} \frac{dx}{(1 + x^2)^{k/2}}$$

The substitution $x = \tan\theta$, $(dx = \sec^2\theta\,d\theta)$ transforms it to

$$2K \int_0^{\pi/2} \frac{\sec^2\theta}{\sec^k\theta}\,d\theta = 2K \int_0^{\pi/2} \cos^{k-2}\theta\,d\theta = K\left[B\left(\frac{k-1}{2}, \frac{1}{2}\right)\right]$$

Therefore

$$f(x) = \frac{1}{B\left[\dfrac{k-1}{2}, \dfrac{1}{2}\right]} \cdot \frac{1}{(1 + x^2)^{k/2}} \qquad k = 1, 2, \ldots \tag{7.2.17}$$

represents a probability function and is called the *Cauchy probability function.* The special case where $k = 2$ makes Eq. (7.2.17)

$$f(x) = \frac{1}{\pi(1 + x^2)} \qquad x \text{ real} \tag{7.2.18}$$

whose graph is shown in Fig. 7.3. The Cauchy p.f. is interesting in that it has no finite moments and hence no finite variance. Consequently, the central limit theorem cannot be applied to it.

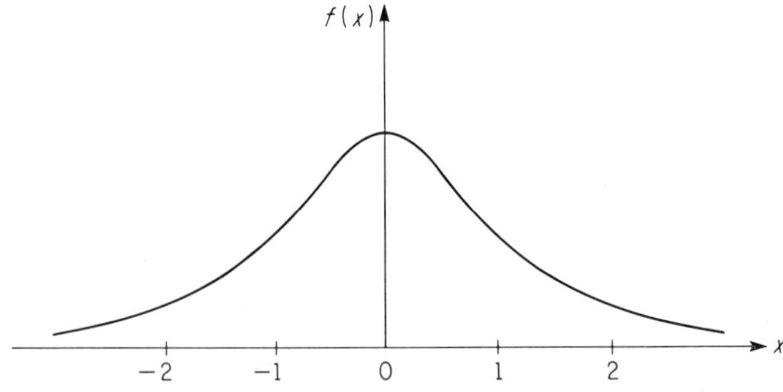

FIG. 7.3. The Cauchy probability function.

(c) Student's p.f. The Student's t probability function is a special case of Eq. (7.2.17). Setting $x = \dfrac{t}{\sqrt{n}}$ and $k = n + 1$ reduces Eq. (7.2.17) to

$$f(t) = \frac{1}{\sqrt{n}\,B\left[\frac{n}{2}, \frac{1}{2}\right]} \frac{1}{\left[1 + \frac{t^2}{n}\right]^{(n+1)/2}} \qquad t \text{ real}$$

$$= \frac{\Gamma\left(\frac{n+1}{2}\right)}{\sqrt{n\pi}\,\Gamma\left(\frac{n}{2}\right)} \left[1 + \frac{t^2}{n}\right]^{-(n+1)/2} \tag{7.2.19}$$

Since n is the independent parameter, the symbol $f(t:n)$ will be used to refer to Eq. (7.2.19). That is,

$$f(t:n) = \frac{\Gamma\left(\frac{n+1}{2}\right)}{\sqrt{n\pi}\,\Gamma\left(\frac{n}{2}\right)} \left[1 + \frac{t^2}{n}\right]^{-(n+1)/2} \tag{7.2.20}$$

which is called the Student's t p.f. or Student's t distribution. Let us note that for the special case $n = 1$ Eq. (7.2.20) reduces to Eq. (7.2.18).

Since $f(t:n) = f(-t:n)$, the t distribution is symmetric about the line $t = 0$, which implies that its mean is zero. However, for $n = 1$, the first moment about the origin does not exist. The variance of the t p.f. is equal to $\dfrac{n}{n-2}$ provided $n > 2$. The parameter n is called the *degrees of freedom* and is usually abbreviated d.f. To be consistent with conventional notation, we should have designated degrees of freedom by ν (nu) since they represent parameters. This departure from conventionality does not present any difficulty because from the context it is clear that we have reference to parameters.

Table V in the Appendix contains representative values of $f(t:n)$. The symbolism $t_{\alpha;n}$ will be used to denote the value of t with n d.f. for which $\int_{t_{\alpha;n}}^{\infty} f(t;n)\,dt = \alpha$. For example, $t_{.05;7}$ represents the value of t with 7 d.f. above (to the right of) which 5 percent of the area is located. From Table V we obtain $t_{.05;7} = 1.895$.

Since $f(t;n)$ is symmetric about $t = 0$,

$$\int_{-\infty}^{-t} f(t:n)\,dt = \int_{t}^{\infty} f(t:n)\,dt \tag{7.2.21}$$

Therefore, Table V provides values also for the left-hand side integral of Eq. (7.2.21). Referring to the above example $t = -1.895$ is the value below (to the left of) which 5 percent of $f(t:7)$ is distributed.

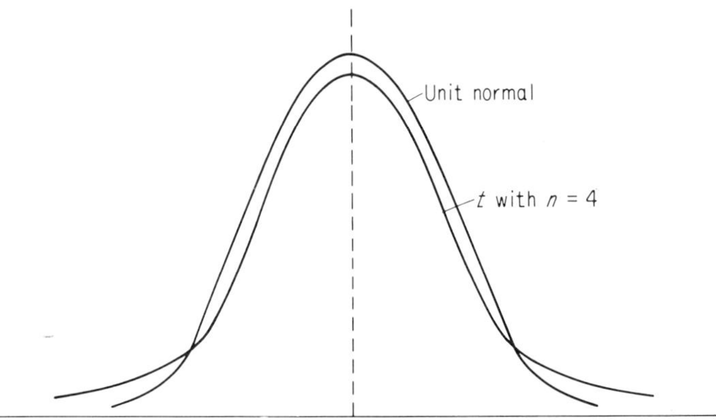

FIG. 7.4. Unit normal and t with $n = 4$.

The t and the unit normal probability functions are very similar. Figure 7.4 represents the unit normal and the t with 4 d.f.

It can be shown that as $n \longrightarrow \infty$ the t probability function approaches the unit normal. (See Prob. 7.2.10.) The following theorem provides the basis for using the t probability function for the sampling distribution of the mean when the population variance is unknown. The proof of the theorem is postponed and given in Prob. 7.3.13.

Theorem 7.2.1 Let $\bar{x}$ represent the mean of a random sample of size n taken from a normal population with mean μ and variance σ^2. Then the random variable $\dfrac{\bar{x} - \mu}{s/\sqrt{n}}$ is distributed according to the Student's t probability function with $n - 1$ degrees of freedom.

Theorem 7.2.1 is not as general as the central limit theorem because it assumes random samples from a normal population. However, for large n, Theorem 7.2.1 can be applied to any probability function with finite variance. There is no definite rule as to what constitutes a large sample size. Usually $n \geq 30$ is accepted to represent a large n. If the normality assumption cannot be asserted, Theorem 7.2.1 can be applied provided $n \geq 30$.

The following example illustrates the notions under discussion.

Example 7.2.3 A random sample of size 16 from a normal population yields a mean value of 87.5 and a standard deviation of 9.35. Find the probability that the population mean does exceed 90.

Solution: Since $E[\bar{x}] = \mu_x$, we assume $\mu_x = 87.5$. The distribution of $\bar{x}$'s obey the t probability function with 15 degrees of freedom. Hence

$$\frac{90 - 87.5}{\dfrac{9.35}{\sqrt{16}}} = \frac{10}{9.35} = 1.06$$

From Table V we note that for 15 d.f. $F(1.341) = .90$ and $F(.691) = .75$. Using linear interpolation we find $F(1.06) = .835$. Therefore the probability that the population mean exceeds 90 is .165.

PROBLEMS

7.2.1 Sketch the beta probability functions (a) $f(x:2,2)$, (b) $f(x:2,4)$, and (c) $f(x:4,2)$.

7.2.2 Given
$$f(x) = kx^5(1-x)^7 \qquad 0 < x < 1$$
$$= 0 \qquad \text{otherwise}$$

(a) For what value of k is $f(x)$ a probability function?
(b) Find the mean of this probability function.
(c) Find the variance of the probability function.
(d) Evaluate $F(.35)$ (distribution function) for this probability function.

7.2.3 Show that for the incomplete beta function

$$I_x(n,m) = 1 - I_{1-x}(m,n)$$

and hence

$$\sum_{x=k}^{N} \binom{N}{k} p^x q^{N-x} = 1 - I_q(N - k + 1, k)$$

7.2.4 Using the result of Prob. 7.2.3, evaluate $F(.65)$ for the probability function of Prob. 7.2.2.

7.2.5 Show that the Cauchy p.f. has no finite mean.

7.2.6 Using Table V in the Appendix, find the values of x such that

$$\text{(a)} \int_0^x f(t:6)\,dt = .45 \qquad \text{(b)} \int_0^x f(t:10)\,dt = .475$$

7.2.7 Using the moment definition of the mean, show that the mean of the t p.f. is zero for $n > 1$.

7.2.8 Find the variance of the t p.f. for $n = 3$.

7.2.9 Show that the variance of the t p.f. for $n > 2$ is $\dfrac{n}{n-2}$.

7.2.10 Show that as $n \to \infty$ the Student's t p.f., Eq. (7.2.20), approaches the unit normal. *Hints:*

$$\lim_{n\to\infty} \left[1 + \frac{t^2}{n}\right]^{-n} = e^{-t^2}$$

$$\Gamma(n+1) \cong \sqrt{2\pi n}\, n^n e^{-n} \quad \text{(Sterling's formula)}$$

$$\Gamma\left(\frac{n+1}{2}\right) = \Gamma\left(\frac{n-1}{2} + 1\right)$$

$$\Gamma\left(\frac{n}{2}\right) = \Gamma\left(\frac{n-2}{2} + 1\right)$$

7.2.11 A random sample of size 36 is taken from a population which yields a mean of 104 and standard deviation of 15. What is the probability that the population mean will be in the interval $103 \le \bar{x} \le 105$?

7.3 SAMPLING DISTRIBUTION OF THE VARIANCE

(a) The Chi-Square Probability Function. Consider the function

$$f(x:n) = kx^{(n/2)-1}e^{-x/2} \qquad x > 0, \quad n > 0$$
$$= 0 \quad \text{otherwise} \tag{7.3.1}$$

Since

$$\int_0^\infty x^{(n/2)-1}e^{-x/2}\,dx = 2^{n/2}\,\Gamma\left(\frac{n}{2}\right) \tag{7.3.2}$$

(see Prob. 7.3.1), $k = \dfrac{1}{2^{n/2}\,\Gamma\left(\frac{n}{2}\right)}$ makes the expression (7.3.1) a probability function, which is called the *chi-square* probability function. The parameter n is called the degrees of freedom of the chi-square p.f. and assumes only integer values. The symbol commonly associated with chi-square is χ^2, where χ is the Greek letter chi. Sometimes Eq. (7.3.1) is equivalently defined as

$$f(\chi^2) = \frac{1}{2^{n/2}\,\Gamma\left(\frac{n}{2}\right)}(\chi^2)^{(n/2)-1}e^{-\chi^2/2} \qquad n > 0$$

Next let us find the mean and the variance of the chi-square p.f. The student is requested to show that the moment-generating function of chi-square p.f. is

$$M_x(t) = E[e^{tx}] = (1 - 2t)^{-n/2} \tag{7.3.3}$$

(See Prob. 7.3.2.) Differentiating this expression we get

$$\frac{d}{dt}M_x(t) = n(1 - 2t)^{-(n/2)-1}$$

which when evaluated at $t = 0$ reduces to n. Therefore,

$$\frac{d}{dt}M_x(0) = n \tag{7.3.4}$$

is the mean of the chi square p.f. Similarly,

$$\frac{d^2}{dt^2}M_x(0) = n^2 + 2n$$

can be readily obtained and hence

$$\sigma^2 = n^2 + 2n - (n)^2 = 2n \tag{7.3.5}$$

represents the variance of the chi-square p.f. That is, the parameter n is the mean and twice its value is the variance of a given chi-square p.f.

Figure 7.5 shows the sketch for chi-square probability functions with 2 and 6 d.f. The former is interesting in that it reduces to the exponential

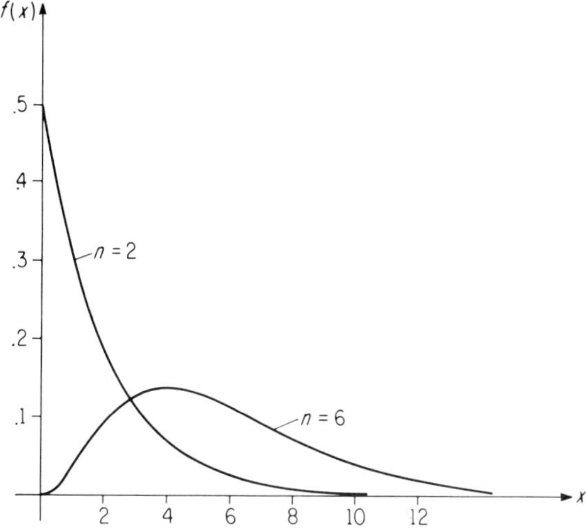

FIG. 7.5. Chi-square p.f. for $n = 2$ and $n = 6$.

p.f. $f(x) = \frac{1}{2} e^{-x/2}$. Now according to Eqs. (4.3.9) and (4.3.10) the mean and the variance for this exponential p.f. are 2 and 4 respectively. This agrees with Eqs. (7.3.4) and (7.3.5). Furthermore, Eqs. (4.3.8) and (7.3.3) agree in that the moment-generating function is $(1 - 2t)^{-1}$ for this case.

Chi-square probability functions with $n > 2$ attain their maximum (mode) at $n - 2$. For example, if $n = 6$, the function has its mode at 4. (See Prob. 7.3.3.)

Table VI in the Appendix contains representative values of the chi-square p.f. We shall use the notation $\chi^2_{\alpha;n}$ to define

$$\int_{\chi^2_{\alpha;n}}^{\infty} f(x:n)\,dx = \alpha$$

where n is the d.f. associated with the chi-square. For instance, we note that $\chi^2_{.05;10} = 18.3070$ and $\chi^2_{.90;20} = 12.4426$.

The chi-square and the unit normal probability functions are related and the following theorem indicates the relation between them.

Theorem 7.3.1 Let x represent a random sample of size 1 from the unit normal. Then the random variable $y = x^2$ has the chi-square p.f. with 1 d.f.

Proof: Problem 4.4.5 is a proof of this theorem. However, let us present another approach here:

$$f(x) = \frac{1}{\sqrt{2\pi}} e^{-x^2/2}$$

represents the unit normal p.f. The moment-generating function of y is

$$\begin{aligned} M_y(t) = M_{x^2}(t) &= E[e^{tx^2}] \\ &= \int_{-\infty}^{\infty} e^{tx^2} \frac{1}{\sqrt{2\pi}} e^{-x^2/2}\,dx \\ &= \frac{1}{\sqrt{2\pi}} \int_{-\infty}^{\infty} e^{-(x^2/2)(1-2t)}\,dx \end{aligned}$$

The substitution $z = x\sqrt{1 - 2t}$ in the last integral yields

$$\begin{aligned} M_y(t) &= \frac{1}{\sqrt{2\pi}} \frac{1}{\sqrt{1-2t}} \int_{-\infty}^{\infty} e^{-z^2/2}\,dz \\ &= (1 - 2t)^{-1/2} \end{aligned} \tag{7.3.6}$$

Applying Theorem 4.2.1 (the uniqueness theorem of m.g.f.) to this result and recalling that the moment-generating function of the chi-square p.f. is $(1 - 2t)^{-n/2}$, then clearly the expression (7.3.6) is the moment-generating function of the chi-square p.f. with 1 degree of freedom.

Theorem 7.3.1 is a special case of the following theorem.

Theorem 7.3.2 Let $x_1, x_2, \ldots, x_k$ represent a random sample of size k from the unit normal probability function. Then, the random variable $y = \sum_{i=1}^{k} x_i^2$ has a chi-square p.f. with k degrees of freedom.

Proof: From Theorem 7.3.1, each of the x_i's has $(1 - 2t)^{-1/2}$ for a moment-generating function. Since x_i's are stochastically independent, (random sample from an infinite population)

$$M_{x_1^2+x_2^2+\cdots+x_k^2}(t) = M_{x_1^2}(t)M_{x_2^2}(t)\cdots M_{x_k^2}(t)$$

holds true by Eq. (6.5.3) and therefore

$$\begin{aligned} M_{\sum_{i=1}^{k} x_i^2}(t) &= [(1 - 2t)^{-1/2}]^k \\ &= (1 - 2t)^{-k/2} \end{aligned}$$

Again by the uniqueness theorem y is chi-square distributed with k degrees of freedom.

A theorem describing the distribution of variances determined by random samples from a normal population is presented next.

Theorem 7.3.3 If x is a random variable whose probability function is normal with mean μ and variance σ^2, $(\sigma \neq 0)$, and s^2 is the variance of a random sample of size n, then the random variable $(n - 1)\dfrac{s^2}{\sigma^2}$ is χ^2 distributed with $(n - 1)$ degrees of freedom.

Proof: Consider the identity

$$(n - 1)s^2 = \sum_{i=1}^{n} (x_i - \bar{x})^2$$

$$= \sum_{i=1}^{n} [(x_i - \mu) - (\bar{x} - \mu)]^2$$

Dividing both sides of this equality by σ^2, we get

$$(n - 1)\frac{s^2}{\sigma^2} = \sum_{i=1}^{n} \left[\left(\frac{x_i - \mu}{\sigma}\right) - \left(\frac{\bar{x} - \mu}{\sigma}\right)\right]^2$$

The right-hand side of the last expression is equivalent to

$$\sum_{i=1}^{n} \left(\frac{x_i - \mu}{\sigma}\right)^2 - n\left(\frac{\bar{x} - \mu}{\sigma}\right)^2$$

and therefore

$$(n - 1)\frac{s^2}{\sigma^2} + \left(\frac{\bar{x} - \mu}{\frac{\sigma}{\sqrt{n}}}\right)^2 = \sum_{i=1}^{n} \left(\frac{x_i - \mu}{\sigma}\right)^2$$

The random variable $\dfrac{x_i - \mu}{\sigma}$ obeys the unit normal p.f. and according to Theorem 7.3.2 the right-hand side is chi-square distributed with n degrees of freedom. From Theorem 7.1.3 (central limit theorem) the random variable $\dfrac{\bar{x} - \mu}{\frac{\sigma}{\sqrt{n}}}$ approaches the unit normal and hence $\left(\dfrac{\bar{x} - \mu}{\frac{\sigma}{\sqrt{n}}}\right)^2$ is chi-square distributed with 1 d.f. It can also be shown that $\bar{x}$ and s^2 are stochastically independent. Therefore the random variable $(n - 1)\dfrac{s^2}{\sigma^2}$ is chi-square distributed with $(n - 1)$ degrees of freedom.

The following example illustrates a chi-square p.f. related notion.

Example 7.3.1 A random sample of size 16 from a normal population yields a variance of 22.1. Find the probability that the population variance is greater than 30.

Solution: According to Theorem 7.3.3 the quantity $\dfrac{(16 - 1)(22.1)}{\sigma^2} = \chi^2$ with 15 d.f. If $\sigma^2 > 30$, we need to locate the value of χ^2 with 15 d.f. such that $15(22.1)/30 = 11.05 < \chi^2$. From Table VI in the Appendix we note that $\chi^2_{.75;15}$ is 11.0365. Therefore the required probability is approximately $1 - 0.75 = 0.25$.

In most applications σ^2 is unknown and s^2 is obtained from a random sample. The sample variance will be used to make an inference regarding the population variance. This concept will be presented in Sec. 7.4.

(b) The *F* Probability Function. The chi-square probability function was shown to be related to the distribution of sample variances from a given population. Often in actual practice we also encounter the problem of comparing the variances of two independent samples to determine whether or not they come from populations having equal variances. One method to verify that the variances are equal would be to show that their ratio is approximately equal to one. If their ratio is either very large or very small the equality of the variances would be doubtful. A probability function, called the F p.f., will be presented next which describes the sampling distribution of the ratio of two independent sample variances. Consider the function

$$\begin{aligned} f(x:\alpha,\beta) &= kx^{\alpha-1}(1+x)^{-(\alpha+\beta)} \qquad \alpha > 0, \beta > 0, x > 0 \\ &= 0 \qquad \text{otherwise} \end{aligned} \tag{7.3.7}$$

We wish to find the value of k for which Eq. (7.3.7) would become a probability function. That is, the value of k is obtained from

$$k\int_0^\infty x^{\alpha-1}(1+x)^{-\alpha-\beta}\cdot dx = 1$$

and in order to integrate this expression we make the substitution $x = \dfrac{y}{1-y}$ which yields

$$k\int_0^1 \left(\frac{y}{1-y}\right)^{\alpha-1}\left(1+\frac{y}{1-y}\right)^{-\alpha-\beta}\frac{dy}{(1-y)^2}$$

$$= k\int_0^1 y^{\alpha-1}(1-y)^{\beta-1}\,dy = k\,B[\alpha,\beta]$$

Consequently $k = \dfrac{1}{B[\alpha,\beta]}$ is the desired constant. That is,

$$f(x:\alpha,\beta) = \frac{1}{B[\alpha,\beta]}x^{\alpha-1}(1+x)^{-\alpha-\beta} \qquad \alpha > 0, \beta > 0, x > 0 \tag{7.3.8}$$

represents a probability function. Suppose we consider the transformation where $\alpha = m/2$, $\beta = n/2$, and $x = \dfrac{m}{n}F$. Since $dx = \dfrac{m}{n}dF$, Eq. (7.3.8) becomes

$$f(F:m,n) = \frac{1}{B\left[\dfrac{m}{2},\dfrac{n}{2}\right]}\left(\frac{m}{n}F\right)^{(m/2)-1}\left(1+\frac{m}{n}F\right)^{-(m+n)/2}\frac{m}{n} \tag{7.3.9}$$

$$f(F{:}m,n) = \frac{1}{B\left[\frac{m}{2}, \frac{n}{2}\right]} \left(\frac{m}{n}\right)^{m/2} F^{(m/2)-1} \left(1 + \frac{m}{n} F\right)^{-(m+n)/2} \tag{7.3.10}$$

Furthermore, since $B[\alpha, \beta] = \dfrac{\Gamma(\alpha)\,\Gamma(\beta)}{\Gamma(\alpha + \beta)}$, Eq. (7.3.10) can be expressed as

$$f(F{:}m,n) = \frac{\Gamma\left(\frac{m+n}{2}\right)}{\Gamma\left(\frac{m}{2}\right)\Gamma\left(\frac{n}{2}\right)} \left(\frac{m}{n}\right)^{m/2} F^{(m/2)-1} \left(1 + \frac{m}{n} F\right)^{-(m+n)/2} \tag{7.3.11}$$

The expression (7.3.11) is called the F p.f. where m and n are its two independent parameters. m and n are called the numerator and denominator degrees of freedom of the F p.f., respectively. For instance, if $m = n = 2$, $f(F) = (1 + F)^{-2}$ and if $m = n = 4$, $f(F) = 6F(1 + F)^{-4}$. The graphs of these two probability functions are presented in Fig. 7.6.

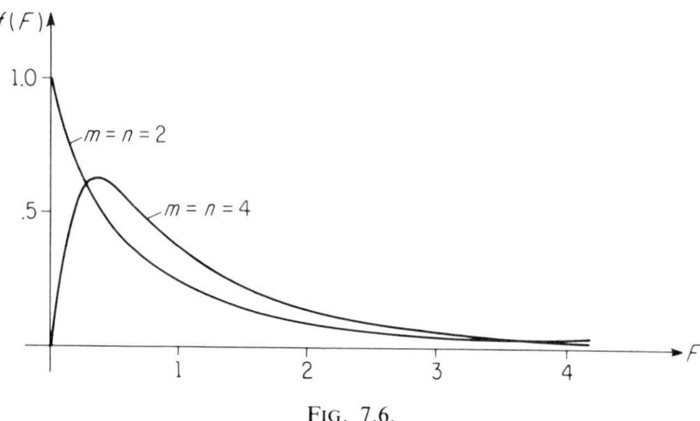

FIG. 7.6.

The student is requested to verify that the particular cases $m = n = 2$, and $m = n = 4$ do represent probability functions. Also, he should review Prob. 4.4.3.

The mean of the F probability function is given by

$$\mu_F = \frac{n}{n-2} \qquad \text{if } n > 2 \tag{7.3.12}$$

It is interesting to note that μ_F depends only on the parameter n. Also,

$$\lim_{n \to \infty} \frac{n}{n-2} = 1 \tag{7.3.13}$$

The variance of the F p.f. is given by the expression

$$\sigma_F^2 = \frac{2n^2(m + n - 2)}{m(n-2)^2(n-4)} \qquad \text{if } n > 4 \tag{7.3.14}$$

Since this probability function has two independent parameters m and n, a three-way table is necessary to tabulate the different probabilities and values of m and n. Table VII in the Appendix gives selected values of $F_{\alpha;m,n}$ where

$$\alpha = \int_{F_{\alpha;m,n}}^{\infty} f(F)\,dF = P[F > F_{\alpha;m,n}] \tag{7.3.15}$$

For example, $F_{.05;4,4} = 6.39$ and $F_{.025;8,10} = 3.85$.

One of the most important applications of the F probability function is described by the following theorem.

Theorem 7.3.4 If χ_1^2 and χ_2^2 are two stochastically independent chi-square distributed random variables with m and n degrees of freedom respectively, then the random variable

$$F = \frac{\dfrac{\chi_1^2}{m}}{\dfrac{\chi_2^2}{n}} \tag{7.3.16}$$

has the F p.f. with m and n degrees of freedom.

From Theorem 7.3.3 we know that the random variables $(n_1 - 1)\dfrac{s_1^2}{\sigma_1^2}$ and $(n_2 - 1)\dfrac{s_2^2}{\sigma_2^2}$ are chi-square-distributed with $n_1 - 1$ and $n_2 - 1$ d.f. Therefore, applying Theorem 7.3.4 to these expressions, we get

$$\begin{aligned} F &= \frac{(n_1 - 1)\dfrac{s_1^2}{\sigma_1^2}}{n_1 - 1} \div \frac{(n_2 - 1)\dfrac{s_2^2}{\sigma_2^2}}{n_2 - 1} \\ &= \frac{s_1^2}{\sigma_1^2} \div \frac{s_2^2}{\sigma_2^2} = \frac{s_1^2}{s_2^2} \cdot \frac{\sigma_2^2}{\sigma_1^2} \end{aligned} \tag{7.3.17}$$

Now if the two samples are taken from identical normal populations, or two independent normal populations with equal variances, σ_1^2 would be equal to σ_2^2 and hence Eq. (7.3.17) becomes

$$F = \frac{s_1^2}{s_2^2} \tag{7.3.18}$$

That is, if s_1^2 and s_2^2 are the variances of two random samples of size m and n, respectively, taken from normal populations with equal variances, the ratio of s_1^2 to s_2^2 is F distributed with $m - 1$ and $n - 1$ d.f. Equations (7.3.17) and (7.3.18) represent important relationships and will be used in subsequent discussions.

Next let us note that according to Theorem 7.3.4 the quantity

$$\frac{1}{F} = \frac{\chi_2^2/n}{\chi_1^2/m}$$

is also F distributed with n and m degrees of freedom. The relationship that holds true in general is

$$F_{(1-\alpha);n,m} = \frac{1}{F_{\alpha;m,n}} \tag{7.3.19}$$

To show Eq. (7.3.19) holds true, we recall that

$$\begin{aligned} \alpha = P[F > F_{\alpha;m,n}] &= 1 - P[F < F_{\alpha;m,n}] \\ &= 1 - P\left[\frac{1}{F} > \frac{1}{F_{\alpha;m,n}}\right] \end{aligned} \tag{7.3.20}$$

Therefore

$$P\left[\frac{1}{F} > \frac{1}{F_{\alpha;m,n}}\right] = 1 - \alpha$$

holds true and Eq. (7.3.19) is verified.

For example, $F_{.05;5,10} = 3.3258$, applying (7.3.19) we obtain

$$F_{.95;10,5} = \frac{1}{3.3258} = .3006$$

In words, for the F p.f. with numerator d.f. = 10 and denominator d.f. = 5, 95 percent of the probability is above (to the right of) the point F = .3006.

As a matter of convenience in notation, *we shall use F and F(m,n) interchangeably*, where m and n are the numerator and denominator degrees of freedom of the associated probability function.

(c) Relationship Between the Normal, Chi-Square, Student *t*, and *F* Probability Functions. From Theorem 7.3.1 we note that the square root of the chi-square p.f. with 1 d.f. is the unit normal. Referring to Table VI in the Appendix, we find, for instance, that $\chi^2_{.05;1} = 3.841$ whose square root yields the values + 1.96 and − 1.96. Now from Table IV we note that for the unit normal

$$\int_{-\infty}^{-1.96} f(z)\,dz = \int_{1.96}^{\infty} f(z)\,dz = .025$$

and hence $P[\,|z| > 1.96] = .05$. Since χ^2 is always positive, while the unit normal symmetric about zero, we need to note that

$$P[\,|z| > z_{\alpha/2}] = \alpha \tag{7.3.21}$$

Therefore

$$\chi^2_{\alpha;1} = (z_{\alpha/2})^2 \tag{7.3.22}$$

The F p.f. with 1 and n d.f. according to Eq. (7.3.11) yields

$$f(F) = \frac{\Gamma\left(\frac{1+n}{2}\right)}{\Gamma\left(\frac{1}{2}\right)\Gamma\left(\frac{n}{2}\right)}\left(\frac{1}{n}\right)^{1/2} F^{-1/2}\left(1 + \frac{F}{n}\right)^{-(n+1)/2} \qquad (7.3.23)$$

Recalling that $\Gamma\left(\frac{1}{2}\right) = \sqrt{\pi}$ and making the transformation $F = t^2$ in Eq. (7.3.23), we obtain

$$f(t;n) = \frac{\Gamma\left(\frac{n+1}{2}\right)}{\sqrt{\pi n}\,\Gamma\left(\frac{n}{2}\right)}\left(1 + \frac{t^2}{n}\right)^{-(n+1)/2} \qquad (7.3.24)$$

which is the t probability function. Therefore, the F p.f. with 1 and n d.f. is related to the square of the t p.f. with n d.f. In this relationship also, we note that F is always positive while t is symmetric about zero; hence,

$$P[\,|\,t\,| > t_{\alpha/2}] = \alpha \qquad (7.3.25)$$

and

$$F_{\alpha;1,n} = t^2_{\alpha/2,n} \qquad (7.3.26)$$

For example, $F_{.10;1,5} = 4.06$ and $t_{.05,5} = 2.015$. Clearly, $(2.015)^2 = 4.06$. Again, $F_{.05;1,10} = 4.96$ and $t_{.025,10} = 2.228$ and $(2.228)^2 = 4.96$.

Lastly, in Prob. 7.2.10 it was shown that the t p.f. approaches the unit normal as n approaches ∞. Therefore, the t p.f. with ∞ d.f. is equal to the unit normal. That is,

$$t_{\alpha;\infty} = z_\alpha \qquad (7.3.27)$$

For instance, $t_{.025;\infty} = 1.96$ and $z_{.025} = 1.96$.

PROBLEMS

7.3.1 Show that

$$\int_0^\infty x^{(n/2)-1} e^{-x/2}\,dx = 2^{n/2}\,\Gamma\left(\frac{n}{2}\right)$$

[*Hint:* Use the transformation $y = \frac{x}{2}$.]

7.3.2 Show that the moment-generating function of the chi-square p.f. is

$$(1 - 2t)^{-n/2}$$

7.3.3 Show that the chi-square p.f. has its maximum at $n - 2$ for $n > 2$.

7.3.4 Given the uniform p.f.

$$f(x) = 1 \quad 0 < x < 1$$
$$= 0 \quad \text{otherwise}$$

show that the random variable $y = -2 \log x$ has a chi-square p.f. with 2 d.f.

7.3.5 Find the mode (relative maximum) of the F p.f. with $m = n = 5$.

7.3.6 Find the mode of the F probability function for the general case.

7.3.7 By actually carrying out the necessary integration(s), verify that Eq. (7.3.14) holds true for $f(F:5,5)$.

7.3.8 Using Table VII in the Appendix, evaluate (a) $F_{.95;6,6}$, (b) $F_{.975;7,12}$, (c) $F_{.99;12,15}$.

7.3.9 Evaluate (a) $F_{.10;1,15}$, (b) $F_{.20;1,15}$, by using the Student's t distribution values given in Table V of the Appendix.

7.3.10 Evaluate $\chi^2_{.03,1}$ and $\chi^2_{.04,1}$ by using the tabulated values of the unit normal probability function.

7.3.11 The chi-square with degrees of freedom greater than 30 is approximately related to the unit normal by the formula

$$\chi^2_{\alpha,n} = \frac{1}{2}\,[z_\alpha + \sqrt{2n - 1}]^2$$

Using this relationship evaluate (a) $\chi^2_{.05,100}$, (b) $\chi^2_{.95,100}$, (c) $\chi^2_{.975,40}$, and compare with the tabulated values in Table VI.

7.3.12 The cumulative Poisson distribution and the cumulative chi-square distribution are related by the formula

$$F(m - 1) = \sum_{x=0}^{m-1} \frac{e^{-k}k^x}{x!} = 1 - F(\chi^2) = \int_{\chi^2}^{\infty} \frac{1}{2^{n/2}\,\Gamma\!\left(\frac{n}{2}\right)} (x)^{(n/2)-1} e^{-x/2}\,dx$$

or equivalently

$$F(\chi^2) = 1 - F(m - 1) = \sum_{x=m}^{\infty} \frac{e^{-k}k^x}{x!}$$

where $\frac{n}{2} = m$ and $\frac{\chi^2}{2} = k$. Using this relationship and Table III in the Appendix evaluate: (a) $F(15)$ for the chi-square with 8 d.f., (b) $F(26)$ for the chi-square with 20 d.f.

7.3.13 Prove Theorem 7.2.1. [*Hint:* Show that the ratio of $\left(\frac{\bar{x} - \mu}{\sigma/\sqrt{n}}\right)^2$ and s^2/σ^2 is F distributed with 1 and $n - 1$ numerator and denominator degrees of freedom and apply Eq. (7.3.26) to the result.]

7.4 CONFIDENCE INTERVALS FOR POPULATION PARAMETERS

Now that the p.f.'s of the sample mean and the sample variance are established, one could make inductive inferences regarding the population mean and the population variance, based upon the statistics obtained from a sample. We shall pursue this concept next.

(a) Inference About the Mean

1. σ^2 Known. In the introduction to this chapter we made reference to the mean of the random sample of 25 height measurements. Let us examine this illustration further. We have calculated the mean of these measurements to be 68.64 inches. (See Example 5.4.1.) What inductive inference can we make regarding the population mean? In order to apply Theorem 7.1.3 (the central limit theorem) we need to know the variance of the population from which the sample was taken. This information might be available or assumed to have a given value. For instance, height measurements of college students, Army personnel, factory workers, etc., might have been studied and the information that the variance of college students' height measurement is 9, or the variance of Army personnel is 7.4, might be known. Since in the example under investigation the sample is from college students, use the information $\sigma^2 = 9$ and note that the random variable

$$\frac{(\overline{x} - \mu)\sqrt{25}}{3}$$

is distributed according to the unit normal p.f. Consequently,

$$P\left[-z_{\alpha/2} < \frac{(\overline{x} - \mu)5}{3} < z_{\alpha/2}\right] = 1 - \alpha$$

which is equivalent to

$$P\left[\overline{x} - z_{\alpha/2}\left(\frac{3}{5}\right) < \mu < \overline{x} + z_{\alpha/2}\left(\frac{3}{5}\right)\right] = 1 - \alpha$$

If it is further assumed that $\alpha = .05$, we get

$$P\left[\overline{x} - z_{.025}\left(\frac{3}{5}\right) < \mu < \overline{x} + z_{.025}\left(\frac{3}{5}\right)\right] = .95$$

$$P\left[\overline{x} - 1.96\left(\frac{3}{5}\right) < \mu < \overline{x} + 1.96\left(\frac{3}{5}\right)\right] = .95$$

$$P[\overline{x} - 1.18 < \mu < \overline{x} + 1.18] = .95$$

This last expression states: The probability that the interval $\overline{x} - 1.18$ and $\overline{x} + 1.18$ inches contains the population mean is 95 percent. To state it differently, we are 95 percent confident that μ, the population mean, is located between $\overline{x} - 1.18$ and $\overline{x} + 1.18$ inches. Notice that there is 5 percent uncertainty in this statement and we are taking this risk of being incorrect, which is inevitable in any inductive inference.

In general, if the variance of the population is known, and a sample of size n is taken, we have

$$P\left[-z_{\alpha/2} < \frac{\overline{x} - \mu}{\frac{\sigma}{\sqrt{n}}} < z_{\alpha/2}\right] = P\left[|\overline{x} - \mu| < z_{\alpha/2}\frac{\sigma}{\sqrt{n}}\right] = 1 - \alpha \quad (7.4.1)$$

which is equivalent to

$$P\left[\overline{x} - z_{\alpha/2}\frac{\sigma}{\sqrt{n}} < \mu < \overline{x} + z_{\alpha/2}\frac{\sigma}{\sqrt{n}}\right] = 1 - \alpha \quad (7.4.2)$$

This is a very useful relationship in making inferences about the mean of any population. The quantities $\bar{x} - z_{\alpha/2} \dfrac{\sigma}{\sqrt{n}}$ and $\bar{x} + z_{\alpha/2} \dfrac{\sigma}{\sqrt{n}}$ are called the *lower* and *upper limits* of the $100(1 - \alpha)$ percent confidence interval respectively. In the above illustration $\bar{x} - 1.18$ inches is the lower and $\bar{x} + 1.18$ inches is the upper limit of the 95 percent confidence interval.

Let us remark that the population mean μ is either in the computed confidence interval or it is not. A more precise interpretation perhaps would be the following. If a sample of size n is taken repeatedly and the sample means calculated, assuming σ^2 is known, we can expect $100 \times (1 - \alpha)$ percent of the time to find the population mean in the intervals

$$\left(\bar{x}_i - z_{\alpha/2} \frac{\sigma}{\sqrt{n}}, \bar{x}_i + z_{\alpha/2} \frac{\sigma}{\sqrt{n}}\right)$$

This is illustrated in Fig. 7.7 below.

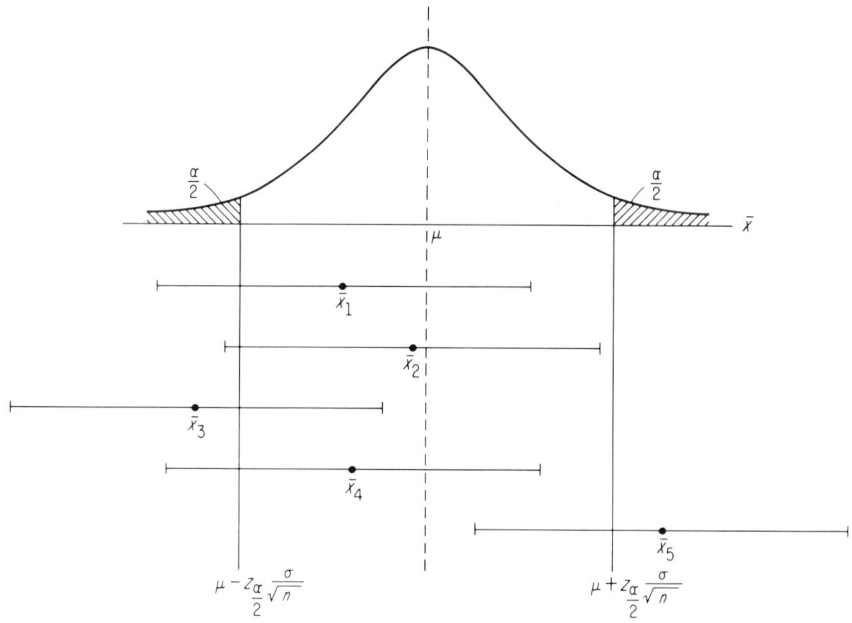

FIG. 7.7.

In Fig. 7.7 we note that the confidence interval associated with $\bar{x}_1$, $\bar{x}_2$, and $\bar{x}_4$ include the population mean μ, while the confidence interval associated with $\bar{x}_3$ and $\bar{x}_5$ do not. If samples of size n were taken repeatedly, we would expect $100(1 - \alpha)$ percent of the confidence intervals to include the population mean.

Example 7.4.1 The distance between the index finger and the thumb in humans is assumed to be normally distributed with a standard deviation of 1.4 inches. If 16 measurements yield a mean value of 7.5 inches, find the 90 percent confidence interval around the population mean.

Solution: $1 - \alpha = .90$ and hence $\alpha = .10$. From Table IV in the Appendix we obtain $z_{.05} = 1.645$. Therefore

$$7.5 - \frac{1.645\,(1.4)}{4} < \mu < 7.5 + \frac{1.645\,(1.4)}{4} \quad \text{yields } 6.92 < \mu < 8.08$$

Example 7.4.2 IQ scores are usually accepted to be normally distributed with a variance of 225. Nine students in a certain high school were randomly selected and tested. Their IQ scores were as follows. 93, 95, 98, 100, 105, 109, 110, 123, and 130. Find the 95 percent confidence interval around the population mean.

Solution:

$$\bar{x} = \frac{\sum_{i=1}^{9} x_i}{9} = 107 \qquad \text{and} \qquad z_{.025} = 1.96$$

Therefore

$$107 - 1.96\frac{\sqrt{225}}{\sqrt{9}} < \mu < 107 + 1.96\frac{\sqrt{225}}{\sqrt{9}}$$

$$107 - 9.8 < \mu < 107 + 9.8$$

That is, we can be 95 percent confident or certain that the population mean IQ score is between 97.20 and 116.80.

Let us refer to Eq. (7.4.1) and note that $z_{\alpha/2}\dfrac{\sigma}{\sqrt{n}}$ is a constant in a given experiment. This constant represents the difference between population and sample means which is merely the magnitude of the error. Letting $z_{\alpha/2}\dfrac{\sigma}{\sqrt{n}}$ equal to the error, designated by E, and solving for n, we get

$$n = \left[\frac{z_{\alpha/2}\,\sigma}{E}\right]^2 \tag{7.4.3}$$

This last expression could be used to predetermine the sample size necessary for a specified magnitude of error.

Example 7.4.3 Refer to Example 7.4.2. How large a sample should one take in order to be 95 percent confident that the sample mean does not differ from the population mean by more than 4 IQ points?

Solution: In this case, $E = 4$, $z_{\alpha/2} = 1.96$ and $\sigma = 15$. Using Eq. (7.4.3) we get

$$n = \left[\frac{1.96\,(15)}{4}\right]^2$$

$$= (.49 \cdot 15)^2 = 54.025$$

Since n cannot be a fractional number, and the problem requires the error not to exceed 4 IQ points, the next large integer should be used. Therefore, a sample of size 55 is required.

2. σ^2 *Unknown*. In many experiments the population variance would not be known and no information be available as to an approximate value. For this reason, the central limit theorem cannot be used to make inferences about the population mean. However, we can use Theorem 7.2.1 and analogously apply the Student's t probability function to compute confidence intervals for population means. There is an obvious drawback to this procedure. Theorem 7.2.1 requires the underlying population to be normally distributed and hence it is less general than the central limit theorem. This condition is not as restrictive as it may seem because the t p.f. is known to approach the unit normal as $n \longrightarrow \infty$ (see Prob. 7.2.10). In general, if $n \geq 30$, Theorem 7.2.1 can be applied to any population just as the central limit theorem was used when σ^2 was known. When samples of size less than 30 are considered the normal distribution of the underlying population cannot be ignored. The computation of confidence intervals is practically identical to the method discussed in part (1) above. That is,

$$P\left[-t_{\alpha/2,n-1} \leq \frac{\bar{x} - \mu_x}{\dfrac{s}{\sqrt{n}}} \leq t_{\alpha/2,n-1}\right] \tag{7.4.4}$$

$$= P\left[\,|\,\bar{x} - \mu_x\,| \leq t_{\alpha/2,n-1}\frac{s}{\sqrt{n}}\right] \tag{7.4.5}$$

$$= P\left[\bar{x} - t_{\alpha/2,n-1}\frac{s}{\sqrt{n}} \leq \mu \leq \bar{x} + t_{\alpha/2,n-1}\frac{s}{\sqrt{n}}\right] = 1 - \alpha \tag{7.4.6}$$

Comparing Eqs. (7.4.2) and (7.4.6) we note that the basic difference in the later is the calculation of s. The following example should illustrate the notion under discussion.

Example 7.4.4 Repeat Example 7.4.2 with the restriction that one cannot assume the population variance to be 225.

Solution: Since the population variance is not known, one needs to calculate s_x^2. For the data given,

$$s_x^2 = \frac{\sum_{i=1}^{9}(x_i - 107)^2}{8} = 159$$

Therefore

$$P\left[\,|\bar{x} - \mu_x| \leq t_{\alpha/2,8}\frac{\sqrt{159}}{\sqrt{9}}\right] = 1 - \alpha$$

For $\alpha = 0.05$ and $n = 8$ we find $t_{\alpha/2,8} = 2.30$, which reduces to

$$|107 - \mu_x| \leq 9.66$$

or equivalently

$$97.34 \leq \mu \leq 116.66$$

A remark should be made regarding the magnitude of the difference between the sample mean and the population mean in this case. In the expression $t_{\alpha/2,n-1}\dfrac{s}{\sqrt{n}}$ both $t_{\alpha/2,n-1}$ and s depend on n and hence $E = t_{\alpha/2,n-1}\dfrac{s}{\sqrt{n}}$ cannot be solved for n explicitly. However, sometimes it might be assumed that s has a fixed value and then a trial and error method may be used to determine the value of n.

Example 7.4.5 How large a sample size should one take if the magnitude of the difference between sample and population means should not exceed 4 units? Assume $s = 10$ and $\alpha = .05$.

Solution:

$$t_{\alpha/2,n-1} = \sqrt{n}\,\frac{E}{s}$$

Trial 1. Let $n = 25$:

$$2.064 = t_{.025,24} \stackrel{?}{=} \sqrt{25}\,\frac{4}{10} = 2.00$$

$$2.064 > 2.00$$

Trial 2. Let $n = 27$:

$$2.056 = t_{.025,26} \stackrel{?}{=} \sqrt{27}\,\frac{4}{10} = 2.07$$

$$2.056 < 2.07$$

Therefore the required value of n is between 25 and 27, i.e., 26.

(b) Inference about the Population Variance

1. Normally Distributed Population. As in the case of the mean, we need to know the p.f. of the sample variance in order to make inductive inferences about the population variance. From Theorem 7.3.3 we know that the random variable $(n - 1)\dfrac{s^2}{\sigma^2}$ is χ^2 distributed with $n - 1$ d.f. Consequently,

$$P\left[\chi^2_{1-\alpha/2,n-1} < (n-1)\frac{s^2}{\sigma^2} < \chi^2_{\alpha/2,n-1}\right] = 1 - \alpha \tag{7.4.7}$$

which is equivalent to

$$P\left[\frac{(n-1)\,s^2}{\chi^2_{\alpha/2,n-1}} < \sigma^2 < \frac{(n-1)\,s^2}{\chi^2_{1-\alpha/2,n-1}}\right] = 1 - \alpha \tag{7.4.8}$$

Equation (7.4.8) represents the 100 $(1 - \alpha)$ percent confidence interval around the population variance from which we obtain

$$P\left[s\sqrt{\frac{n-1}{\chi^2_{\alpha/2,n-1}}} < \sigma < s\sqrt{\frac{n-1}{\chi^2_{1-\alpha/2,n-1}}}\right] = 1 - \alpha \tag{7.4.9}$$

for the $100(1 - \alpha)$ percent confidence interval around the population standard deviation.

The following example illustrates the concept under consideration.

Example 7.4.6 Write the 95 percent confidence interval around the population standard deviation for the data of Example 7.4.4.

Solution: In Example 7.4.4 it was given that $s^2 = 159$ and $n = 9$. From Table VI we note $\chi^2_{.025,8} = 17.535$ and $\chi^2_{.975,8} = 2.179$. Applying Eq. (7.4.9) we get

$$\sqrt{159}\sqrt{\frac{8}{17.535}} < \sigma < \sqrt{159}\sqrt{\frac{8}{2.179}}$$

$$8.4 < \sigma < 24.1$$

2. *Underlying Distribution Unknown and Sample Size Large.* The random variable $\chi^2 = (n-1)\dfrac{s^2}{\sigma^2}$ assumed that the underlying distribution of the population is normal. However, if n is greater than 30, then it can be shown that the sampling distribution of standard deviation approaches the normal with mean σ and variance $\dfrac{\sigma^2}{2n}$. (See Prob. 7.4.9.) That is, the random variable

$$z = \frac{s - \sigma}{\dfrac{\sigma}{\sqrt{2n}}}$$

is approximately normal distributed. Consequently,

$$P\left[-z_{\alpha/2} < \frac{s-\sigma}{\dfrac{\sigma}{\sqrt{2n}}} < +z_{\alpha/2}\right] = 1 - \alpha \tag{7.4.10}$$

which is equivalent to

$$P\left[\frac{s}{\left(1+\frac{z_{\alpha/2}}{\sqrt{2n}}\right)} < \sigma < \frac{s}{\left(1-\frac{z_{\alpha/2}}{\sqrt{2n}}\right)}\right] = 1-\alpha \qquad (7.4.11)$$

Equation (7.4.11) may be used whenever the sample size is large otherwise Eq. (7.4.8) is the correct expression to apply.

Example 7.4.7 Refer to Example 5.6.2. Write a 99 percent confidence interval around the population standard deviation based on the sample size 40.

Solution: From Example 5.6.2 we get $s = 11.94$. Furthermore $z_{.005} = 2.58$ is obtained from the unit normal table. Applying Eq. (7.4.11) one gets

$$\frac{11.94}{1+\frac{2.58}{\sqrt{80}}} < \sigma < \frac{11.94}{1-\frac{2.58}{\sqrt{80}}}$$

$$9.27 < \sigma < 16.77$$

(c) Inference about the Proportion of a Binomial. In Sec. 4.6 it was shown that the binomial p.f. $f(x:n,p)$ can be approximated by the normal p.f. with mean np and variance npq. That is, the random variable $z = \frac{x-np}{\sqrt{npq}}$ is approximately normally distributed with mean zero and variance one. Therefore

$$P\left[-z_{\alpha/2} < \frac{x-np}{\sqrt{npq}} < z_{\alpha/2}\right] = 1-\alpha \qquad (7.4.12)$$

or equivalently,

$$P\left[\left|\frac{x}{n}-p\right| < z_{\alpha/2}\sqrt{\frac{pq}{n}}\right] = 1-\alpha \qquad (7.4.13)$$

We cannot use the last expression for estimating p because the parameter p is needed to estimate p. However, using $\frac{x}{n} = \bar{p}$ as an estimate of p and substituting it in Eq. (7.4.13) we get

$$P\left[|\bar{p}-p| < z_{\alpha/2}\sqrt{\frac{\bar{p}\bar{q}}{n}}\right] = 1-\alpha$$

which can be expressed as

$$P\left[\bar{p} - z_{\alpha/2}\sqrt{\frac{\bar{p}\bar{q}}{n}} < p < \bar{p} + z_{\alpha/2}\sqrt{\frac{\bar{p}\bar{q}}{n}}\right] = 1-\alpha \qquad (7.4.14)$$

This expression represents 100 $(1 - \alpha)$ percent confidence interval around the population parameter p.

Example 7.4.8 One hundred registered voters were selected at random from the registration lists of whom 43 were Republicans and 57 Democrats. Write a 95 percent confidence interval around the true (population) proportion for the Republicans.

Solution: Here $x = 43$, $n = 100$ and $z_{\alpha/2} = 1.96$. Using these values in Eq. (7.4.14) yields

$$P[.43 - .196\sqrt{.43(.57)} < p < .43 + .196\sqrt{.43(.57)}] = .95$$

$$P[.430 - .097 < p < .430 + .097] = .95$$

$$P[.333 < p < .527] = .95$$

That is, it could be stated with 95 percent certainty that the exact proportion of registered Republicans in this community is in the derived interval.

Let us emphasize that Eq. (7.4.14) represents an approximate probability and not an exact expression. If an exact probability were required, one encounters two major difficulties. First, since the binomial is a noncontinuous p.f. we cannot in general find values x_1 and x_2 such that

$$\sum_{x=0}^{x_1} f(x:n,p) = \frac{\alpha}{2} \quad \text{and} \quad \sum_{x=x_2}^{n} f(x:n,p) = \frac{\alpha}{2}$$

Therefore one needs to locate the largest integer x_1 and smallest integer x_2 in order to satisfy simultaneously the relationships

$$\sum_{x=0}^{x_1} f(x:n,p) \leq \frac{\alpha}{2} \tag{7.4.15}$$

and

$$\sum_{x=x_2}^{n} f(x:n,p) = 1 - \sum_{x=0}^{x_2-1} f(x:n,p) \leq \frac{\alpha}{2} \tag{7.4.16}$$

Second, the unknown parameter p, which we wish to estimate or make an inference about, is one of the dependent variables of the binomial p.f. These obstacles are partly removed if one uses Eq. (7.4.14), provided n is large.

An exact method for finding confidence intervals around the parameter p of the binomial distribution utilizes the relationship

$$1 - F(x:n,p) = P\left[F(m_1, m_2) < \frac{n-x}{x+1}\,\frac{p}{1-p}\right] \tag{7.4.17}$$

where $m_1 = 2(x + 1)$ and $m_2 = 2(n - x)$. The proof of Eq. (7.4.17) is based upon the relationship of the cumulative binomial and the incomplete beta function, but will not be given.

Substituting Eq. (7.4.17) for Eq. (7.4.15), we get

$$P\left[F(m_1, m_2) < \frac{n-x}{x+1}\frac{p}{1-p}\right] = 1 - \frac{\alpha}{2} \tag{7.4.18}$$

To obtain the value of F explicitly we set

$$F_{\alpha/2;m_1,m_2} = \frac{n-x}{x+1}\frac{p}{1-p} = \frac{n-x}{x+1}\left[\frac{1}{1-p} - 1\right] \tag{7.4.19}$$

which when solved for p reduces to

$$p = 1 - \frac{n-x}{n-x+(x+1)F_{\alpha/2;m_1,m_2}} \tag{7.4.20}$$

or equivalently

$$1 - p = q = \frac{n-x}{n-x+(x+1)F_{\alpha/2;m_1,m_2}} \tag{7.4.21}$$

Equation (7.4.20) represents the upper limit of 100 $(1 - \alpha)$ percent confidence interval around the parameter p. To obtain the lower limit, consider

$$F_{\alpha/2;m_1,m_2} = \frac{1}{F_{1-\alpha/2;m_2,m_1}}$$

and applying it to Eq. (7.4.16) we get

$$1 - \frac{\alpha}{2} = \sum_{x=0}^{x_2-1} f(x{:}n,p) = P\left[\frac{1}{F_{\alpha/2;m_2,m_1}} < \frac{n-x}{x+1}\frac{p}{1-p}\right] \tag{7.4.22}$$

since the summation in the last expression has an upper limit $x_2 - 1$, the degrees of freedom are $m_1 = 2[(x-1)+1] = 2x$ and $m_2 = 2[n - (x-1)] = 2(n-x+1)$. Therefore, following similar steps as above, we get

$$\frac{1}{F_{\alpha/2;m_2,m_1}} = \frac{n-x+1}{x}\frac{p}{1-p} \tag{7.4.23}$$

which when solved for $1 - p$ reduces to

$$1 - p = q = \frac{(n-x+1)F_{\alpha/2;m_2,m_1}}{x+(n-x+1)F_{\alpha/2;m_2,m_1}} \tag{7.4.24}$$

and is the lower limit of 100 $(1 - \alpha)$ percent confidence interval for p.

Example 7.4.9 A random sample of 18 registered voters were selected from the registration list for jury duty. If 7 of the selected persons were Democrats, write a 98 percent confidence interval around the true population proportion p.

Solution: In this problem $n = 18$, $x = 7$, and $\alpha = 2$ percent. Therefore, for the upper confidence limits we have

$$m_1 = 2(7+1) = 16 \quad \text{and} \quad m_2 = 2(18-7) = 22$$

Applying these values to Eq. (7.4.21) we get

$$q = \frac{11}{11 + 8(F_{.01;16,22})} = \frac{11}{11 + 8(2.95)} = \frac{11}{34.60}$$
$$= .317$$

and $p = .683$.

Similarly, applying $m_1 = 2x = 14$ and

$$m_2 = 2(n - x + 1) = 2(18 - 7 + 1) = 24$$

to Eq. (7.4.24) we get for lower 98 percent confidence limit

$$q = \frac{12\,(F_{.01;24,14})}{7 + 12\,(F_{.01;24,14})} = \frac{12\,(3.43)}{7 + 12\,(3.43)} = \frac{41.16}{48.16} = .855$$

and $p = .145$. That is,

$$.145 < p < .683$$

or

$$.317 < q < .855$$

PROBLEMS

7.4.1 From past experience it is known that the standard deviation of the lives of certain type of tires is 1,000 miles. If two sets of four tires had to be replaced after 18,500, 17,600, 19,100, 18,000, 16,800, 17,400, 18,300, and 16,600 miles, write a 95 percent confidence interval for the true population mean.

7.4.2 Under the assumptions of Prob. 7.4.1, how large a sample should one take in order that the difference between the true and the sample mean does not exceed 250 miles?

7.4.3 Repeat Prob. 7.4.1 if the population standard deviation is unknown. Assume that the population is normal.

7.4.4 A new automobile engine is designed and the prototype on 12 test runs yields a mean of 27.3 miles per gallon and 1.2 for the standard deviation. (a) write a 95 percent confidence interval for the population mean. (b) write a 95 percent confidence interval around the population variance.

7.4.5 In a random sample of size 16 the sums of squares, $[\Sigma\,(x_i - \overline{x})^2]$, is equal to 416.3. Write a 99 percent confidence interval around the population standard deviation.

7.4.6 In a random sample of size 128 the sums of squares was 2,946.7. Write a 95 percent confidence interval around the population standard deviation.

7.4.7 In a random sample of 1,000 articles taken from a day's production, 40 percent were defective. Write a 95 percent confidence interval for the true proportion of defective articles.

7.4.8 Suppose the sample size in Prob. 7.4.7 is 20. Obtain an exact confidence interval for the true proportion of defectives.

7.4.9 Show that

$$z = \frac{s - \sigma}{\dfrac{\sigma}{\sqrt{2n}}}$$

is approximately normal distributed. [*Hint:* Refer to Prob. 7.3.11 and solve for z in the relationship

$$(n - 1)\frac{s^2}{\sigma^2} = \frac{1}{2}[z + \sqrt{2(n - 1) - 1}]^2$$

assuming $\sqrt{2n - 2}$ and $\sqrt{2n - 3}$ are approximately equal to $\sqrt{2n}$ for large n.]

7.4.10 A random sample of 33 observations is taken from a large population which yields 16.8 for the standard deviation. Write a 90 percent confidence interval around the population standard deviation.

7.4.11 A random sample of 36 is taken from an infinite population having a standard deviation of 9. What is the probability that the mean of the sample will differ from the mean of the population by more than 3 units of measurement?

7.4.12 In Prob. 7.4.11, how large a sample should one take in order to be 90 percent confident that the mean of the sample does not differ from the mean of the population by more than two units of measurement?

7.4.13 Verify Eq. (7.4.17) for the special case $n = 5$, $x = 2$, and $p = \frac{1}{2}$.

7.5 SOME DESIRABLE CHARACTERISTICS OF STATISTICS

In the discussion of the sampling distribution of the mean and the variance, it was mentioned that the primary objective of taking samples from a population, (random variable, say x), was for the purpose of estimating the unknown parameters associated with the probability function of x. For instance, if μ and σ^2 were the unknown parameters of the distribution of x, and a random sample of 25 observations were taken from the population in question, the sample mean, $\bar{x}$, and the sample variance, s^2, represent estimates of the parameters μ and σ^2, respectively. However, it should be recalled that $\bar{x}$ is not necessarily equal to μ, nor s^2 equal to σ^2. In general, we cannot assert that the sample mean equals the population mean, nor the sample variance equals the population variance, unless the whole population is taken for a sample. Since it is not practical to take for a sample the whole population, we need to establish some characteristics of a "good" estimate (statistic) and secondly, to determine criteria on which to decide whether a given estimate is "better" than another. In this section we shall present a few criteria for "good" estimates.

(a) Unbiased Statistic. Suppose π represents a parameter of the probability function of the random variable x. Let $x_1, x_2, \ldots, x_n$ be a random sample of x and $\hat{p}$ a function of the sample. If

$$E[\hat{p}] = \pi \tag{7.5.1}$$

then $\hat{p}$ is said to be an *unbiased* statistic. In words, whenever the expected value of a statistic is equal to the actual value of a parameter, the statistic is said to be unbiased or unbiased estimator.

Since $E[\bar{x}] = \mu_x$ [Eq. (7.1.9)], the sample mean is an unbiased statistic. Not all statistics are unbiased. For example, when the sample variance was defined in Sec. 5.3 by Eq. (5.3.6) it was stated that for technical reasons the denominator was $n - 1$ and not n. Let us next prove that the sample variance defined by Eq. (5.3.6) is an unbiased estimator of σ_x^2:

$$E[s^2] = E\left[\frac{1}{n-1}\sum_{i=1}^{n}(x_i - \bar{x})^2\right] = E\left[\frac{1}{n-1}\sum_{i=1}^{n}[(x_i - \mu) - (\bar{x} - \mu)]^2\right]$$

$$= E\left[\frac{1}{n-1}\Sigma(x_i - \mu)^2\right] - E\left[2(\bar{x} - \mu)\frac{1}{n-1}\Sigma(x_i - \mu)\right] + E\left[\frac{1}{n-1}\Sigma(\bar{x} - \mu)^2\right]$$

$$= E\left[\frac{n}{n-1}\frac{\Sigma(x_i - \mu)^2}{n}\right] - E\left[2(\bar{x} - \mu)\frac{n}{n-1}\frac{\Sigma(x_i - \mu)}{n}\right] + E\left[\frac{n}{n-1}(\bar{x} - \mu)^2\right]$$

$$= \frac{n}{n-1}\sigma^2 - E\left[2(\bar{x} - \mu)\frac{n}{n-1}(\bar{x} - \mu)\right] + \frac{n}{n-1}E[(\bar{x} - \mu)^2]$$

$$= \frac{n}{n-1}\sigma^2 - \frac{n}{n-1}E[(\bar{x} - \mu)^2]$$

However, $E[(\bar{x} - \mu)^2] = \sigma_{\bar{x}}^2$ which is equal to σ^2/n. Therefore, we get

$$E[s^2] = \frac{n}{n-1}\sigma^2 - \frac{n}{n-1}\frac{\sigma^2}{n}$$

$$= \frac{n}{n-1}\left[1 - \frac{1}{n}\right]\sigma^2 = \sigma^2 \qquad (7.5.2)$$

which verifies that the sample variance is an unbiased statistic of the population variance.

Let us remark at this juncture that the sample standard deviation is a *biased* statistic. That is, $E[s] \neq \sigma$.

(b) Consistent Statistic. Suppose π represents a parameter of the probability function of the random variable x and $\hat{p}$ is an estimate of π based on a random sample of size n. $\hat{p}$ is said to be a *consistent statistic* if the probability of $\hat{p}$ approaching π becomes one as n increases without bound. In symbols,

$$\lim_{n\to\infty} P[\,|\hat{p} - \pi| < \epsilon] = 1 \qquad \text{for all } \epsilon > 0 \qquad (7.5.3)$$

defines a consistent statistic. For example the sample mean $\bar{x}$ satisfies the criterion of consistency. Recalling that Eq. (7.1.12) can be expressed equivalently as

$$P[\,|\bar{x} - \mu| < \epsilon] > 1 - \frac{\sigma^2}{\epsilon^2 n} \qquad (7.5.4)$$

and letting $n \longrightarrow \infty$, we obtain

$$\lim_{n \to \infty} P[\,|\bar{x} - \mu| < \epsilon] = 1$$

and hence Eq. (7.5.3) is satisfied.

(c) Relative Efficiency. Suppose $\hat{p}_1$ and $\hat{p}_2$ are two unbiased statistics estimating the parameter π and if

$$E[(\hat{p}_1 - \pi)^2] < E[(\hat{p}_2 - \pi)^2] \qquad (7.5.5)$$

then $\hat{p}_1$ is said to be *more efficient relative to* $\hat{p}_2$. Some authors refer to $\hat{p}_1$ as more reliable than $\hat{p}_2$.

For instance, let us consider the sample mean and the sample median from a normal population. From the central limit theorem the sampling distribution of the mean is asymptotically normal with mean μ and variance σ^2/n. It can be shown that the sampling distribution of the median is asymptotically normal with mean μ and variance $\frac{\pi}{2}\frac{\sigma^2}{n}$. That is,

$$E[(\bar{x} - \mu_x)^2] = \frac{\sigma^2}{n}$$

and

$$E[(\tilde{x} - \mu_x)^2] = \frac{\pi}{2}\frac{\sigma^2}{n}$$

Since $\frac{\sigma^2}{n} < \frac{\pi}{2}\frac{\sigma^2}{n}$, the sample mean relative to the median is more efficient.

Unbiasedness, consistency, and relative efficiency are desirable characteristics of estimates. However, it is not always feasible to find statistics that have all these qualities.

In Fig. 7.8 the sampling distribution of two estimates $\hat{p}_1$ and $\hat{p}_2$ are given; $\hat{p}_1$ is unbiased estimate of π, while $\hat{p}_2$ is biased.

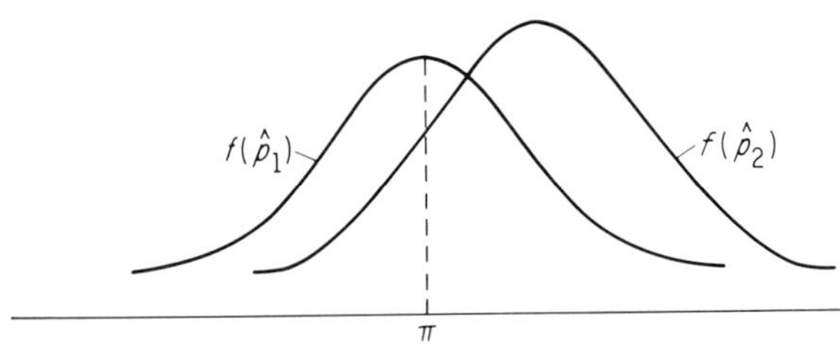

FIG. 7.8.

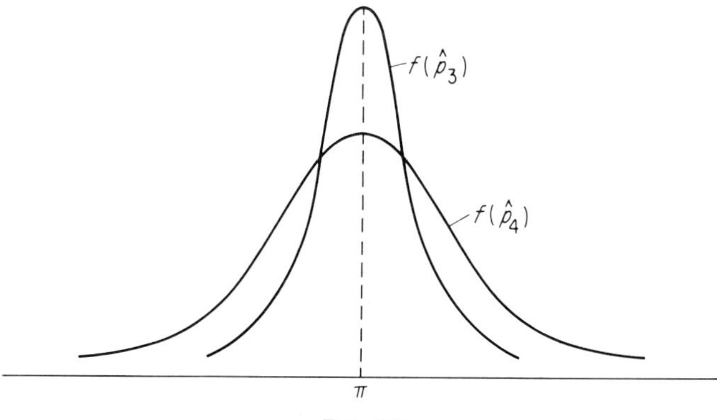

FIG. 7.9.

In Fig. 7.9 the sampling distribution of two other unbiased estimates $\hat{p}_3$ and $\hat{p}_4$ are given, for which $\hat{p}_3$ is more efficient relative to $\hat{p}_4$.

PROBLEMS

7.5.1 Suppose that p_1, p_2 and p_3 are three estimates of the population parameter π. If $E[p_1] = .9\pi$, $E[p_2] = \pi$, and $E[p_3] = 1.2\pi$ is the only information given, which estimate would you prefer? Why?

7.5.2 Suppose p_1, p_2 and p_3 are three unbiased estimates of π. If $\sigma_{p_1}^2 = .25\pi^2$, $\sigma_{p_2}^2 = .09\pi^2$ and $\sigma_{p_3}^2 = .04\pi^2$, which estimate would you prefer? Explain why.

7.5.3 Show that the expectation of the sample proportions taken from a binomial population is equal to the parameter p. That is, $E\left[\frac{x}{n}\right] = p$, where n represents the sample size and x the number of successes.

7.5.4 Show that the sample proportions taken from a binomial population is a consistent statistic. [*Hint:* Refer to Eq. (7.1.13).]

7.5.5 Show that the sample standard deviation s_x is a biased statistic—that is, $E[s_x] \neq \sigma$.

8

Statistical Decision Making

8.0 INTRODUCTION

While discussing random sampling in Chapter 5, it was mentioned that the fundamental notion underlying statistical theory is to obtain information regarding a population by examining a random sample from that population. The information thus obtained is generally used for making decisions. For example, suppose a box contains 24 items. Two are randomly selected and tested for defect. (See Example 2.6.2.) Should one accept or reject the assumption that all items in this box are nondefective, if the two inspected articles prove nondefective? Again, if the mean height of 25 randomly selected students from College A is 68.64 inches, should one accept or reject the assumption that the exact average height of students at College A is 68.75 inches? In the discussion on confidence intervals in Sec. 7.4, we considered probabilities associated with such assumptions. That is, it was stated with certain confidence (probability) that the population parameter lies in a given interval of values. Often, however, one might not be interested in whether the parameter is or is not in an interval, but rather an assumption made about a parameter is correct or incorrect. In this chapter we shall consider statistical assumptions or hypotheses and with a prescribed decision rule determine whether to accept or to reject the hypothesis in question.

8.1 ACCEPTANCE SAMPLING PLAN

An ideal manufacturing process requires that all items produced during production be nondefective. This of course occurs very seldom. In most manufacturing situations it is anticipated and perhaps expected that certain portion of the manufactured articles be defective; but the objective still remains to keep the quantity of defectives as "low" as possible. What might be a low percentage in one situation could be considered very high for another process. For example, if the production items are 1-inch nails, the process might be considered good when 10 percent of the nails turn defective. However, if the items manufactured are automobiles, a single defective article (a car) might prove costly and present a poor and unacceptable process. Therefore, the producers try to maximize the number of nondefective articles, which also intuitively implies the maximizing of the probability that a purchaser will accept the product as satisfactory.

Manufacturers commonly use a method called the *acceptance sampling plan* to determine whether the number of defective articles is small in a shipment or not. Basically, acceptance sampling requires that a predetermined number of items when randomly selected and inspected prove to be nondefective. The sampling method described in Example 2.6.2 illustrates a typical acceptance sampling plan. In that illustration a box contained 24 articles, a random sample of size 2 was taken, and both had to prove to be nondefective during inspection.

In general, in an acceptance sampling plan a lot contains N items, a random sample of size n is taken from it and if k out of n prove to be nondefective the manufacturer ships the lot as quality merchandise. Otherwise, the lot is retained at the plant for possible replacing of defective items. Whenever defective items are replaced by nondefective articles, the new lot is said to be *rectified*.

The reader might ask the question, "Why does the manufacturer not test or inspect all items in the lot and do away with acceptance sampling plans?" It should be recalled that some tests or inspection methods are self-destructive by nature. For instance, when a light bulb is given a simulated life test it is obviously destroyed at the end of the experiment. Another reason for not testing each item individually is that the testing process is usually very costly. Expensive instruments, extra plant space required, and labor cost make inspecting all manufactured items prohibitive.

The number k in an acceptance sampling is called the *allowable number* of the sampling plan. It should be clear from the discussion and reference to Example 2.6.2 that there are three independent parameters in the acceptance sampling—the lot size, the sample size, and the actual number of defective items in the lot. The allowable number k is a random variable $k = 0, 1, \ldots, k \leq n$. [Refer to Eq. (2.6.2).] However, in this situation the actual number of defective items are not known, hence Np represents the number of defective items in the lot, where p represents the proportion of defective items. Therefore the hypergeometric p.f. applied to this case becomes

$$f(x:N,Np,n) = \frac{\binom{Np}{x}\binom{N-Np}{n-x}}{\binom{N}{n}} \qquad x = 0, 1, \ldots, k \leq n \qquad (8.1.1)$$

Designating the probability of accepting a lot as quality merchandise by $P(A)$, we note that

$$P(A) = F(k:N,Np,n) \qquad (8.1.2)$$

[Refer to Eq. (2.3.1).]

Assuming that for a given acceptance sampling plan the quantities N, n and k are fixed, we designate by $P(A \mid p)$ the probability of accepting a lot given p and express Eq. (8.1.2) as

$$P(A \mid p) = F(k:N,Np,n) \tag{8.1.3}$$

Let us illustrate the concepts under consideration.

Example 8.1.1 In a manufacturing process lots are comprised of 10 items per box. The acceptance sampling plan to be pursued is the following. A random sample of 2 items are selected and the lot is accepted (shipped to the consumer) if both items prove nondefective. Find the probabilities of accepting the lot if $p = 0, 0.1, \ldots, 1$.

Solution: The statement "Both items to prove nondefective" is equivalent to "No item in the sample is defective." Hence, for this case Eq. (8.1.3) reduces to $f(k\colon N,Np,n)$ and we obtain

$$P(A \mid 0) = f(0\colon\ 10, 0, 2) = \frac{\binom{0}{0}\binom{10}{2}}{\binom{10}{2}} = 1$$

$$\begin{aligned}
P(A \mid .1) &= f(0\colon 10,1,2) = .80 \\
P(A \mid .2) &= f(0\colon 10,2,2) = .62 \\
P(A \mid .3) &= f(0\colon 10,3,2) = .466 \\
P(A \mid .4) &= f(0\colon 10,4,2) = .333 \\
P(A \mid .5) &= f(0\colon 10,5,2) = .222 \\
P(A \mid .6) &= f(0\colon 10,6,2) = .133 \\
P(A \mid .7) &= f(0\colon 10,7,2) = .066 \\
P(A \mid .8) &= f(0\colon 10,8,2) = .022 \\
P(A \mid .9) &= f(0\colon 10,9,2) = 0 \\
P(A \mid 1) &= f(0\colon 10,10,2) = 0
\end{aligned}$$

In Sec. 2.6 the relationship between the hypergeometric and the binomial p.f. was considered. It was shown that for large N the hypergeometric p.f. approaches the binomial p.f. [Refer to Eq. (2.6.9).] Therefore, if the lot is large, the binomial p.f. provides a close approximation for Eq. (8.1.3). That is,

$$P(A \mid p) = F(k:N,Np,n) \cong F(k:n,p) \tag{8.1.4}$$

and the values of $P(A \mid p)$ can be obtained from the cumulative binomial tables (Table II in the Appendix). The following example illustrates this notion.

Example 8.1.2 Repeat Example 8.1.1 using the corresponding binomial p.f. as an approximation to the indicated acceptance sampling plan.

Solution: From Eq. (8.1.4) we note that $F(0{:}10,10p,2) \cong F(0{:}2,p)$. Recalling that $F(0{:}2,p) = f(0{:}2,p)$, we obtain, using Table II in the Appendix,

$f(0{:}2,0) = 1$	$f(0{:}2,.6) = .16$
$f(0{:}2,.1) = .81$	$f(0;2,.7) = .09$
$f(0{:}2,.2) = .64$	$f(0{:}2,.8) = .04$
$f(0{:}2,.3) = .49$	$f(0{:}2,.9) = .01$
$f(0{:}2,.4) = .36$	$f(0{:}2,1.0) = 0$
$f(0{:}2,.5) = .25$	

If the values of $P(A \mid p)$ are plotted with $P(A \mid p)$ as ordinate and p as the abscissa, the resulting sketch is called the *operating characteristic curve* of the acceptance sampling plan. Commonly, the operating characteristic curves are referred to as *OC* curves. The *OC* curves for the acceptance sampling plans of Example 8.1.1 and 8.1.2 are shown in Fig. 8.1.

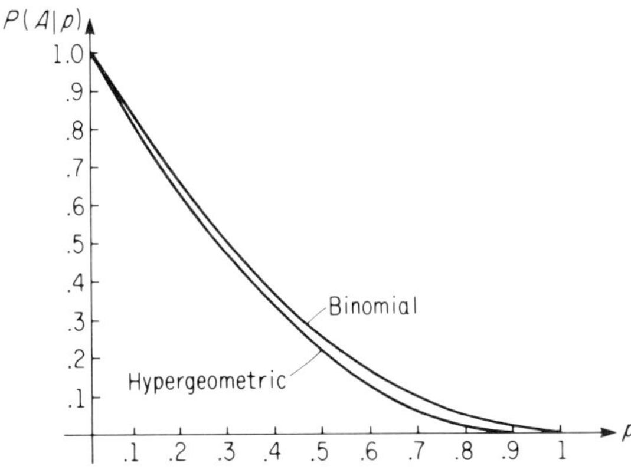

FIG. 8.1. *OC* curves.

For the *OC* curves of the acceptance sampling plans described above, it was assumed that $0 \leq p \leq 1$. However, in most manufacturing processes only a small fraction of the articles produced are defective; say 10 percent or less. Under such conditions acceptance sampling plans are conveniently calculated by the Poisson probability function. In Sec. 2.5 it was shown that the Poisson p.f. is related to the binomial p.f. and therefore

$$P(A \mid p) \cong F(k{:}\mu) \tag{8.1.5}$$

where μ is equal to $n \cdot p$ and represents the single parameter of the Poisson probability function. [Equations (3.2.5) and (3.2.19).] The following example illustrates an acceptance sampling based on the Poisson probability function.

Example 8.1.3 An acceptance sampling plan requires that 10 percent of the articles in a lot of 1,000 be inspected. The lot is accepted if at most two of the inspected items are defective. Graph the *OC* curve for this plan.

Solution: In this illustration $n = 1000(0.1) = 100$ and hence the parameter for the Poisson p.f. is $\mu = 100p$. Using Table III in the Appendix, the distribution function of the Poisson probability function, we get for increments of .01 of p,

$$
\begin{aligned}
P(A \mid 0) &= F(2{:}0) = 1 \\
P(A \mid .01) &= F(2{:}1) = .92 \\
P(A \mid .02) &= F(2{:}2) = .677 \\
P(A \mid .03) &= F(2{:}3) = .423 \\
P(A \mid .04) &= F(2{:}4) = .238 \\
P(A \mid .05) &= F(2{:}5) = .125 \\
P(A \mid .06) &= F(2{:}6) = .062 \\
P(A \mid .07) &= F(2{:}7) = .030 \\
P(A \mid .08) &= F(2{:}8) = .014 \\
P(A \mid .09) &= F(2{:}9) = .006
\end{aligned}
$$

The *OC* curve for this acceptance sampling plan is given in Fig. 8.2.

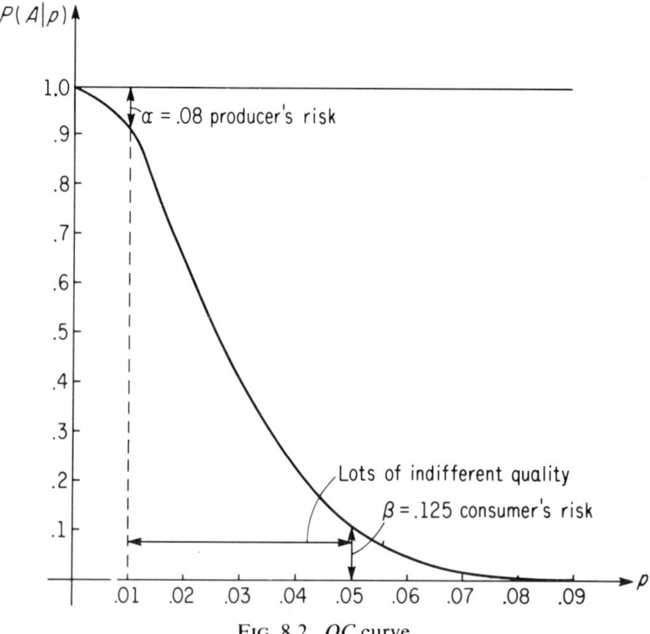

FIG. 8.2. *OC* curve.

The *OC* curves of various acceptance sampling plans describe the mutual risks that manufacturers and the purchasers agree to take. The manufacturer takes the risk that a good lot will be rejected by the purchaser. This risk is called the *producer's risk*, and the probability associated with it is designated by α. The buyer takes the risk that a bad lot will be accepted from the producer which risk is referred to as *consumer's risk*. The probability associated with the consumer's risk is designated by β.

Table 8.1.1 summarizes the two types of risks being discussed.

TABLE 8.1.1

	Lot Is Good	Lot Is Bad
Accept the lot	Correct choice $P(A \mid p) = (1 - \alpha)$	Incorrect choice consumer's risk, β
Reject the lot	Incorrect choice producer's risk, α	Correct choice, $(1 - \beta)$

Let us illustrate how an *OC* curve determines the two types of risks. Suppose a producer is using the acceptance sampling plan given in Example 8.1.3. Furthermore, the producer realizes that at least 1 and at most 5 percent of the manufactured items are defective, and informs the purchaser of these conditions. The buyer agrees to this plan with the stipulation to reject (return) the lot only if $p > .05$. The lots for which $.01 < p < .05$ are called *lots of indifferent quality*. From Example 8.1.3 we notice that $P(A \mid .01) = .92$ and hence $\alpha = .08$. That is, the probability is 8 percent for a good lot being rejected by this sampling plan, and represents the risk which the producer takes. On the other hand, $P(A \mid .05) = .125$, represents the consumer's risk because it corresponds to accepting a bad lot. These risks along with the interval for lots of indifferent quality are shown in Fig. 8.2.

The *OC* curves can be used conversely as well. That is, for any specified values of α and β the interval corresponding to the lots of indifferent quality can be located. For instance, from Fig. 8.2 we note that for $\alpha = .20$ and $\beta = .25$, the interval for lots of indifferent quality is approximately $.015 < p < .04$.

Next let us consider acceptance sampling plans where each rejected lot is rectified. If the exact (true) proportion of defective articles produced by the manufacturer is p, then α, the producer's risk represents the proportion of lots that has to be rectified and $(1 - \alpha)$ the proportion that will be outgoing without rectification. Clearly, $(1 - \alpha) = P(A \mid p)$ (see Table 8.1.1), and hence the average proportion of outgoing articles without rectification will be

$$p(1 - \alpha) = p[P(A \mid p)] \qquad (8.1.6)$$

and is referred to as the *average outgoing quality*. The average outgoing quality is designated by AOQ. In general, $P(A \mid p)$ is calculated for an acceptance sampling plan and therefore the computation of AOQ does not require new information. Since Eq. (8.1.6) is a function of p, it may be represented as

$$f(p) = AOQ(p) = p[P(A \mid p)] \tag{8.1.7}$$

The graph of $f(p)$ with p as the abscissa is called *the AOQ curve*. The following example illustrates the notions being discussed.

Example 8.1.4 Graph the AOQ curve for the data of Example 8.1.1.
Solution: The product of the various values of p with $P(A \mid p)$ yields

$$\begin{aligned}
f(0) &= (0)(1) &&= 0\\
f(.1) &= (.1)(0.80) &&= .080\\
f(.2) &= (.2)(0.62) &&= .124\\
f(.3) &= (.3)(0.466) &&= .140\\
f(.4) &= (.4)(0.333) &&= .133\\
f(.5) &= (.5)(0.222) &&= .111\\
f(.6) &= (.6)(0.133) &&= .080\\
f(.7) &= (.7)(0.066) &&= .046\\
f(.8) &= (.8)(0.022) &&= .018\\
f(.9) &= (.9)(0) &&= 0\\
f(1) &= (1)(0) &&= 0
\end{aligned}$$

whose graph is shown in Fig. 8.3.

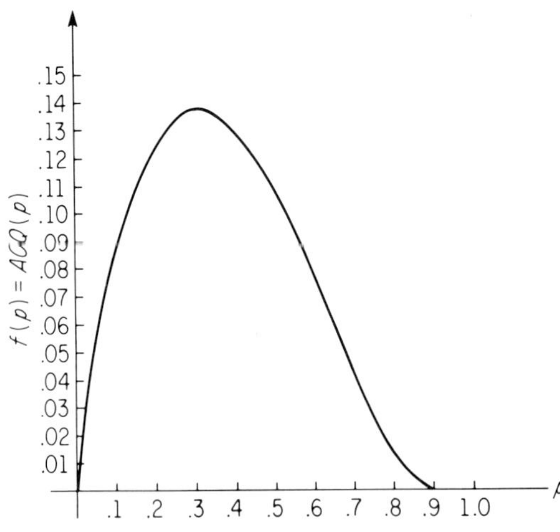

FIG. 8.3. AOQ curve.

In Fig. 8.3 we note that in the neighborhood of $p = .3$ the AOQ curve attains its maximum. This value of AOQ is called the *average outgoing quality limit* and is denoted by $AOQL$. In Example 8.1.4 $AOQL$ is approximately .14 and represents the maximum of the average proportions of defective articles not rectified under the given acceptance sampling plan. To state differently, the particular acceptance sampling plan will rectify at least 86 percent of the outgoing lots. In general then, at least $(1 - AOQL)100$ percent of outgoing articles will be rectified by a particular acceptance sampling plan.

PROBLEMS

8.1.1 In a manufacturing process 60 articles are placed in each box. The acceptance sampling plan is to test 3 items from each box and if all 3 prove nondefective the articles are shipped as quality merchandise; otherwise, the lot is rectified. (a) Graph the OC curve for this acceptance sampling plan, using $p = 0$, .1, .2, . . . , 1; (b) If $.1 < p < .4$ represents the interval for lots of indifferent quality, find the purchaser's and consumer's risks; (c) Graph the AOQ curve for this plan and find the value of $AOQL$.

8.1.2 Repeat Prob. 8.1.1 with 30 articles in each box instead of 60.

8.1.3 Repeat Prob. 8.1.1 using the binomial approximation to the hypergeometric.

8.1.4 Repeat Prob. 8.1.2 using the binomial approximation to the hypergeometric.

8.1.5 A wholesaler buys 6,000 flashlight bulbs. The acceptance sampling plan is to test 1 percent of the merchandise and accept the lot if at most 4 bulbs prove defective. Assuming that no more than 10 percent of the articles are defective, use the Poisson approximation to (a) graph the OC curve for this sampling plan; (b) graph the AOQ curve for this plan; (c) find the producer's and consumer's risks, if the interval for lots of indifferent quality is $.01 < p < .03$; (d) estimate p from the OC curve if the producer's risk is .02; (e) estimate p from the OC curve if the consumer's risk is .05.

8.2 TESTING STATISTICAL HYPOTHESES

In a broad sense, a statistical hypothesis is an assumption or a statement concerning the distribution and the parameters of a population. For instance, the statement "The breakdown voltage of a certain type of transistors is log-normal distributed." is an example of a statistical hypothesis. [Refer to Eq. (4.7.1).] Again, "The mean breakdown voltage of these transistors is 5.7 microvolts." is a statistical hypothesis because the mean of the log-normal p.f. is a function of its parameters α and β. [Refer to Eq. (4.7.11).] Likewise, "The variance of the log-normal p.f. describing the breakdown voltages of the transistors is greater than 4 microvolts^2." is a statistical hypothesis since according to Eq. (4.7.14) the variance is a function of the parameters α and β also.

Each of these statistical hypotheses is either true or false and we must decide whether to accept the hypothesis as correct or incorrect. The statement regarding the mean should either be accepted as valid or invalid. How does one decide between these two states of affairs? Usually information obtained from samples constitute the basis for deciding whether to accept or to reject a statistical hypothesis. Such a decision rule or procedure is called a *statistical test.* Let us illustrate a few statistical tests and show how they are applied to hypotheses.

Suppose we assume that x, the distance between the index finger and the thumb in stretched position, is normally distributed with mean 7 inches and variance 1.21. Then the probability function of the random variable is, according to Eq. (4.5.9), $f(x:7, 1.21)$. Now suppose that the procedure which we shall adopt for our decision is as follows. Toss a coin once. Observe the outcome. If heads show accept the assumption as true, if tails show reject the hypothesis. This certainly is a testing procedure but we hasten to admit that is not a "good" test to base decisions upon. A more desirable test might be the following procedure. Select at random n persons, say 10, and measure the distance between their index finger and thumb. Compute the mean and the standard deviation of this particular sample. If the mean is between 6 and 8 inches and the standard deviation between 1 and 1.5 inches, accept the hypothesis as true. Otherwise, reject. The distributions of the mean and the variance of the sample then, are used as the *test statistics.*

We shall refer to a statistical hypothesis as *compound* whenever more than one hypothesis is tested to accept or reject its validity. The illustration just cited is an example of compound statistical hypothesis because three different hypotheses are to be examined separately; namely,

1. x is normally distributed.
2. $\mu_x = 7$—that is, the mean of the p.f. is 7.
3. $\sigma_x = 1.1$ or the variance is 1.21.

These hypotheses have to be tested individually, but not necessarily in that order. Usually, a compound hypothesis is tested by examining each hypothesis separately and assuming during the process that the remaining hypotheses are true. In the example under discussion, to test the hypothesis $\mu_x = 7$ it would be assumed that hypotheses 1 and 3 are correct.

Next let us assume that the available information to be used for the test indicates to reject a hypothesis. This implies that we really accept an *alternative* hypothesis. For instance, if we reject $\mu_x = 7$ we must of necessity accept one or the other of the following hypotheses. (a) $\mu_x > 7$ (b) $\mu_x < 7$. More specifically in (a) we may state $\mu_x = 8$ or $\mu_x = 7.7$ or $\mu_x = 9.24$ and in (b) $\mu_x = 6.5$ or $\mu_x = 5.93$, *ad infinitum.*

Conventionally, the symbols H_0 and H_1 are used to signify a hypothesis and its alternative, respectively. H_0 is read "the null hypothesis" and H_1 "the alternative hypothesis."

1. $H_0: \mu_x = 50$
   $H_1: \mu_x \neq 50$
2. $H_0: \mu = 50$
   $H_1: \mu > 50$
3. $H_0: \mu = 50$
   $H_1: \mu < 50$

are examples of the three major categories of null and alternative hypotheses. We shall discuss these types of hypotheses in detail in subsequent sections.

Each correct and incorrect decision can be partitioned into two parts. Hence there are four possible choices in a given double dichotomy (accept or reject) and (H_0 is true or H_1 is true). Table 8.2.1 shows the choice possibilities under consideration.

From Table 8.2.1 one notes that there are two possibilities for correct and two possibilities for incorrect decisions or conclusions. The two correct and the two incorrect choices are not the same and in general the incorrect decisions are of concern. To distinguish between these two kinds of errors, commonly they are referred to as *type I* and *type II errors*. Type I and type II errors are indicated in Table 8.2.1.

TABLE 8.2.1

	H_0 Is the True Assumption	H_1 Is the True Assumption
Accept H_0	H_0 is true and we accept H_0. Correct decision.	H_1 is true and we accept H_0. Incorrect decision—type II error.
Reject H_0	H_0 is true and we reject H_0. Incorrect decision—type I error.	H_1 is true and we reject H_0. Correct decision.

To illustrate the difference between type I and type II errors, let us consider a nonstatistical situation. A jury is to decide whether a defendant is guilty of murder or not. The two state of affairs are either he is a murderer or he is not, and the jury can offer two verdicts, guilty or not guilty. Table 8.2.2 is an analogous table to Table 8.2.1.

Type I error on the part of the jury permits the murderer to go free, while type II error condemns an innocent person. Which of these two types of errors is of more serious consequence? Since this is a nonstatistical illustration, we let nonstatisticians answer the question and we return to statistical hypotheses.

Both type I and type II errors may be expressed as probabilities and conventionally the symbols α and β are associated with them, respectively.

TABLE 8.2.2

	The Defendant Is a Murderer	The Defendant Is Not a Murderer
Accept: he is murderer.	Correct verdict of guilty.	Incorrect verdict of guilty. Type II error.
Reject: he is murderer.	Incorrect verdict of not guilty. Type I error.	Correct verdict of not guilty.

That is,

$$P[(H_0 \text{ is true}) \cap (\text{reject } H_0)] = \alpha \tag{8.2.1}$$

$$P[(H_1 \text{ is true}) \cap (\text{accept } H_0)] = \beta \tag{8.2.2}$$

Consequently,

$$P[(H_0 \text{ is true}) \cap (\text{accept } H_0)] = 1 - \alpha \tag{8.2.3}$$

$$P[(H_1 \text{ is true}) \cap (\text{reject } H_0)] = 1 - \beta \tag{8.2.4}$$

α, the probability of rejecting a true hypothesis, is commonly referred to as the *significance level* of the test. Usually, values of .10 or less are associated with α. The most often used values of α are .05 and .01 and conventionally $\alpha = .05$ is called *significant* and $\alpha = .01$ *highly significant.* $(1 - \alpha)$ represents the probability of accepting the null hypothesis when the null hypothesis is actually true.

β, the probability of accepting a hypothesis when the alternative is true, is called the *operating characteristic* of a test, and is designated by *OC*. $(1 - \beta)$ is referred to as the *power* of a test, which represents the probability of rejecting a hypothesis when the alternative hypothesis is true. The magnitudes of the *OC* and the power depend on the alternative hypothesis and the parameters of the distribution in question. Let us illustrate the computations of *OC* and power for a particular situation. Referring once again to the example of the distribution of the distances between index finger and thumb, suppose we wish to test the hypothesis $H_0: \mu = 7$ inches against the alternative $H_1: \mu \neq 7$ inches. In all tests of hypotheses the significance level must be stated *a priori.* Hence we further assume that $\alpha = .10$. Since H_1 has infinitely many alternatives, we choose a single alternative and consider it. Arbitrarily say $\mu = 8$. Therefore $H_0: \mu = 7$, $H_1: \mu = 8$ and $\alpha = .1$. The statistic which we shall use as a basis for our decision is the sample mean. Recalling that the sampling distribution of the mean is normal with mean μ and variance σ^2/n, we need to decide next on the sample size. Assuming arbitrarily $n = 4$, we obtain

$$P\left[|\bar{x} - 7| < 1.645 \sqrt{\frac{1.21}{4}}\right] = 1 - .1 = 0.90$$

which reduces to $P[6.1 < \bar{x} < 7.9] = .90$. [Refer to Eq. (7.4.1).]

The interval $6.1 < \bar{x} < 7.9$ represents the values of the test statistic which lead to the acceptance of the null hypothesis. The interval of test statistics that leads to the acceptance of a hypothesis is called the *acceptance interval* or *acceptance region.* The values of test statistics that lead to the rejection of a hypothesis is called the *rejection interval* or *rejection region.* These regions are shown in Fig. 8.4, under the H_0 curve.

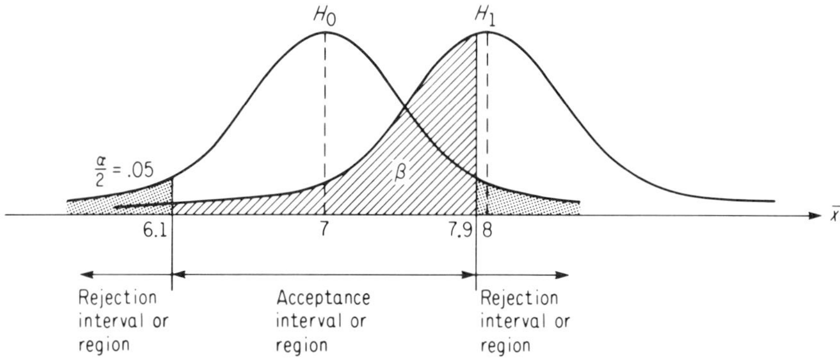

FIG. 8.4.

The dark-shaded area in Fig. 8.4 represents $\alpha = .1$, which is equally divided into two portions, and corresponds to the probability associated with the rejection intervals.

The curve under H_1 represents the assumptions made in the alternative hypothesis. The assumptions made under H_0 and H_1 are identical except for the means, which is shown in Fig. 8.4.

To compute β, the operating characteristic of the test, we refer to the definition. β corresponds to that area under H_1 which covers the interval $6.1 < \bar{x} < 7.9$. This portion is represented by the light-shaded area in Fig. 8.4 and is computed by

$$\beta = P\left[\frac{(6.1 - 8)2}{\sqrt{1.21}} < z < \frac{(7.9 - 8)2}{\sqrt{1.21}}\right]$$
$$= P[-3.46 < z < -.18] = .42831$$

The power of the test, therefore, is equal to $1 - .42831 = .57169$. Let us remark in passing that α, the probability of type I error, is analogous to α, the producer's risk, and β, the probability of type II error, analogous to β, the consumer's risk discussed in Sec. 8.1.

In general, we have $H_0{:}\mu = \mu_0$ and $H_1{:}\mu = \mu_1$, where μ_0 and μ_1 are the hypothesized values and specific values of α, σ^2 and n are given. These assumptions are shown in Fig. 8.5.

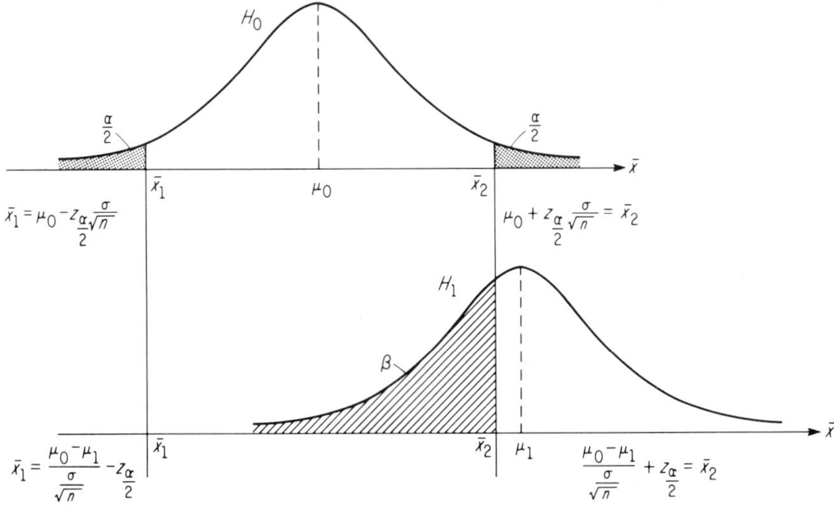

FIG. 8.5.

From Fig. 8.5 it should be evident that

$$\beta = P\left[\frac{\mu_0 - \mu_1}{\frac{\sigma}{\sqrt{n}}} - z_{\alpha/2} < z_\beta < \frac{\mu_0 - \mu_1}{\frac{\sigma}{\sqrt{n}}} + z_{\alpha/2}\right] \tag{8.2.5}$$

The following example illustrates the use of Eq. (8.2.5) in computing the power of a test.

Example 8.2.1 Is is assumed that the weekly wages of college students are normally distributed with variance 64. Calculate the power of the test under the assumptions,

$$H_0:\mu = \$50$$
$$H_1:\mu = \$51$$
$$\alpha = 0.1 \text{ and sample size } n = 25$$

Solution: Using Eq. (8.2.5) we get

$$\beta = P\left[\frac{50 - 51}{\frac{8}{5}} - 1.645 < z_\beta < \frac{50 - 51}{\frac{8}{5}} + 1.645\right]$$
$$= P[-2.27 < z_\beta < 1.02] = .87333$$

Hence the power of the test is $1 - .87333 = .12667$.

From the above discussion and illustrations it should be evident that β not only depends on the parameters of the distribution but on α also. Holding all other parameters fixed, as α increases β decreases and hence

$(1 - \beta)$, the power of a test, increases. The objective in testing statistical hypotheses is to choose a test among all tests having the same significance level α such that β is minimum or the power is maximum. If β is considered a function of the parameter in question, say $\beta = g(\theta)$, then we seek to find the minimum of $g(\theta)$. For instance, in the above illustrations the parameter considered has been the mean, hence, from Eq. (8.2.5) we obtain

$$\beta = g(\mu) = P\left[\frac{(\mu_0 - \mu)\sqrt{n}}{\sigma} - z_{\alpha/2} < z_\beta < \frac{(\mu_0 - \mu)\sqrt{n}}{\sigma} + z_{\alpha/2}\right]$$

In order to find the minimum of $g(\mu)$, we consider all possible values of μ (alternative choices for the hypothesis) and graph $g(\mu)$. This graph is called the *operating characteristic curve* of the test, and is commonly designated by *OC* curve.

The power curve is related to the *OC* curve, since power $= 1 - \beta$.

Example 8.2.2 Sketch the *OC* curve and the power curve for the data of Example 8.2.1.

Solution: Using different values of μ in Example 8.2.1 and graphing the resulting values of β and $1 - \beta$, we obtain the results as in Figs. 8.6*a* and 8.6*b*.

Referring to Eq. (8.2.5) once again, we notice that β depends on the sample size, n, also. [σ and z_α are constants for a given problem.] Therefore, both the *OC* curve and the power curve given in Figs. 8.6*a* and 8.6*b* represent a single member of a family of such curves. (See Prob. 8.2.1.)

Basically there are two types of tests: (1) one-sided, (2) two-sided. The one-sided are further categorized by left-sided and right-sided test. Some authors refer to sides as *tails*. As the names might suggest, a one-sided test is a test in which the alternative hypothesis has associated with itself only one rejection region, either the left or the right, but not both. A one-sided test is used when we wish to reject a hypothesis if the test statistic yields either large or small values, but not both.

$$H_0:\mu = 75$$
$$H_1:\mu > 75$$

is an example of right-sided hypothesis and

$$H_0:\mu = 100$$
$$H_1:\mu < 100$$

is an example of left-sided hypothesis.

A two-sided test is a test in which the alternative hypothesis has associated with itself two rejection regions, one in the left side and the other on the right. The two-sided tests are used when one wants to reject a hypothesis if the test statistic yields either too large or too small a value. In

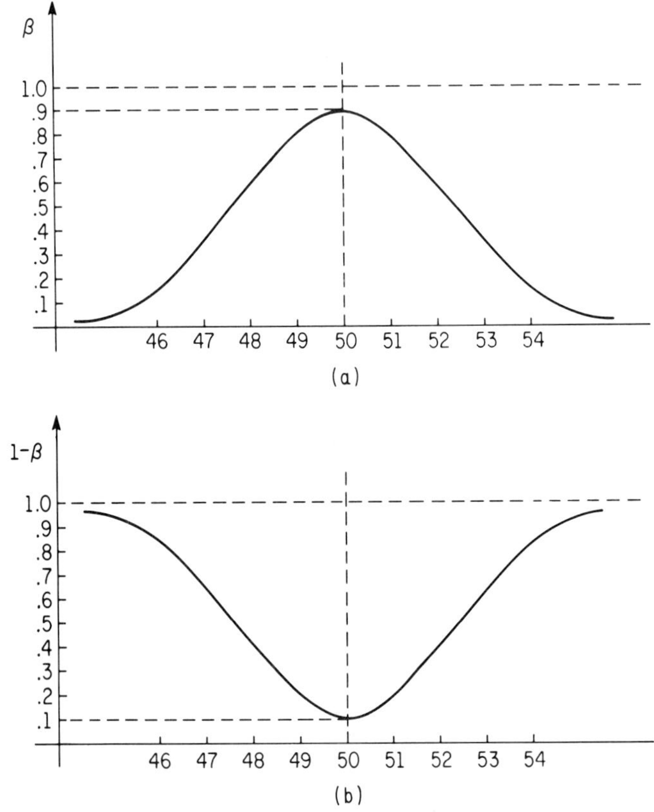

FIG. 8.6. (a) *OC* curve for two-sided test. (b) Power curve for two-sided test.

all our discussions thus far we have considered two-sided tests. (See Figs. 8.4 and 8.5.)

The computation of β, and hence of $(1 - \beta)$, for a one-sided test follows similar steps as the derivation of Eq. (8.2.5). In the one-sided case the totality of α is either in the right or in the left side. Therefore

$$\beta = P\left[z_\beta < \frac{(\mu_0 - \mu_1)\sqrt{n}}{\sigma} + z_\alpha\right] \qquad (8.2.6)$$

for the right-sided test and

$$\beta = P\left[z_\beta > \frac{(\mu_0 - \mu_1)\sqrt{n}}{\sigma} - z_\alpha\right] \qquad (8.2.7)$$

for the left-sided test.

Example 8.2.3 Refer to Example 8.2.1. Sketch the *OC* curve and the power curve if the hypotheses are

(a)

$$H_0: \mu = 50$$
$$H_1: \mu > 50$$

(b)

$$H_0: \mu = 50$$
$$H_1: \mu < 50$$

Solution: (a) This is a right-sided test and hence using Eq. (8.2.6) we get

$$\beta = g(\mu) = P\left[z_\beta < \frac{(50 - \mu)5}{8} + 1.28\right]$$

which yields

$$\begin{aligned} g(49) &= P[z_\beta < 1.905] = .972 \\ g(50) &= P[z_\beta < 1.28] = .900 \\ g(51) &= P[z_\beta < .655] = .744 \\ g(52) &= P[z_\beta < .03] = .516 \\ g(53) &= P[z_\beta < -.60] = .274 \end{aligned}$$

Graphing these results with β as the ordinate we obtain the graph of Fig. 8.7*a*. Similarly, graphing $1 - \beta$ we get the graph of Fig. 8.7*b*. Note that $g(50) = .9$ and $1 - \beta = 1 - g(50) = .1$ which corresponds to the significance level α of the test.

(b) This is a left-sided test and therefore using Eq. (8.2.7) we obtain

$$\beta = g(\mu) = P\left[z_\beta > \frac{(50 - \mu)5}{8} - 1.28\right]$$

Evaluating this expression for various values of μ and graphing we get the result as shown in Fig. 8.8*a*. Also, graphing $1 - \beta$ we obtain the sketch of Fig. 8.8*b*.

It can be readily seen that the *OC* curves of $H_1: \mu < \mu_0$ and $H_1: \mu > \mu_0$ under identical assumptions are symmetric about the line $\mu_0 = \mu$. Similarly, the power curves of these two alternative hypotheses are symmetric about the line $\mu_0 = \mu$. Refer to Figs. 8.7*a* and 8.8*a* as well as Figs. 8.7*b* and 8.8*b*.

Finally, let us remark that the primary purpose in testing hypotheses is to disprove the null hypothesis, in order to accept with high probability the alternative hypothesis. In other words, an important objective in testing hypotheses is to maximize the power.

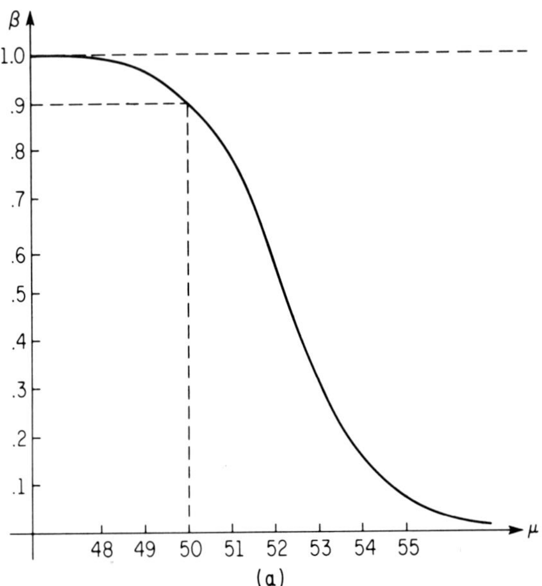

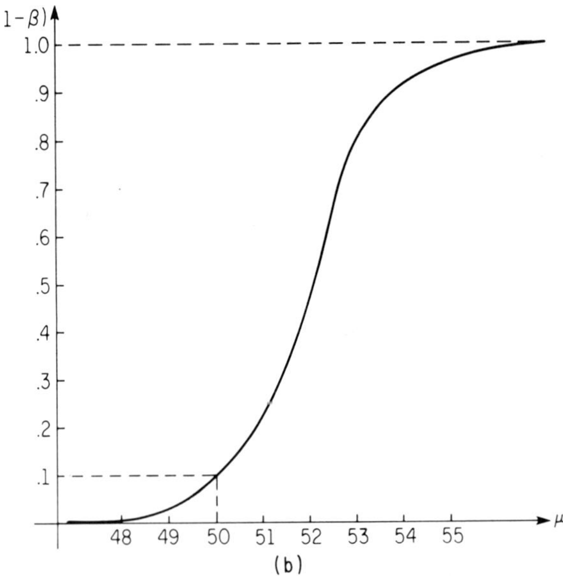

FIG. 8.7. (a) *OC* curve $H_1: \mu > 50$. (b) Power curve $H_1: \mu > 50$.

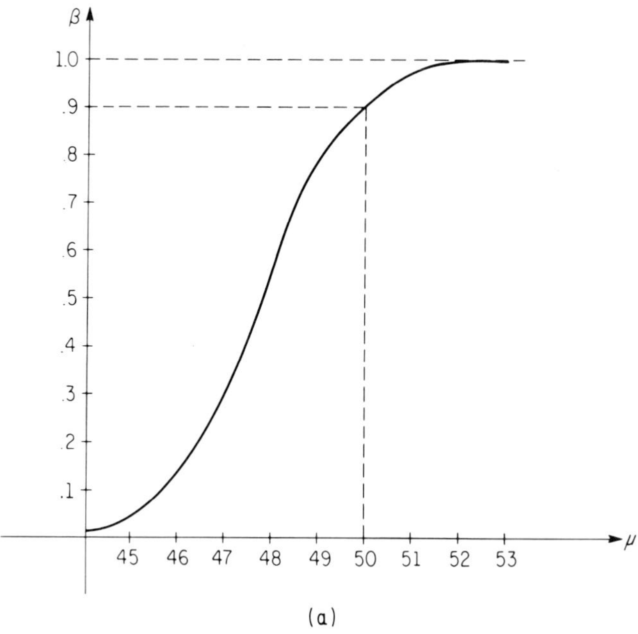

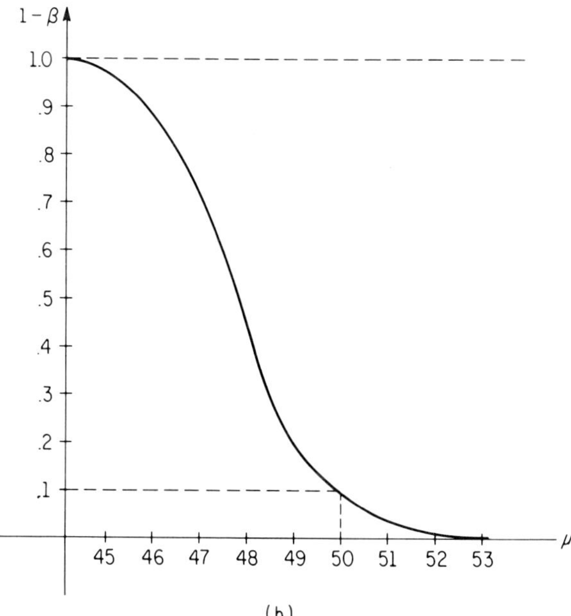

FIG. 8.8. (a) *OC* curve H_1: $\mu < 50$. (b) Power curve H_1: $\mu < 50$.

PROBLEMS

8.2.1 Sketch the OC and the power curves for the data of Example 8.2.1 with (a) $n = 16$, (b) $n = 36$ and (c) $n = 64$. Compare these results with Figs. 8.6*a* and 8.6*b*.

8.2.2 Repeat Prob. 8.2.1 assuming the alternative hypotheses, (a) $\mu > 50$, (b) $\mu < 50$.

8.2.3 Referring to the results of Example 8.2.2 and Prob. 8.2.1, choose the nearest suitable value of n so that β will be less than or equal to 0.20 in the alternative hypothesis $\mu = 52$.

8.2.4 In a two-sided test for the mean of a normal population with variance σ^2, values of α and β are specified. Show that sample sizes

$$\left[\frac{(z_\beta + z_{\alpha/2})\sigma}{\mu_0 - \mu_1}\right]^2 \leq n$$

satisfy simultaneously the specified probabilities of α and β. [*Hint:* Refer to Eq. (8.2.5).]

8.2.5 Using the result of Prob. 8.2.4, determine the minimum sample size if the conditions, $\alpha = .05$, $\beta = .1$, $\mu_0 = 60$, $\mu_1 = 62$ and $\sigma = 5$ are to be satisfied simultaneously.

8.2.6 Obtain the expressions analogous to that in Prob. 8.2.4, if the test is (a) right-sided, (b) left-sided.

8.2.7 Given a right-sided test, determine the minimum sample size if the conditions given in Prob. 8.2.5 are to be satisfied simultaneously.

8.3 TEST OF HYPOTHESES CONCERNING POPULATION MEANS

As mentioned in the previous section, most hypotheses occur in compound form and the procedure followed is to test each hypothesis separately, assuming the rest to be true. In this section we will consider hypotheses pertaining to the population mean making various assumptions regarding the variance and the distribution of the random variable. The two broad classifications of assumptions are (a) variance known, and (b) variance unknown. We shall examine these categories in detail and state further assumptions regarding the distributions.

(a) Variance Known

1. One Population Involved. In this situation the statistician has knowledge concerning the population variance σ^2 either through previous experience or information provided from other sources. For instance, he may have information about the variance of the distribution of certain IQ examination scores. Again, he might be given information about the life of certain electric light bulbs, such as $\sigma = 1{,}600$ hours, etc.

Next, let us recall the central limit theorem. According to this theorem, the random variable

$$z = \frac{\bar{x} - \mu}{\frac{\sigma}{\sqrt{n}}} = \frac{(\bar{x} - \mu)\sqrt{n}}{\sigma} \tag{8.3.1}$$

will be normally distributed with mean zero and variance 1. If $\mu = \mu_0$, then the random variable $z = \frac{(\bar{x} - \mu_0)\sqrt{n}}{\sigma}$ will have the same distribution and may be used for a test statistic.

The rejection interval or the *critical region* of a test statistic should be stated as an assumption prior to testing a hypothesis. For example, in a two-sided test the rejection interval may be $|z| > 1.5$ or $|z| > 1.96$ etc., and in a one-sided test $z < -1.75$ or $z > 2$ might have been selected. From the Unit Normal Table in the Appendix we can find the probabilities associated with each stated z value. That is, $P[|z| > 1.5] = .13362$, $P[|z| > 1.96] = .05$, $P[z < -1.75] = .04006$ and $P[z > 2] = .02275$.

Conversely, for any specified probability associated with the significance level of a test, we can locate the corresponding z value. Therefore, we can find the critical region(s) whenever the significance level of a test is stated. The following are the necessary steps to describe the procedure for testing hypotheses.

1. State the hypothesis and its alternative(s) precisely.
2. Specify significance level α *a priori.*
3. Indicate test statistic or rule for decision to be used.
4. Find the rejection region(s) corresponding to assumed α.
5. Calculate the statistic from the sample under investigation.
6. Decide on the basis of result obtained whether to accept or to reject the hypothesis.

Example 8.3.1 During the last 10 years a college professor taught a calculus course every semester. He has given parallel forms (equivalent in content both quality and quantity wise) of a test and has obtained an overall mean of 75 points. Last semester he taught a class of 36 students and the mean for this particular class was 79. Assuming that the population variance for this test is 25, is there statistical evidence on the 10 percent significance level to believe that the last semester's students were different from the previous years' students?

Solution:

1. H_0: Population mean μ is equal to specified mean 79—that is, $\mu = 79$.

H_1: Population mean μ is not equal to the specified mean 79—that is, $\mu \neq 79$.

2. $\alpha = 10$ percent. (This is arbitrarily specified.)

3. Statistic to be used is $z = \dfrac{\overline{x} - \mu}{\dfrac{\sigma}{\sqrt{n}}}$.

4. Critical region: $|z| > 1.645$.

5. $z = \dfrac{79 - 75}{\dfrac{\sqrt{25}}{\sqrt{36}}} = \dfrac{4 \times 6}{5} = 4.8$.

6. Reject the hypothesis. That is, we have sufficient reason to believe at 10 percent significance level that the students of last semester have different mean compared to the students of previous years.

Example 8.3.2 Refer to Example 8.3.1, and answer the following question. Is there statistical evidence on the 5 percent significance level to believe that last semester's students had a larger mean than the previous years' students?

Solution: First we express the question in the form of a statistical hypothesis. Let

1. $H_0: \mu = 75$.
   $H_1: \mu > 75$.
2. $\alpha = .05$.
3. Test statistic: $z = \dfrac{\overline{x} - 75}{\dfrac{\sigma}{\sqrt{n}}}$.
4. Critical region: $z > 1.645$.
5. $z = \dfrac{79 - 75}{\dfrac{5}{6}} = 4.8$.
6. Therefore, reject H_0 and accept H_1. Which is to say, there is statistical evidence to believe that the students of last semester had a greater mean than the students of previous years.

2. *Two Populations Involved.* Often one needs to decide between hypotheses of the types

$$H_0:\theta_1 = \theta_2 \qquad H_0:\theta_1 = \theta_2 \qquad H_0:\theta_1 = \theta_2$$
$$H_1:\theta_1 \neq \theta_2 \qquad H_1:\theta_1 < \theta_2 \qquad H_1:\theta_1 > \theta_2$$

where θ_1 is a parameter from a population and θ_2 is the corresponding parameter in another population.

Suppose, for instance, a professor who teaches two different sections of the same class wishes to test the statement

$$H_0:\mu_1 = \mu_2$$
$$H_1:\mu_1 \neq \mu_2$$

where μ_1 and μ_2 are the average scores in the respective sections. How can he proceed to test these hypotheses? The main difficulty lies in securing a test statistic, which we wish to establish next.

Let x_1 represent the scores in the first class and x_2 the scores in the second. Applying the central limit theorem to these random variables we get

$$f\left(\bar{x}_1:\mu_1, \frac{\sigma_1^2}{n_1}\right) \quad \text{and} \quad f\left(\bar{x}_2:\mu_2, \frac{\sigma_2^2}{n_2}\right)$$

where μ_1 and μ_2 are the means of the two distributions of scores, σ_1^2 and σ_2^2 the variances, and n_1 and n_2 represent the sample sizes taken from the respective populations.

Assuming that $\bar{x}_1$ and $\bar{x}_2$ are stochastically independent, we get applying Theorem 7.1.1 to the random variable $w = \bar{x}_1 - \bar{x}_2$,

$$\mu_w = \mu_{\bar{x}_1} - \mu_{\bar{x}_2}$$

and

$$\sigma_w^2 = \sigma_{\bar{x}_1 - \bar{x}_2}^2 = \frac{\sigma_1^2}{n_1} + \frac{\sigma_2^2}{n_2} \tag{8.3.2}$$

Therefore, again by the central limit theorem, we have

$$f\left(\bar{x}_1 - \bar{x}_2:\mu_1 - \mu_2, \frac{\sigma_1^2}{n_1} + \frac{\sigma_2^2}{n_2}\right)$$

or equivalently the random variable

$$z = \frac{(\bar{x}_1 - \bar{x}_2) - (\mu_1 - \mu_2)}{\sqrt{\dfrac{\sigma_1^2}{n_1} + \dfrac{\sigma_2^2}{n_2}}} \tag{8.3.3}$$

is distributed according to $f(z:0, 1)$, the unit normal. As in the case of one population, here also the test statistic is the random variable z.

Some special cases of Eq. (8.3.3) are worth noting.

1. If $\mu_1 = \mu_2$, we get

$$z = \frac{\bar{x}_1 - \bar{x}_2}{\sqrt{\dfrac{\sigma_1^2}{n_1} + \dfrac{\sigma_2^2}{n_2}}} \tag{8.3.4}$$

2. If $\sigma_1^2 = \sigma_2^2 = \sigma^2$, we get

$$z = \frac{(\bar{x}_1 - \bar{x}_2) - (\mu_1 - \mu_2)}{\sigma\sqrt{\dfrac{1}{n_1} + \dfrac{1}{n_2}}} \tag{8.3.5}$$

3. If $n_1 = n_2 = n$, we get

$$z = \frac{(\bar{x}_1 - \bar{x}_2) - (\mu_1 - \mu_2)}{\dfrac{\sqrt{\sigma_1^2 + \sigma_2^2}}{\sqrt{n}}} \tag{8.3.6}$$

In general, assuming that $\mu_1 - \mu_2 = \delta$, we have the statistic

$$z = \frac{(\bar{x}_1 - \bar{x}_2) - \delta}{\sigma_{\bar{x}_1 - \bar{x}_2}} \tag{8.3.7}$$

which can be used to test hypotheses concerning the means of two independent random variables.

Example 8.3.3 Identical mathematics examinations were given to eighth- and ninth-grade pupils. There were 50 students in each group and the means were 70 and 76 respectively. Assuming that the respective variances were 137 and 63, is there statistical evidence at the 5 percent significance level to reject the hypothesis that the mean of the ninth-graders exceeds that of the eighth-graders by more than 9 points?

Solution: Let μ_n and μ_e represent the population means of the ninth- and eighth-graders:

1. H_0: $\mu_n - \mu_e = 9$.
   H_1: $\mu_n - \mu_e < 9$.
2. $\alpha = .05$.
3. The test statistic is $z = \dfrac{\bar{x}_1 - \bar{x}_2 - 9}{\sqrt{\dfrac{\sigma^2_{x_1}}{n_1} + \dfrac{\sigma^2_{x_2}}{n_2}}}$.
4. Rejection or critical interval: $z < -1.64$.
5. Calculations: $z = \dfrac{(76 - 70) - 9}{\sqrt{\dfrac{137}{50} + \dfrac{63}{50}}} = \dfrac{-3}{\sqrt{4}} = -1.5$.
6. Accept H_0. That is, we accept the hypothesis that the mean of the ninth-graders exceeds the mean of the eighth-graders by 9 points.

(b) Variance Unknown

1. One Population Involved. The discussion in this situation will be analogous to that of case (a), except for the fact that the population variance σ^2 is unknown. This is the more common case because only in very limited circumstances the population variance will be known. Hence, both the mean and the variance of the sample will be needed to test statistical hypotheses concerning the population mean. It has been indicated already that the random variable $t = \dfrac{\bar{x} - \mu}{\dfrac{s}{\sqrt{n}}}$ is distributed according to the Student's t p.f. with $(n - 1)$ degrees of freedom. If the hypothesis $\mu = \mu_0$ is true the random variable $t = \dfrac{\bar{x} - \mu_0}{\dfrac{s}{\sqrt{n}}}$ will also be distributed according to the Student's t p.f. with $(n - 1)$ degrees of freedom. Therefore the random variable

$$t = \frac{\bar{x} - \mu_0}{\frac{s}{\sqrt{n}}} \tag{8.3.8}$$

may be used for a test statistic.

The rejection interval will be

$$|t| > t_{\alpha/2, n-1}$$

for the two-sided test, $t < -t_{\alpha, n-1}$ for the left-sided test, and $t > t_{\alpha, n-1}$ for the right-sided test. The prescribed steps given in case (a) should be followed here also.

Example 8.3.4 In a certain company the mean annual vacation period is 15 days. Twelve employees were randomly chosen from this company and their vacation days inquired. The following data were obtained:

5, 12, 4, 8, 15, 14, 10, 12, 9, 17, 12, 11.

Is there statistical evidence on the 5 percent significance level to believe that the company mean vacation days is not 15 days?

Solution:

1. H_0: company mean vacation days $\mu = 15$ days.
   H_1: company mean vacation days $\mu \neq 15$ days.
2. $\alpha = 5$ percent.
3. Test statistic: $t = \dfrac{\bar{x} - \mu_0}{\frac{s}{\sqrt{n}}}$.
4. Critical region: $|t| > t_{.025, 11} = 2.2010$.
5. $t = \dfrac{10.75 - 15}{\frac{3.84}{\sqrt{12}}} = -3.8$.

6. Reject H_0. That is, with this particular sample size of 12, we do have statistical evidence to question (reject) the statement that the company mean vacation period is 15 days.

Example 8.3.5 For the data of Example 8.3.4, test the hypothesis that the company mean vacation is less than 15 days.

Solution:

1. H_0: Company mean vacation days = 15.
   H_1: Company mean vacation days > 15.
2. $\alpha = .05$.
3. Test statistic: $t = \dfrac{\bar{x} - \mu}{\frac{s}{\sqrt{n}}}$.

4. Critical region: $t > t_{.05,11} = 1.7959$.
5. $t = -3.8$.
6. Accept H_0—that is, we have no statistical evidence to disprove the null hypothesis.

2. *Two Populations Involved.* For this case our discussion will parallel that of the case (a) presented above. The main difficulty in the current situation is to find a random variable that could be used for a test statistic in comparing two means from independent populations. These populations could have either equal or unequal variances and hence the two possibilities should be examined separately.

1. Equal Variances $\sigma_1^2 = \sigma_2^2$. In Sec. 7.5 it was shown that the sample variance is an unbiased estimate of the population variance. Therefore, weighted or pooled variances would be an unbiased estimate also. That is, if s_1^2 and s_2^2 are two sample variances from two independent populations with equal variances, and k_1 and k_2 are any constants, then

$$S_w^2 = \frac{k_1 s_1^2 + k_2 s_2^2}{k_1 + k_2} \tag{8.3.9}$$

is an unbiased estimate. The proof is straightforward:

$$\begin{aligned} E[S_w^2] &= \frac{1}{k_1 + k_2} E[k_1 s_1^2 + k_2 s_2^2] \\ &= \frac{1}{k_1 + k_2} E[k_1 s_1^2] + E[k_2 s_2^2] \\ &= \frac{k_1 \sigma_1^2 + k_2 \sigma_2^2}{k_1 + k_2} \end{aligned} \tag{8.3.10}$$

However, since $\sigma_1^2 = \sigma_2^2$ by assumption, the last expression becomes

$$E[S_w^2] = \sigma^2 \tag{8.3.11}$$

If n_1 and n_2 represent random sample sizes from two independent populations, then for $k_1 = n_1 - 1$ and $k_2 = n_2 - 1$, Eq. (8.3.9) becomes

$$S_w^2 = \frac{(n_1 - 1)s_1^2 + (n_2 - 1)s_2^2}{n_1 + n_2 - 2} \tag{8.3.12}$$

Which as shown above is an unbiased estimator of the unknown variance of the populations.

Applying Theorem 7.3.3 to Eq. (8.3.12) we note that the quantity

$$(n_1 + n_2 - 2)\frac{S_w^2}{\sigma^2} = \frac{(n_1 - 1)s_1^2 + (n_2 - 1)s_2^2}{\sigma^2}$$

is chi-square distributed with $n_1 + n_2 - 2$ d.f. Next, applying Theorem 7.3.1 to Eq. (8.3.5) we deduce that the expression

$$z^2 = \frac{[(\bar{x}_1 - \bar{x}_2) - (\mu_1 - \mu_2)]^2}{\sigma^2\left(\dfrac{1}{n_1} + \dfrac{1}{n_2}\right)}$$

is chi-square distributed with 1 d.f. Hence by Theorem 7.3.4 the random variable

$$t^2 = \frac{[(\bar{x}_1 - \bar{x}_2) - (\mu_1 - \mu_2)]^2}{S_w^2\left(\frac{1}{n_1} + \frac{1}{n_2}\right)}$$

is F distributed with 1 and $n_1 + n_2 - 2$ numerator and denominator degrees of freedom, respectively. Therefore, from Eq. (7.3.26) we note that the random variable

$$t = \frac{(\bar{x}_1 - \bar{x}_2) - (\mu_1 - \mu_2)}{S_w\sqrt{\frac{1}{n_1} + \frac{1}{n_2}}} \tag{8.3.13}$$

obeys the Student's p.f. with $n_1 + n_2 - 2$ degrees of freedom. In general, letting $\mu_1 - \mu_2 = \delta$ and substituting the value of S_w, Eq. (8.3.13) becomes

$$t = \frac{(\bar{x}_1 - \bar{x}_2) - \delta}{\sqrt{(n_1 - 1)s_1^2 + (n_2 - 1)s_2^2}}\sqrt{\frac{n_1 n_2(n_1 + n_2 - 2)}{n_1 + n_2}} \tag{8.3.14}$$

Either Eq. (8.3.13) or Eq. (8.3.14) is the test statistic to use for testing hypotheses concerning the means of two independent random variables with equal variances.

Example 8.3.6 9 men and 16 women are selected at random from a large list of applicants, and given a vocabulary examination. The mean number of correct answers for the groups were 15.7 and 21.8 respectively. The sample variances were calculated to be 18.0 for the man and 21.0 for the women. Is there reason to doubt at the 5 percent significance level that vocabulary achievements are the same for both groups?

Solution: We assume that the distribution of vocabulary achievement scores for men and women have equal variances and hypothesize

1. H_0: $\mu_1 = \mu_2$.
   H_1: $\mu_1 \neq \mu_2$.

or equivalently

   H_0: $\mu_1 - \mu_2 = 0$.
   H_1: $\mu_1 - \mu_2 \neq 0$.

2. $\alpha = .05$.
3. Test statistic, expression (8.3.13) with 23 degrees of freedom.
4. Rejection interval: $|t| > t_{.05,23} = 2.069$.
5. Calculations: $t = \dfrac{15.7 - 19.8}{\sqrt{8(18) + 15(21)}}\sqrt{\dfrac{9(16)\,23}{25}}$

$$= (15.7 - 19.8)\,\frac{3 \times 4}{5}\sqrt{\frac{23}{459}}$$

$$= -2.23.$$

6. Therefore, the null hypothesis is rejected.

2. Unequal Variances $\sigma_1^2 \neq \sigma_2^2$. If the underlying distributions of the two random variables are normal, and $\sigma_1^2 \neq \sigma_2^2$, then the random variable

$$t = \frac{(\bar{x}_1 - \bar{x}_2) - (\mu_1 - \mu_2)}{\sqrt{\dfrac{s_1^2}{n_1} + \dfrac{s_2^2}{n_2}}} \tag{8.3.15}$$

may be used for a test statistic to test the hypothesis $\mu_1 = \mu_2$ and it approximately obeys Student's t p.f. with

$$\frac{\left[\dfrac{s_1^2}{n_1} + \dfrac{s_2^2}{n_2}\right]^2}{\dfrac{(s_1^2 / n_1)^2}{n_1 - 1} + \dfrac{(s_2^2 / n_2)^2}{n_2 - 1}} \tag{8.3.16}$$

degrees of freedom. The degrees of freedom computed by Eq. (8.3.16) is not necessarily an integer and hence the next smaller integer should be used.

When n_1 and n_2 are both greater than 25, one can assume $s_1 = \sigma_1$, $s_2 = \sigma_2$ and use the test statistic (8.3.3) instead of (8.3.15) as an approximation.

3. *Paired Samples.* Often it is convenient to select $n_1 = n_2$ or circumstances may naturally lead to $n_1 = n_2$. For instance, suppose a final examination is composed of two parts. Let μ_1 and μ_2 represent the means of the respective parts. How could we test the hypothesis that $\mu_1 - \mu_2 = 15$? Let x_1 and x_2 represent the scores on parts 1 and 2 respectively. The random variable $d = x_1 - x_2$ can be tested by applying Eq. (8.3.8). That is, the random variable

$$t = \frac{\bar{d} - (\mu_1 - \mu_2)}{\dfrac{s_d}{\sqrt{n}}} \tag{8.3.17}$$

is distributed according to Student's t p.f. with $(n - 1)$ degrees of freedom.

Another example of paired samples might be a pre-test and post-test situation in an experiment. For instance, ten 2-ft-long rods of a certain alloy is to be subjected to a tensile test before and after a hardening treatment. One method of pairing would be to divide the 2-ft rod into two 1-ft rods and match the two parts. Hypotheses regarding the mean difference of the paired observations can be tested by Eq. (8.3.17). Table 8.3.1 summarizes the discussion of this section and is presented here for convenience.

TABLE 8.3.1
Summary Table For Testing Hypotheses Concerning Means

Variance Known				Variance Unknown				
One Population		Two Populations		One Population		Two Populations		
						$\sigma_1^2 = \sigma_2^2$		$\sigma_1^2 \neq \sigma_2^2$
I. Test Statistic Eq. (8.3.1)		II. Test Statistic Eq. (8.3.3)		III. Test Statistic Eq. (8.3.8)		IV. Test Statistic Eq. (8.3.13) or (8.3.14)		V. Test Statistic Eq. (8.3.15) with degrees of freedom given by Eq. (8.3.16)
H_0: $\mu = \mu_0$		H_0: $\mu_1 - \mu_2 = \delta$		H_0: $\mu = \mu_0$		H_0: $\mu_1 - \mu_2 = \delta$		H_0: $\mu_1 - \mu_2 = \delta$
H_1:	Critical Region	H_1:	Critical Region	H_1:	Critical Region	H_1:	Critical Region	Same as in IV
$\mu \neq \mu_0$	$\lvert z \rvert > z_{\alpha/2}$	$\mu_1 - \mu_2 \neq \delta$	$\lvert z \rvert > z_{\alpha/2}$	$\mu \neq \mu_0$	$\lvert t \rvert > t_{\alpha/2,n-1}$	$\mu_1 - \mu_2 \neq \delta$	$\lvert t \rvert > t_{\alpha/2,n_1+n_2-2}$	
$\mu < \mu_0$	$z < -z_\alpha$	$\mu_1 - \mu_2 < \delta$	$z < -z_\alpha$	$\mu < \mu_0$	$t < -t_{\alpha,n-1}$	$\mu_1 - \mu_2 < \delta$	$t < -t_{\alpha,n_1+n_2-2}$	
$\mu > \mu_0$	$z > z_\alpha$	$\mu_1 - \mu_2 > \delta$	$z > z_\alpha$	$\mu > \mu_0$	$t > t_{\alpha,n-1}$	$\mu_1 - \mu_2 > \delta$	$t > t_{\alpha,n_1+n_2-2}$	
				If $n > 30$, Test Statistic I may be used instead of III.		If n_1 and n_2 are both greater than 25, Test Statistic II may be used instead of IV.		

PROBLEMS

8.3.1 At a certain college the average monthly salary offer to 50 graduating seniors is \$680. Assuming that the starting salaries for graduating seniors are normally distributed with mean \$640 and standard deviation of \$100, is there reason to believe that these 50 seniors are offered better than average salaries? Use $\alpha = .05$ for significance level.

8.3.2 The tensile strength of a certain type of steel cables is assumed to be normally distributed with mean 15 and standard deviation of .4 ton. If the average tensile strength of 9 randomly tested cables of this type is 14.2 tons, does this result cast doubt on the claim? Use $\alpha = .01$.

8.3.3 In a manufacturing process the average daily production of articles is 144 on machine A with a standard deviation of 15. The average daily production of articles on machine B is 152 with a standard deviation of 20. Two new empolyees are randomly assigned to operate machines A and B on a trial basis for 4 working days. The average yield for machine A was 120 and for B 140. Test the hypothesis that the two operators are equally qualified. State whatever other assumptions are needed to make the test.

8.3.4 A certain type of flashlight battery is advertised to last 90 hours or more in continuous use. Eight such batteries when tested lasted for 103, 90, 75, 112, 94, 83, 91, and 87 hours. Does this data contradict the claim on the .05 significance level?

8.3.5 Thirty-six randomly selected students from the freshman class were randomly divided into two equal groups. The first group was taught calculus by the lecture, textbook and homework method while the other students were taught by showing movies covering the same topics. On identical examinations given to the two groups after the learning period, the first group had a mean of 80 with a standard deviation of 10 and the second group had a mean of 83 with a standard deviation of 7. What conclusions can be drawn from these results? Make the necessary assumptions to carry out the analysis.

8.3.6 The management of a company decides to replace the tires on all company-owned cars. Two brands of tires are available for the same price. Forty-eight tires of brand A and 72 tires of brand B were purchased. Brand A tires lasted on the average 20,500 miles with a standard deviation of 1,500 miles while brand B tires lasted on the average 22,000 miles with a standard deviation of 2,000 miles. Are the qualities of the two brands significantly different? Use $\alpha = .05$.

8.3.7 Two types of diets are used for one month to determine the weight gains in calves. The result of the experiment is as follows:

	Group A	Group B
Mean gain....................	44 lb	35 lb
Variance......................	62.5	53.8
Number in group............	9	11

Is there significant difference between the mean gains in weight for the two groups? Use $\alpha = .01$.

8.3.8 A teacher claims that his students perform similarly on the midterm and the final examinations. To test his claim he randomly chose to examine the records of 9 students from last semester and the following is the data.

Student	Midterm	Final
1	67	81
2	77	82
3	80	73
4	83	84
5	92	87
6	54	68
7	89	97
8	67	76
9	85	94

Assuming $\alpha = 0.05$ he tests the hypothesis

$$H_0: \quad \mu_{\text{midterm}} = \mu_{\text{final}}$$
$$H_1: \quad \mu_{\text{midterm}} \neq \mu_{\text{final}}$$

What conclusion did he draw?

8.3.9 Repeat Prob. 8.3.8 and use the test statistic for paired observations, Eq. (8.3.17).

8.4 TESTS OF HYPOTHESES CONCERNING POPULATION VARIANCE

It was indicated in the previous section that most hypotheses occur in compound form and each hypothesis in a given compound situation should be tested separately. With different assumptions regarding the underlying distributions and their variances, we tested hypotheses concerning population means. Now we wish to test hypotheses concerning the population variances, and shall present three cases: (a) one population involved, (b) two populations involved, and (c) more than two populations involved.

(a) One Population Involved. The student may recall from the discussion in Chapter 5 that the variance of a random variable x is not altered by a transformation of the type $y = x \pm k$. [See Eq. (5.4.3).] Hence the variances of the random variables x and $y = x - \mu_x$ are equal. Therefore in testing hypotheses concerning variances one usually does not make any assumption regarding the population mean.

However, we need to make an assumption regarding the distribution of the random variable under question. If one makes the assumption that the underlying distribution is normal, then Theorem 7.3.3 can be immediately applied and used as a test statistic for the population variance. That is, the random variable $(n - 1)\dfrac{s_x^2}{\sigma_x^2}$ is chi-square distributed with $n - 1$ degrees of freedom. The steps to be followed in testing a hypothesis here are identical to those of Sec. 8.3. Let us illustrate by an example the concept under discussion.

Example 8.4.1 A test producer claims that his mathematics achievement test has a variance of 20. A high school mathematics teacher administers to 31 of his pupils this examination and obtains a variance of 26.8. Is there statistical evidence on the 5 percent significance level that the claim is not true?

Solution:

H_0: $\sigma^2 = 20.$

H_1: $\sigma^2 \neq 20.$

$\alpha = 0.05.$

Test statistic: $(n - 1)\dfrac{s^2}{\sigma^2}$.

Critical region: $\chi^2 < 16.791$ or $\chi^2 > 46.979.$

Calculations: $\chi^2 = (31 - 1)\dfrac{26.8}{20} = 40.2.$

Accept the null hypothesis, since the calculated value of the χ^2 is not in the critical region. Hence we have no significant evidence to reject the claim that $\sigma^2 = 20$.

As in testing hypotheses of population means, here also there are three types of alternative hypotheses which lead to either a two-sided test or a one-sided test. These alternative hypotheses are

$$H_1\colon \sigma^2 \neq \sigma_0^2 \qquad H_1\colon \sigma^2 > \sigma_0^2 \qquad \text{and} \qquad H_1\colon \sigma^2 < \sigma_0^2$$

Furthermore, the objective still remains to disprove the stated null hypothesis which implies that the critical regions are in the direction of the alternative hypotheses.

Example 8.4.2 In Example 8.4.1, test the hypothesis that $\sigma^2 \leq 20$.

Solution:

H_0: $\sigma^2 = 20.$

H_1: $\sigma^2 > 20.$

$\alpha = .05.$

Test statistic: $(n - 1)\dfrac{s^2}{\sigma^2}$.

Critical region: $\chi^2 > 43.77.$

Calculations: $\chi^2 = 40.2.$

The stated null hypothesis is accepted.

(b) Two Populations Involved. In this case also we make the assumption that the two populations are normally distributed, and look for a test statistic which utilizes the two parameters. Recalling that the statistic

$$F = \frac{s_1^2}{s_2^2} \cdot \frac{\sigma_2^2}{\sigma_1^2}$$

obeys the F probability function with $n_1 - 1$ and $n_2 - 1$ numerator and denominator degrees of freedom [refer to Eq. (7.3.17)], we can use this expression as a test statistic and proceed to test hypotheses concerning the variances of two populations.

Again, there are three types of alternative hypotheses which one needs to examine:

$$H_1\colon \sigma_1^2 \neq \sigma_2^2 \qquad H_1\colon \sigma_1^2 < \sigma_2^2 \qquad \text{and} \qquad H_1\colon \sigma_1^2 > \sigma_2^2$$

An example should suffice to illustrate the notion discussed in this case.

Example 8.4.3 Two machines in a factory produce the same kind of yarn. The random variable in question is shrinkage. A random sample of 21 from the first and 11 from the second machine yield variances of 40 and 30, respectively. Is there statistical evidence on the 10 percent significance level that the population variances of the machines are different?

Solution:

$H_0\colon \sigma_1^2 = \sigma_2^2$.

$H_1\colon \sigma_1^2 \neq \sigma_2^2$.

$\alpha = .10$.

Test statistic: $F = \dfrac{s_1^2}{s_2^2}$.

Critical region: $F > 2.7740 \quad \text{or} \quad F < \dfrac{1}{2.35} = .426.$

Calculations: $F = \dfrac{40}{30} = 1.333.$

Accept H_0. There is no sufficient evidence to reject H_0 on the .1 significance level.

Let us observe that if we always associate with the numerator the larger of the two sample variances, the resulting F-ratio will be greater than one and hence the step of finding the left rejection interval in a two-sided test can be eliminated. However, care should be taken in associating the correct degree of freedoms with the numerator and the denominator.

(c) More Than Two Populations Involved. When there are more than two populations whose variances are to be tested, one could take all different possible combinations of two variances and test each pair by the method of case (b). For instance, if there are three populations, the possible combinations of null hypotheses would be $H_0\colon \sigma_1^2 = \sigma_2^2$, $H_0\colon \sigma_1^2 = \sigma_3^2$, and $H_0\colon \sigma_2^2 = \sigma_3^2$. Clearly, each of these three hypotheses may be tested by the method discussed for two populations, when independent samples are taken from populations 1 and 2, populations 1 and 3, and populations 2 and 3 respectively. However, this approach has a serious disadvantage. Assuming that the significance levels are .1 in each test and since the tests are independent, the probability that one will make a Type I error in all three is $(.1)^3 = .001$, the probability that two Type I errors will be made is $3(.1)^2(.9) = 0.027$ and the probability that one type I error will be made is $3(.1)(.9)^2 = .243$. It should be evident from this illustration that two independent samples must be taken from each of the three populations and the probability of making at least one type I error is equal to .271. Therefore, other more efficient methods should be utilized.

1. Bartlett's Test. A test statistic commonly used to test hypotheses of the type

$$H_0\colon \sigma_1^2 = \sigma_2^2 = \cdots = \sigma_k^2$$

$$H_1\colon \text{At least two variances are unequal}$$

is called *Bartlett's test*, which we state next.

Theorem 8.4.1 If random samples of size $n_1, n_2, \ldots, n_k$ are taken from k normally distributed populations and if H_0 is true (all variances are equal), then the random variable B,

$$B = \frac{2.3026}{C}\left[\sum_{i=1}^{k}(n_i - 1)\log S_w^2 - \sum_{i=1}^{k}(n_i - 1)\log s_i^2\right] \tag{8.4.1}$$

$$= \frac{1}{C}\left[\sum_{i=1}^{k}(n_i - 1)\ln S_w^2 - \sum_{i=1}^{k}(n_i - 1)\ln s_i^2\right] \tag{8.4.2}$$

is approximately χ^2 distributed with $k - 1$ degrees of freedom, where

$$S_w^2 = \frac{\sum_{i=1}^{k}(n_i - 1)s_i^2}{\sum_{i=1}^{k}(n_i - 1)}$$

and

$$C = 1 + \frac{1}{3(k-1)}\left[\sum_{i=1}^{k}\frac{1}{n_i - 1} - \frac{1}{\sum_{i=1}^{k}(n_i - 1)}\right] \tag{8.4.3}$$

Notice that S_w^2 is the weighted or the pooled variance of the k sample variances. The statistic defined by Eq. (8.4.1) or Eq. (8.4.2) will be small whenever the variances are nearly equal. Specifically, if all the variances are equal, B will be zero. Hence, we reject the null hypothesis if the test statistic B yields a large value. That is, the rejection interval for this statistic is $\chi^2 > \chi^2_{\alpha\, k-1}$. Let us illustrate Bartlett's test by an example.

Example 8.4.4 Suppose random samples of size 7, 9, and 11 are taken from three normally distributed populations and the respective variances are 30, 40, and 50. Test the homogeneity of the variances of the underlying populations.

Solution:

H_0: $\sigma_1^2 = \sigma_2^2 = \sigma_3^2$.

H_1: At least two variances are unequal.

$\alpha = 0.05$.

Critical region: $\chi^2 > \chi^2_{2:0.95} = 5.99$.

Computation of test statistic:

$$C = 1 + \frac{1}{3(3-1)}\left[\frac{1}{6} + \frac{1}{8} + \frac{1}{10} - \frac{1}{24}\right] = 1.059$$

$$S_w^2 = \frac{6(30) + 8(40) + 10(50)}{24} = 41.666$$

$$B = \frac{2.3026}{1.059}[24 \log 41.666 - (6 \log 30 + 8 \log 40 + 10 \log 50)]$$

$$= .4422.$$

Accept H_0. There is no statistical evidence to reject the hypothesis of equality of the three population variances.

2. *Hartley's Test.* Another test statistic to test the homogeneity of variances, provided all sample sizes are equal, is due to Hartley. The test statistic has two independent parameters k and n, where k is the number of variances to be tested and n the sample size in each sample. This test considers the ratio of the largest sample variance to the smallest and is designated by $F_{\max}$. That is,

$$F_{\max} = \frac{s^2_{\max}}{s^2_{\min}} \tag{8.4.4}$$

Table VIII in the Appendix gives the upper 5 and 1 percent points of the $F_{\max}$ probability function. The degrees of freedom associated with $F_{\max}$ are $n - 1$. Since

$$s^2 = \frac{\text{sums of squares}}{n-1} = \frac{\sum_{i=1}^{n}(x_i - \overline{x})^2}{n-1}$$

$$= \frac{ss}{n-1} \tag{8.4.5}$$

$$F_{\max} = \frac{ss_{\max}}{ss_{\min}} \tag{8.4.6}$$

The following illustration shows the use of Hartley's test.

Example 8.4.5 A random sample of size 9 is taken from each of three independent normally distributed populations and the respective variances are 51.6, 39.4, and 28.7. Test the equality of the variances for three populations.

Solution: Because the sample sizes are equal, we use Hartley's test.

TABLE 8.4.1
SUMMARY TABLE FOR TESTING HYPOTHESES CONCERNING VARIANCES
All Variances from Normally Distributed Populations

One Population Involved		Two Populations Involved		More Than Two Populations	
Test Statistic: $\chi^2 = (n-1)\dfrac{s^2}{\sigma^2}$		Test Statistic: $F = \dfrac{s_L^2}{s^2}$ where s_L^2 is the larger of s_1^2 or s_2^2		Test Statistic: 1. If all sample sizes are unequal B given by Eq. 8.4.1 2. If sample sizes are all the same $F_{max} = \dfrac{s_{max}^2}{s_{min}^2}$	
H_0: $\sigma^2 = \sigma_0^2$		H_0: $\sigma_1^2 = \sigma_2^2$		H_0: $\sigma_1^2 = \sigma_2^2 = \cdots = \sigma_k^2$	
H_1:	Critical Region	H_1:	Critical Region	H_1:	Critical Region
$\sigma^2 \neq \sigma_0^2$	$\chi^2 > \chi^2_{\alpha/2,n-1}$ or $\chi^2 < \chi^2_{1-\alpha/2,n-1}$	$\sigma_1^2 \neq \sigma_2^2$	$F > F_{\alpha/2,n_1-1,n_2-1}$	At least two of the variances are unequal	1. $B > \chi^2_{\alpha:k-1}$
$\sigma^2 > \sigma_0^2$	$\chi^2 > \chi^2_{\alpha,n-1}$	$\sigma_1^2 > \sigma_2^2$	$F > F_{\alpha,n_1-1,n_2-1}$		2. $F_{max} > F_{max:k,n-1}$
$\sigma^2 < \sigma_0^2$	$\chi^2 < \chi^2_{1-\alpha,n-1}$	$\sigma_2^2 > \sigma_1^2$	$F > F_{\alpha,n_2-1,n_1-1}$		

H_0: Three variances are equal.

H_1: At least two variances are unequal.

$\alpha = .05.$

Test statistic: $F_{max} = \dfrac{s^2_{max}}{s^2_{min}}$.

Critical region: $F_{max} > F_{max:3,8} = 6.00.$

Calculations: $F_{max} = \dfrac{51.6}{28.7} = 1.78.$

Accept H_0.

Table 8.4.1 is given as a summary reference for this section.

PROBLEMS

8.4.1 Refer to Prob. 8.3.4 and test the hypothesis that the population variance is 49 hours^2, against the alternative that it is (a) less than 49. (b) more than 49. Use $\alpha = .05$.

8.4.2 Refer to Prob. 7.4.1 and 7.4.3. Using the information given there, test the hypothesis that the population standard deviation is 1,000 miles against the alternative that it is less than 1,000 miles. Use $\alpha = .01$.

8.4.3 Using the data in Prob. 7.4.10 test the hypothesis that the population variance is 441. Let $\alpha = .01$.

8.4.4 If the underlying distribution is not normal but the sample size is large, the random variable

$$z = \frac{s - \sigma}{\dfrac{\sigma}{\sqrt{2n}}}$$

is approximately unit normal. [See Prob. 7.4.9.] Repeat Prob. 8.4.3, assuming 33 constitutes a large sample size.

8.4.5 Using the information given in Prob. 8.3.5, test the hypothesis that the variances of the populations are equal. Use $\alpha = .05$.

8.4.6 Refer to Prob. 8.3.6 and test the hypothesis that $\sigma_A^2 = \sigma_B^2$ against the alternative that $\sigma_A^2 < \sigma_B^2$.

8.4.7 Refer to Prob. 8.3.8 and test the hypothesis that the variance of the midterm examination scores equals the variance of the final examination scores.

8.4.8 The weights in pounds of three groups of students are as follows:

A	B	C
121	180	140
129	170	160
125	190	174
120	182	144
144	183	151
140	167	174
136	166	
	172	
	182	

Test the hypothesis of homogeneity of the variances. Use $\alpha = .05$.

8.4.9 Simplify Eq. (8.4.2) and (8.4.3) under the assumption that the sample sizes are equal in each group. That is, $n_i = n; i = 1, 2, \ldots, k$.

8.4.10 Using the result of Prob. 8.4.9 test the homogeneity of the variances for the following data (in pounds.) Let $\alpha = .05$.

A	B	C	D
121	180	140	190
129	170	160	202
125	190	174	208
120	182	144	195
140	167	174	205
136	172	182	200

8.4.11 Test the homogeneity of the variances for the data in Prob. 8.4.10 using the F_{max} test statistic.

8.5 TESTS OF HYPOTHESES CONCERNING PROPORTIONS

Often one needs to make an inference concerning the parameter p (proportion) of a binomial p.f. For instance, in an acceptance sampling plan, one might want to test the hypothesis $p = .07$, or in another situation one might want to compare two proportions, say p_1 and p_2; in general, the testing of a hypothesis of the type $p_1 = p_2 = \cdots = p_k$ needs to be considered. As in all previous discussions regarding tests of hypotheses, here also the primary burden lies in locating a test statistic which describes the distribution of the parameter p. Following a similar pattern as in Sec. 8.4, here we shall examine three cases for proportions. Namely, (a) one population (b) two populations and (c) more than two populations involved.

(a) One Population Involved. In Sec. 7.4 it was shown that the random variable

$$z = \frac{x - np}{\sqrt{npq}} = \frac{\frac{x}{n} - p}{\sqrt{\frac{pq}{n}}} \cong \frac{\bar{p} - p}{\sqrt{\frac{\bar{p}\,\bar{q}}{n}}} \tag{8.5.1}$$

where $\bar{p} = \frac{x}{n}$ and $\bar{q} = 1 - \bar{p}$, is approximately unit normal distributed for large values of n. Also, an exact test statistic concerning population proportion was presented there. [Refer to Eq. (7.4.17)]. The critical or rejection regions for the latter approach are obtained by the use of expressions (7.4.20) and 7.4.24). In some applications, particularly during manufacturing processes of small articles, large sample sizes can be taken without jeopardizing the production and hence the test statistic given by Eq. (8.5.1) could be used to test hypotheses concerning proportions.

The hypothesis that we shall test is H_0: $p = p_0$ against one of the three alternatives H_1: $p \neq p_0$, H_1: $p > p_0$ and H_1: $p < p_0$. The procedure to be followed here are identical to the steps suggested in Secs 8.3 and 8.4. An example will illustrate the notion.

Example 8.5.1 From the production line of a manufacturing process 400 items are inspected after random selection. It was found that 28 of the articles were defective. Does this contradict the hypothesis that more than 10 percent of the articles are defective? Use $\alpha = .05$.

Solution:

H_0: $p = .1$.

H_1: $p < .1$.

$\alpha = .05$.

Test statistic: Eq. (8.5.1).

Rejection or critical region: $z < -1.645$.

Calculations: $z = \dfrac{.07 - .1}{\sqrt{\dfrac{(.07)(.93)}{400}}} = -2.36.$

Reject H_0. That is, the proportion of defective articles does not exceed 10 percent.

(b) Two Populations Involved. Suppose $\bar{p}_1 = x_1/n_1$ represents the sample proportion from one and $\bar{p}_2 = x_2/n_2$ the sample proportion from another population. From Eq. (8.5.1) the random variables

$$z_1 = \frac{\bar{p}_1 - p_1}{\sqrt{\dfrac{\bar{p}_1\bar{q}_1}{n_1}}} \quad \text{and} \quad z_2 = \frac{\bar{p}_2 - p_2}{\sqrt{\dfrac{\bar{p}_2\bar{q}_2}{n_2}}}$$

are approximately unit normal distributed. That is, the random variables $\bar{p}_1$ and $\bar{p}_2$ are normally distributed with means p_1 and p_2: and variances $\bar{p}_1\bar{q}_1/n_1$ and $\bar{p}_2\bar{q}_2/n_2$ respectively. Assuming that $\bar{p}_1$ and $\bar{p}_2$ are stochastically independent and applying Theorem 7.1.1 to their difference, we note that the random variable

$$z = \frac{(\bar{p}_1 - \bar{p}_2) - (p_1 - p_2)}{\sqrt{\dfrac{\bar{p}_1\bar{q}_1}{n_1} + \dfrac{\bar{p}_2\bar{q}_2}{n_2}}} \tag{8.5.2}$$

is also approximately unit normal distributed. If we replace $\bar{p}_1$ and $\bar{p}_2$ by their weighted or pooled average,

$$\bar{p}_w = \frac{n_1\bar{p}_1 + n_2\bar{p}_2}{n_1 + n_2} = \frac{x_1 + x_2}{n_1 + n_2} \tag{8.5.3}$$

Equation (8.5.2) reduces to

$$z = \frac{(\bar{p}_1 - \bar{p}_2) - (p_1 - p_2)}{\sqrt{\bar{p}_w \bar{q}_w \left(\frac{1}{n_1} + \frac{1}{n_2}\right)}} \tag{8.5.4}$$

This last expression is the desired test statistic to test hypotheses of the type H_0: $p_1 - p_2 = \delta$ against one of the three alternatives H_1: $p_1 - p_2 \neq \delta$, H_1: $p_1 - p_2 > \delta$ and H_1: $p_1 - p_2 < \delta$. Let us note that if $\delta = 0$, the null hypothesis becomes $p_1 = p_2$ and the àlternatives are $p_1 \neq p_2$, $p_1 > p_2$ and $p_l < p_2$ respectively.

Example 8.5.2 A random sample of 250 students from College A and 300 from College B were interviewed by a polling agency regarding a controversial issue. 200 students in College A and 200 students in College B indicated favorable response. Test the hypothesis that students at these two colleges have similar points of view on this particular issue.

Solution: Interpreting similar points of view to imply that students on both campuses agree on this issue with equal proportions, we let

H_0: $p_1 = p_2$.

H_1: $p_1 \neq p_2$.

$\alpha = .05$.

Test statistic: Eq. (8.5.4).

Rejection or critical region: $|z| > 1.96$.

Calculations: $$z = \frac{\frac{4}{5} - \frac{2}{3}}{\sqrt{\frac{400}{550}\frac{150}{550}\left(\frac{1}{250} + \frac{1}{300}\right)}} = \frac{\sqrt{110}}{3} = 3.5.$$

Therefore, reject the null hypothesis.

(c) More Than Two Populations Involved. Often one encounters situations where there are more than two populations to compare. For instance, in Example 8.5.2 the opinion of college students regarding a controversial issue was sampled from two colleges. Suppose a polling agency plans to interview students from k different colleges. What statistic could one apply to test the hypothesis $p_1 = p_2 = \cdots = p_k = p$, where p presents the true proportion of students expressing favorable response to the interview question in all colleges?

Let us represent the data from the k samples as

x_1	x_2	$\cdots$	x_k	X
$n_1 - x_1$	$n_2 - x_2$	$\cdots$	$n_k - x_k$	$N - X$
n_1	n_2	$\cdots$	n_k	N

where $X = \sum_{i=1}^{k} x_i$ represents the horizontal marginal corresponding to the total favorable responses and $N = \sum_{i=1}^{k} n_i$ is the total number in the combined samples.

The required test statistic can be readily obtained by applying Theorem 7.3.2 to the random variable

$$z_i = \frac{x_i - n_i p}{\sqrt{n_i pq}} = \frac{\bar{p}_i - p}{\sqrt{\dfrac{pq}{n_i}}}$$

which according to Eq. (8.5.1) is unit normal distributed. Therefore, applying Theorem 7.3.2 to the last expression, we note that the r.v. $\sum_{i=1}^{k} z_i^2$ is χ^2 distributed with k degrees of freedom.

We cannot use this result for a test statistic because p is unknown. As in case (b) above, here also we remove the obstacle by substituting $\bar{p}_w$, the weighted average of $\bar{p}_i$'s for p. However, one degree of freedom is associated with $\bar{p}_w$ and therefore the random variable

$$\sum_{i=1}^{k} z_i^2 \simeq \sum_{i=1}^{k} \left[\frac{\bar{p}_i - \bar{p}_w}{\sqrt{\dfrac{\bar{p}_w \bar{q}_w}{n_i}}} \right]^2 = \sum_{i=1}^{k} \frac{n_i(\bar{p}_i - \bar{p}_w)^2}{\bar{p}_w \bar{q}_w} \tag{8.5.5}$$

is χ^2 distributed with $(k - 1)$ degrees of freedom where

$$\bar{p}_w = \frac{\sum_{i=1}^{k} n_i p_i}{\sum_{i=1}^{k} n_i} = \frac{\sum_{i=1}^{k} x_i}{\sum_{i=1}^{k} n_i} = \frac{X}{N} \tag{8.5.6}$$

The expression (8.5.5) is the desired test statistic for testing hypotheses concerning homogeneity (equality) of proportions, provided no $n_i \bar{p}_i$ is less than 5. To state this restriction positively, there should be at least 5 observations in each group whose population proportion is among the proportions being tested. This restriction is necessary because for too small values of p, the random variable

$$z = \frac{\bar{p}_i - p}{\sqrt{\dfrac{\bar{p}_w \bar{q}_w}{n_i}}}$$

deviates from the normal assumption. The reader is requested at this juncture to refer to Eq. (2.5.3) and examine Sec. 4.6.

Since the null hypothesis will be rejected whenever the magnitude of the differences between the proportions is large, the critical region for the test statistic (8.5.5) is $\chi^2 > \chi^2_{\alpha:k-1}$.

In many special cases it is feasible to take samples of equal size from each of the populations. Equal sample sizes reduce Eq. (8.5.5) to

$$\chi^2 = \sum_{i=1}^{k} z_i^2 \cong n\sum_{i=1}^{k} \frac{(\bar{p}_i - \bar{p}_w)^2}{\bar{p}_w \bar{q}_w} \tag{8.5.7}$$

which simplifies the computation of the test statistic considerably. The following hypothetical example illustrates the concept under discussion.

Example 8.5.3 Four machines are used in the manufacturing of certain articles. A random sample of size 100 is taken from a day's production of each machine. The result is as follows:

	Machine			
	A	B	C	D
Defective	10	30	20	40
Nondefective	90	70	80	60

Is there statistical evidence to disprove the hypothesis that each machine functions equally well?

Solution: We interpret the expression "equally well" to imply that proportions of nondefectives for each machine are equal.

H_0: $p_1 = p_2 = p_3 = p_4$.
H_1: At least 2 proportions are not equal.
$\alpha = .01$.

Test statistic: Eq. (8.5.7).
Critical region: $\chi^2 > \chi^2_{.01:3} = 11.34$.

$$\chi^2 = 100\left[\frac{(.15)^2}{(.75)(.25)} + \frac{(.05)^2}{(.75)(.25)} + \frac{(.05)^2}{(.75)(.25)}\frac{(.15)^2}{(.75)(.25)}\right] = \frac{80}{3} = 26.66$$

Reject H_0. That is, there is statistical evidence on the 1 percent significance level that the proportion of nondefectives are unequal.

Although Eq. (8.5.5) is developed for the case where more than 2 populations are involved, it could be applied for the situation where there are two populations. We illustrate such an example next.

Example 8.5.4 Repeat Example 8.5.2 using Eq. (8.5.5) for test statistic.

Solution:

H_0: $p_1 = p_2$.
H_1: $p_1 \neq p_2$.
$\alpha = 5$ percent.

Test statistic: Eq. (8.5.5).

Critical region: $\chi^2 > \chi^2_{.05:1} = 3.841$.

$$\text{Calculations: } \chi^2 = 250\left[\frac{\left(\frac{4}{5} - \frac{8}{11}\right)^2}{\left(\frac{8}{11}\right)\left(\frac{3}{11}\right)}\right] + 300\left[\frac{\left(\frac{2}{3} - \frac{8}{11}\right)^2}{\left(\frac{8}{11}\right)\left(\frac{3}{11}\right)}\right] = \frac{110}{9}.$$

Reject H_0.

Note that, as expected the results are identical in Examples 8.5.2 and 8.5.4. That is, the random variable z^2 is χ^2 distributed with 1 d.f.

Next let us derive an equivalent expression for Eq. (8.5.5), which may prove less laborious in computation. Recalling that

$$\bar{p}_w = \frac{\sum_{i=1}^{k} x_i}{N}$$

and $1 - \bar{p}_w = \bar{q}_w$, it can be shown that

$$\frac{(x_i - n_i\bar{p}_w)^2}{n_i\bar{p}_w\bar{q}_w} = \frac{(x_i - n_i\bar{p}_w)^2}{n_i\bar{p}_w} + \frac{[(n_i - x_i) - n_i\bar{q}_w]^2}{n_i\bar{q}_w} \tag{8.5.8}$$

(See Prob. 8.5.5.) Therefore, Eq. (8.5.5) can be expressed equivalently:

$$\chi^2 = \sum_{i=1}^{k}\left\{\frac{\left(x_i - n_i\frac{X}{N}\right)^2}{n_i\frac{X}{N}} + \frac{\left[(n_i - x_i) - n_i\left(1 - \frac{X}{N}\right)\right]^2}{n_i\left(1 - \frac{X}{N}\right)}\right\} \tag{8.5.9}$$

This relationship could be notationally simplified further. Letting $x_i = f_{i1}$, $n_i - x_i = f_{i2}$ and $n_i\frac{X}{N} = e_{ij}$, Eq. (8.5.9) reduces to

$$\chi^2 = \sum_{i=1}^{k}\sum_{j=1}^{2}\frac{(f_{ij} - e_{ij})^2}{e_{ij}} \tag{8.5.10}$$

and as in Eq. (8.5.5), χ^2 has $(k - 1)$ d.f.

Example 8.5.5 Repeat Example 8.5.3 using for test statistic (8.5.10).
Solution: We shall calculate only the test statistic:

10	30	20	40	100
90	70	80	60	300
100	100	100	100	400

$$\chi^2 = \frac{(10 - 25)^2}{25} + \frac{(30 - 25)^2}{25} + \frac{(20 - 25)^2}{25} + \frac{(40 - 25)^2}{25}$$

$$+\frac{(90-75)^2}{75}+\frac{(70-75)^2}{75}+\frac{(80-75)^2}{75}+\frac{(60-75)}{75}$$

$$=\frac{1}{25}[15^2+5^2+5^2+15^2]+\frac{1}{75}[15^2+5^2+5^2+15^2]$$

$$=\frac{4}{75}[225+25+25+225]$$

$$=\frac{4[500]}{75}=\frac{2000}{75}=\frac{400}{15}=\frac{80}{3}$$

which as anticipated agrees with the result of Example 8.5.3.

Before we conclude the discussion of testing hypotheses concerning proportions a final remark is in order. In Sec. 4.4, while considering the normal approximation of the binomial, it was indicated that a correction for continuity is necessary to obtain more accurate results in the approximating process. In the derivation of Eq. (8.5.5) and hence in Eq. (8.5.10) if the random variable

$$z_i=\frac{|x_i-n_ip_i|-\frac{1}{2}}{\sqrt{n_ip_iq_i}}$$

were used instead of

$$z_i=\frac{x_i-n_ip_i}{\sqrt{n_ip_iq_i}}$$

we would have obtained for Eq. (8.5.10)

$$\chi^2=\sum_{i=1}^{k}\sum_{j=1}^{2}\frac{\left(|f_{ij}-e_{ij}|-\frac{1}{2}\right)^2}{e_{ij}} \tag{8.5.11}$$

This last expression is referred to as the *adjusted* χ^2.

Table 8.5.1 summarizes the test statistics presented in this section.

TABLE 8.5.1

TESTS OF HYPOTHESES FOR PROPORTIONS

One Population		Two Populations		More than Two Populations	
Test Statistic: $z=\frac{\bar{p}-p}{\sqrt{\frac{\bar{p}\bar{q}}{n}}}$		Test Statistic: $z=\frac{(\bar{p}_1-\bar{p}_2)-(p_1-p_2)}{\sqrt{\bar{p}_w\bar{q}_w\left(\frac{1}{n_1}+\frac{1}{n_2}\right)}}$		Test Statistic: Eq. (8.5.5) or (8.5.9)	
$H_0: p=p_0$		$H_0: p_1-p_2=\delta$		$H_0: p_1=p_2=\cdots=p_k$	
H_1:	Critical Region	H_1:	Critical Region	H_1:	Critical Region
$p\neq p_0$ $p<p_0$ $p>p_0$	$\lvert z\rvert>z_{\alpha/2}$ $z<z_{1-\alpha}$ $z>z_\alpha$	$p_1-p_2\neq\delta$ $p_1-p_2<\delta$ $p_1-p_2>\delta$	$\lvert z\rvert>z_{\alpha/2}$ $z<z_{1-\alpha}$ $z>z_\alpha$	At least two proportions are not equal	$\chi^2>\chi^2_{\alpha;k-1}$

PROBLEMS

8.5.1 It is hypothesized that in Collegetown, U.S.A., 60 percent of the voters are Republicans against the alternative that the percentage is less than 60 percent. If the assumed significance level is .05, and a random sample of 40 is taken from the registration list of voters, what is the largest number of Republicans in this sample that will lead to the rejection of the null hypothesis?

8.5.2 In Prob. 8.5.1, find the type II error of the test if the actual proportion of Republicans is 55 percent.

8.5.3 In a random sample of 500 students at a college 155 were coeds. Test the hypothesis that the proportion of coeds at this institution is .25 against the alternative that the proportion is larger than .25.

8.5.4 In Prob. 8.5.3, sketch the power curve of the test if the actual proportion of coeds at the college is .32.

8.5.5 Verify Eq. (8.5.8).

8.5.6 The president of the Sport Cars Club at a Western University claims that no more than 20 percent of sport cars in use in this country are American made. In a random sample of 100 cars, what is the minimum number of American cars required to prove this claim incorrect?

8.5.7 The effectiveness of a new drug is to be tested on guinea pigs with certain eye desease. Among 15 animals treated with the new and 25 with standard treatment, ten in each group recovered from the disorder. Based on this data, is there evidence that the new drug is superior? Use $\alpha = .05$.

8.5.8 The following is a detailed two way classification of the data in Prob. 8.5.3.

	Freshmen	Sophomores	Juniors	Seniors
Coeds........	50	52	37	16
Males........	150	98	63	34

Test the hypothesis that the proportion of coeds in each class is the same. Use significance level of .01.

8.5.9 A course in mathematical statistics is offered during each term of the academic year at a university. If 63, 82, and 56 students registered for this course during the Fall, Winter, and Spring quarters, test the hypothesis that the number of students registered for the course are equal each term. Use $\alpha = .05$.

8.5.10 An ordinary six-sided die and a coin are tossed 240 times and the following frequencies observed.

Coin \ Die	1	2	3	4	5	6
Heads........	22	17	28	25	16	17
Tails	23	15	26	20	18	13

Using this information and $\alpha = .05$ test the hypotheses that (a) the coin is honest, (b) the die is honest, or (c) the joint events are stochastically independent.

8.6 TEST OF HYPOTHESES CONCERNING DISTRIBUTIONS—GOODNESS OF FIT

Suppose in Example 8.5.2 the polling agency as part of the same experiment wants to examine the response given by freshmen, sophomores, juniors, and seniors in 10 different colleges. Or in Example 8.5.3 that the articles from each machine are classified as nondefective, usable, and defective. What test statistic could one use to make inferences concerning proportions of freshmen, sophomores, juniors, and seniors in each of the ten institutions of the former and proportions of nondefective, usable, and defective in the latter illustration? In general, when there are I classifications and within each of them there are J classifications, we could represent the data in an $I \times J$ two-way table as shown.

	1	2	3	$\cdots$	j	$\cdots$	J	
1	f_{11}	f_{12}	f_{13}	$\cdots$	f_{1j}	$\cdots$	f_{1J}	$f_{1\cdot}$
2	f_{21}	f_{22}	f_{23}	$\cdots$	f_{2j}	$\cdots$	f_{2J}	$f_{2\cdot}$
3	f_{31}	f_{32}	f_{33}	$\cdots$	f_{3j}	$\cdots$	f_{3J}	$f_{3\cdot}$
$\vdots$	$\vdots$	$\vdots$	$\vdots$		$\vdots$		$\vdots$	$\vdots$
i	f_{i1}	f_{i2}	f_{i3}	$\cdots$	f_{ij}	$\cdots$	f_{iJ}	$f_{i\cdot}$
$\vdots$	$\vdots$	$\vdots$	$\vdots$		$\vdots$		$\vdots$	$\vdots$
I	f_{I1}	f_{I2}	f_{I3}	$\cdots$	f_{Ij}	$\cdots$	f_{IJ}	$f_{I\cdot}$
	$f_{\cdot 1}$	$f_{\cdot 2}$	$f_{\cdot 3}$	$\cdots$	$f_{\cdot j}$	$\cdots$	$f_{\cdot J}$	$f_{\cdot\cdot}$

In the table, f_{ij} represents the frequency of observations in the ijth cell and the dot notation for the marginals was defined in Sec. 1.5. Such a representation of data is called a *contingency table.* Assuming stochastic independence, the expected frequency for the ijth cell is

$$e_{ij} = \frac{f_{i\cdot}\, f_{\cdot j}}{f_{\cdot\cdot}} \tag{8.6.1}$$

and as in the derivation of Eq. (8.5.10), it can be shown that the quantity

$$\chi^2 = \sum_{i=1}^{I} \sum_{j=1}^{J} \frac{(f_{ij} - e_{ij})^2}{e_{ij}} \tag{8.6.2}$$

is chi-square-distributed with $(I - 1)(J - 1)$ degrees of freedom. Can you justify the reason for the degrees of freedom associated with Eq. (8.6.2)?

Let us remark that in the special case where $J = 1$ (or $I = 1$), expression (8.6.2) reduces to

$$\chi^2 = \sum_{i=1}^{I} \frac{(f_i - e_i)^2}{e_i} \tag{8.6.3}$$

with $(I - 1)$ degrees of freedom.

The restriction that all frequencies in the two-way classification should be greater than 5 also applies in this situation. If this requirement is not met, usually two or if necessary more than two classifications are combined to yield the desired condition. For instance, suppose the following frequencies exist in a 4×6 classification.

4	10	15	20	1	10
2	10	20	30	4	24
5	12	17	28	13	19
10	2	5	6	11	8

Since the first and fifth columns have some entries with frequencies less than 5, we combine the first column with the second as well as the last column with the fifth, obtaining a 4×4 table

14	15	20	11
12	20	30	28
17	17	28	32
12	5	6	19

Now any inference regarding proportions should be made concerning the combined proportions of first and second columns, third, fourth, and combined fifth and sixth columns.

In this illustration, instead of combining columns one could have combined rows to achieve the same objective. If after combining two columns or rows there still exist cell frequencies with less than 5 observations, the process of combining groups is continued until the stated criterion is satisfied.

Combining various columns or rows in a contingency table is not limited to the situation where cell frequencies are less than five. Sometimes it is advantageous or even necessary to group frequencies prior to testing hypotheses. Suppose the following detailed data represents the responses of the male and female students at four colleges:

	College (Male students)				College (Female students)			
	A	B	C	D	A	B	C	D
Favorable.........	15	23	17	20	20	20	24	26
Not favorable.....	25	25	25	50	40	25	15	30

and a hypothesis concerning the favorable responses of students in general is to be tested. The two separate data could be combined to yield

	College			
	A	B	C	D
Favorable.........	35	43	41	46
Not favorable.....	65	50	40	80

which represents a combined sample of students from the four colleges.

Equation (8.6.2) is the test statistic to test hypotheses of the type

H_0: $p_{i1} = p_{i2} = \cdots = p_{iJ}$
H_1: At least two proportions are unequal

or

H_0: $p_{1j} = p_{2j} = \cdots = p_{Ij}$
H_1: At least two proportions are unequal

To be precise, Eq. (8.6.2) is the test statistic to test hypotheses of stochastic independence for joint events.

Example 8.6.1 Two hundred students from each of the four colleges were selected at random and their year in college observed. The following data was obtained.

	Year in College				
	1	2	3	4	
College A	80	60	30	30	200
College B	100	50	30	20	200
College C	50	50	50	50	200
College D	60	60	40	40	200
	290	220	150	140	800

Test the hypothesis that the proportions of students in each classification are equal in the four colleges. Use $\alpha = 5$ percent.

Solution: We first compute the expected frequencies for each cell entry and obtain

	Year in College			
	1	2	3	4
College A	$\frac{145}{2}$	55	$\frac{75}{2}$	35
College B	$\frac{145}{2}$	55	$\frac{75}{2}$	35
College C	$\frac{145}{2}$	55	$\frac{75}{2}$	35
College D	$\frac{145}{2}$	55	$\frac{75}{2}$	35

H_0: $p_{1\cdot} = p_{2\cdot} = p_{3\cdot} = p_{4\cdot}$.
H_1: At least a pair of proportions are unequal.
$\alpha = 5$ percent.

Test Statistic: Eq. (8.6.2).
Critical region: $\chi^2 > \chi^2_{\alpha;9} = 16.919$.

$$\chi^2 = \sum_{i=1}^{4} \sum_{j=1}^{4} \frac{(f_{ij} - e_{ij})^2}{e_{ij}}$$

$$= \frac{\left(80 - \frac{145}{2}\right)^2}{\frac{145}{2}} + \frac{\left(100 - \frac{145}{2}\right)^2}{\frac{145}{2}} + \frac{\left(50 - \frac{145}{2}\right)^2}{\frac{145}{2}} + \frac{\left(60 - \frac{145}{2}\right)^2}{145}$$

$$+ \frac{(60 - 55)^2}{55} + \frac{(50 - 55)^2}{55} + \frac{(50 - 55)^2}{55} + \frac{(60 - 55)^2}{55}$$

$$+ \frac{\left(30 - \frac{75}{2}\right)^2}{\frac{75}{2}} + \frac{\left(30 - \frac{75}{2}\right)^2}{\frac{75}{2}} + \frac{\left(50 - \frac{75}{2}\right)^2}{\frac{75}{2}} + \frac{\left(40 - \frac{75}{2}\right)^2}{\frac{75}{2}}$$

$$+ \frac{(30 - 35)^2}{35} + \frac{(20 - 35)^2}{35} + \frac{(50 - 35)^2}{35} + \frac{(40 - 35)^2}{35}$$

$$= \frac{1}{290} [15^2 + 55^2 + (-45)^2 + (-25)^2]$$

$$+ \frac{1}{55} [5^2 + 5^2 + 5^2 + 5^2]$$

$$+ \frac{1}{150} [(-15)^2 + (-15)^2 + 25^2 + 5^2]$$

$$+ \frac{1}{35} [5^2 + (-15)^2 + 15^2 + 5^2]$$

$$= \frac{5{,}900}{290} + \frac{100}{55} + \frac{1{,}100}{150} + \frac{500}{35} = 43.7$$

Reject H_0.

It was stated in the opening remarks of Sec. 8.2 that usually a statistical hypothesis is comprised of several different hypotheses and each is to be examined separately. It was mentioned there that the statistical hypothesis: the random variable x is normally distributed with mean μ and variance σ^2, includes three distinct hypotheses, namely, (1) x is normally distributed, (2) $\mu = \mu_0$, and (3) $\sigma^2 = \sigma_0^2$. The last two hypotheses can be tested by the methods presented in Secs. 8.3 and 8.4 respectively. Here we shall consider a method for testing hypotheses of the first type—that is, hypotheses concerning the underlying distribution of a given random variable.

Before developing the necessary test statistic to accomplish the stated objective, let us consider an illustration. Suppose we wish to test the hypothesis that a random variable x is normally distributed with mean 54.5 and variance 144. To test this compound hypothesis a random sample of size 40 yields the following frequency distribution

25–34	1
35–44	9
45–54	12
55–64	11
65–74	5
75–84	2

The mean and the standard deviation of this sample are 53.5 and 11.94 respectively. (See Examples 5.6.1 and 5.6.2.) If one tests the hypotheses $\mu = 54.5$ and $\sigma^2 = 144$ based upon the given sample, both hypotheses are accepted as true on 5 percent significance level. The reader is requested to verify these two assumptions.

Assuming that $\mu = 54.5$ and $\sigma = 12$, a procedure is necessary to test the hypothesis that the random variable x is normally distributed. The normal p.f. with mean 54.5 and standard deviation of 12 is represented in Fig. 8.9, along with the histogram of the given data. We wish to compare the histogram with its associated normal curve. What criteria could be used to accept the hypothesis that these figures are similar? Specifically, what test statistic can be used to test the hypothesis that x is normally distributed? We have already used a statistic of central tendency, the mean, and a statistic of dispersion, the variance, to test two parts of the compound hypothesis; hence some other criteria of comparison are desired. Let us consider as a possible criterion of comparison the corresponding areas of the histogram and the normal curve—that is, the areas associated with the intervals 24.5–34.5, 34.5–44.5, . . . , 74.5–84.5 for the histogram and the normal curve. For instance, the area related with the interval 54.5–64.5 is $11/40 = .275$ for the histogram while the corresponding area for the normal curve is .2976 [$P(0 < z < 5/6)$.] Similarly,

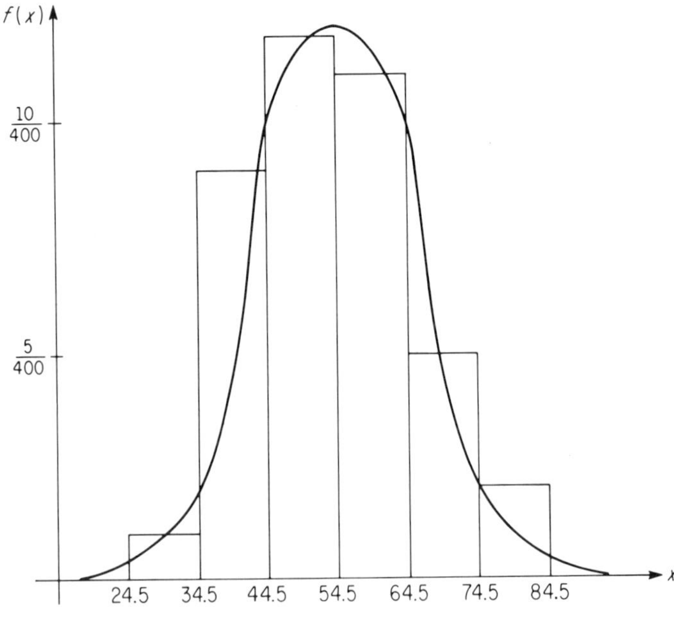

FIG. 8.9.

for the interval 44.5–54.5 the respective areas are $12/40 = .30$ and .2976, and so on. In order to test whether the differences between these proportions are significant, the test statistic given by Eq. (8.6.3) could be applied, which is the desired test statistic.

Let us observe that the frequencies in the first and the last class intervals are both less than 5. Therefore it is necessary to combine them with the adjacent intervals which yields

Class Limit	f_i	e_i
25–44	10	8.1
45–54	12	11.9
55–64	11	11.9
65–84	7	8.1
	40	40.0

The corresponding expected frequencies are $40p_1$, $40p_2$, $40p_3$, and $40p_4$, respectively, where

$$p_1 = P[x < 44.5] = P\left[z < \frac{44.5 - 54.5}{12} = -.833\right] = .2024$$

$$p_2 = P[44.5 \leq x \leq 54.5] = P[-.833 \leq z \leq 0] = .2976$$

and because of symmetry, $p_1 = p_4$ and $p_2 = p_3$. Hence, $40p_1 = 8.1$, $40p_2 = 11.9$, $40p_3 = 11.9$ and $40p_4 = 8.1$ which are listed under the column e_i above.

Calculation of

$$\chi^2 = \sum_{i=1}^{4} \frac{(f_i - e_i)^2}{e_i}$$

yields

$$\chi^2 = \frac{(1.9)^2}{8.1} + \frac{(0.1)^2}{11.9} + \frac{(0.9)^2}{11.9} + \frac{(1.1)^2}{8.1} = .67$$

which is less than $\chi^2_{.05,3}$ and consequently we accept the assumption that the sample "fits" the normal curve. The degrees of freedom in this example is $4 - 1 = 3$, because *one* statistic was used in the computation of the expected frequencies. Namely, the total number of frequencies. This illustration indicates the general approach to test hypotheses concerning distribution of random variables. Because basically we are testing the "fit" of the sample to the hypothesized probability function, tests of this type are called *goodness of fit*.

The statistic to test goodness of fit is Eq. (8.6.3) with varying degrees of freedom. In general the degrees of freedom associated with it is the number of grouped frequencies *minus* the total number of statistics used from the original data in the computation of the expected frequencies. In the goodness of fit for the normal p.f. the degrees of freedom is $(k - 3)$ because three statistics are used in the computation of the expected frequencies. That is, the total number of frequencies, the mean and the variance. For the Poisson p.f. the associated degrees of freedom are $(k - 2)$, since in the Poisson p.f. the mean and the variance are equal to the single parameter.

Example 8.6.2 The number of typographical errors per page prior to proofreading is assumed to obey the Poisson probability function. (Refer to Example 2.5.3.) During the publication of a textbook of 400 pages the following were observed:

Errors per page:	0	1	2	3	4	5	6	7	8
Frequency:	28	58	72	92	75	40	25	8	2

Test the hypothesis that this sample obeys Poisson probability function, using 5 percent significance level.

Solution: The mean number of errors per page for this sample is 1,200/400 = 3.00. Therefore, we test the Poisson p.f. with parameter $k = 3$ for goodness of fit. The Poisson probability function is applicable to a countably infinite random variable. Since there are no errors greater than 8 per page and the frequency for 8 errors per page is less than five, we combine the frequencies of all possible errors of greater than 7 with 7. Using Table III in the Appendix, one obtains the theoretical probabilities associated with the random variable x. That is,

$$f(0:3) = .0498$$

$$f(1:3) = .1494$$

$$f(2:3) = .2240$$

$$f(3:3) = .2240$$

$$f(4:3) = .1680$$

$$f(5:3) = .1008$$

$$f(6:3) = .0504$$

$$f(x \geq 7:3) = .0335$$

Multiplying each of these probabilities by 400, the expected frequencies are obtained, as listed in the adjoining table. Since $\chi^2_{.05:6} = 12.5916$, the hypothesis that the sample fits the Poisson p.f. with $k = 3$ cannot be rejected.

f_i	e_i	$\frac{(f_i - e_i)^2}{e_i}$
28	19.92	3.27743
58	59.76	.05183
72	89.60	3.45714
92	89.60	.06429
75	67.20	.90535
40	40.32	.00254
25	20.16	1.16198
10	13.40	.86268
400	399.96	9.78324

PROBLEMS

8.6.1 Show that for a 2 × 2 contingency table with cell frequencies a, b, c, and d, Eq. (8.6.2) reduces to

$$\chi^2 = \frac{N(ad - bc)}{(a + b)(a + c)(c + d)(b + d)}$$

where $N = a + b + c + d$.

8.6.2 A cereal manufacturing company is planning to put a new product on the market. A random sample of adults and children yields the following result.

	Women	Men	Children
Like..........	14	25	96
Dislike.......	43	15	14
Indifferent ...	13	40	40

Using $\alpha = .01$, test the hypotheses that (a) adults and children have similar preferences, (b) men and women have similar preferences, (c) the proportions of likes and dislikes are equal—assuming indifference to the product implies dislike.

8.6.3 The manager of a beauty salon keeps records of the number of requests for coloring hair. During a certain week, among 120 women 65 requested the hair color blonde, 35 brunette, and 20 red. Test the hypothesis that the hair color preference of the ladies are in the ratio of 4:3:1.

8.6.4 The following data represents a random sample:

15–19 6

20–24 9

25–29 9

30–34 6

Test the hypothesis that the underlying population is normally distributed.

8.6.5 Repeat Prob. 8.6.4 if it is known that $\mu = 22$ and $\sigma = 5$.

8.6.6 The number of accidents per week on a certain highway over a period of two years has the following frequencies.

Number of Accidents per Week	Frequency
0	34
1	40
2	15
3	6
4	5
5	4

Test the hypothesis that this data obeys the Poisson p.f. State whatever assumptions are needed to make the test.

8.7 NONPARAMETRIC STATISTICS

Nearly in all the discussions in previous sections it was assumed that the distributions of the random variables were known and in most cases assumed to obey the normal probability function. For instance, the sampling distribution of $(n - 1)s^2/\sigma^2$ was shown to be χ^2, provided the random variable was normally distributed. The sampling distribution of the mean for any random variable was shown to be asymptotically normal (central limit theorem). There are occasions when one could not assume that the random variable in question is normally distributed, and there may be evidence to the contrary. For example, it may be known that a given random variable is distributed according to the exponential probability function. In general, however, one has no knowledge about the distribution or the parameters of a population and the question "Are the theorems and concepts of the previous sections applicable in these circumstances?" might be asked. Fortunately the normal distribution assumption, even though the random variable does not obey the normal probability function, leads to accurate conclusions with high probability. Therefore the notions described in Chapters 7 and 8 could be utilized in situations where the normal distribution assumption cannot be made.

In recent years there have been attempts to test hypotheses concerning distributions without specifying its exact distribution or the parameters associated with it. Because the test statistics to be discussed compare distributions and not parameters, they are commonly referred to as *nonparametric statistics*. The goodness-of-fit test presented in Sec. 8.6 is perhaps the most commonly used nonparametric statistic.

In this section we shall present one more nonparametric statistic to test hypotheses concerning proportions, which could be readily extended to test hypotheses concerning means. It is called the *sign test*.

The analysis of the sign test is based on the binomial probability function and perhaps is one of the most simple nonparametric tests. This technique assumes that each member of N pairs of measurements has been obtained under identical conditions by either of two available methods. If the two methods yield equal results, the differences in the observations will include equal number of plus (+) and minus (−) signs. That is, the proportion of + signs equals the proportion of − signs, and either too many + or too many − signs will cause one to reject the hypothesis $p = 1/2$.

Let us designate the N pairs of observations by (x_i, y_i) $i = 1, 2, \ldots, N$ and their differences by d_i. We delete from the set of observations any value of d_i that is equal to zero and let the number of remaining observations equal to n. Let k represent the number of times the less frequent sign occurs and hence $n - k$ represents the number of more frequently occuring sign. If the hypothesis $p = 1/2$ is true, the exact probability of this event happening is obtained from the binomial probability function which yields

$$f\left(k: n, \frac{1}{2}\right) = \binom{n}{k}\left(\frac{1}{2}\right)^k\left(\frac{1}{2}\right)^{n-k} = \binom{n}{k}\left(\frac{1}{2}\right)^n$$

and the probability that k or fewer signs occur is

$$F\left(k: n, \frac{1}{2}\right) = \sum_{x=0}^{k}\binom{n}{x}(0.5)^n$$

Using Table II in the Appendix, we can find the required probability and decide whether to accept the stated null hypothesis or not. The following example illustrates how the sign test is used.

Example 8.7.1 Twelve identical twins are randomly assigned to two kinds of diets and their gains in weight, in pounds, is as shown below.

Pair	Diet I	Diet II
1	3	4
2	1	1
3	0	1
4	2	3
5	2	4
6	1	2
7	4	2
8	5	4
9	3	3
10	3	1
11	1	2
12	3	6

Is there statistical evidence to accept the hypothesis that either diet is superior? Use $\alpha = .05$.

Solution: The difference in the paired diets are $-1, 0, -1, -1, -2, -1, +2, +1, 0, +2, -1, -3$. In this example $N = 12$ and $n = 10$ because two pairs have zero difference. $k = 3$ represents the number of + signs.

H_0: $p = .5$.
H_1: $p \neq .5$.
$\alpha = .05$.
Test statistic: $F(k\colon n, .5)$.
Critical region: $F(k\colon n, .5) < .025$.
Calculations: $F(3\colon 10, .5) = .1719$.

Therefore accept H_0. Notice that either of the diets are superior implies one of the diets must be significantly different.

The sign test has an obvious disadvantage in that n has to be large. If the number of pairs yielding nonzero difference is too small, the sign test fails. For instance, in a two-sided test as in Example 8.7.1, if n were 4 and $k = 0$, the smallest probability associated with the test statistic is $2(.0625) = .125$, and hence it is impossible with $\alpha = .05$ to reject the hypothesis $p = .5$. From Table II in the Appendix we note that the expression $\binom{n}{0}\left(\frac{1}{2}\right)^n < .025$ holds true for $n > 5$. That is, the minimum number of paired measurements with nonzero differences is 6 for a two-sided test with .05 significance level. Likewise, for a two-sided test with $\alpha = .01$ the minimum number of paired measurements with nonzero differences is 8.

On the other hand, for large n the sign test has an advantage. If n is large, the normal approximation to the binomial can be utilized to test the hypothesis $p = 1/2$, which we illustrate next.

Example 8.7.2 Repeat Example 8.7.1 by using the normal approximation to the binomial.

Solution: In this case $\mu = np$ yields $10(.5) = 5$ and $\sigma = \sqrt{npq} = \sqrt{10(.5)(.5)} = \sqrt{2.5}$.

H_0: $p = .5$.
H_1: $p \neq .5$.
$\alpha = .05$.

Test statistic: $z = \dfrac{\bar{x} - np}{\sqrt{npq}}$.

Critical region: $z > 1.96$ or $z < -1.96$.

Calculations: $z = \dfrac{3.5 - 5}{\sqrt{2.5}} = -.95$.

Therefore accept H_0.

PROBLEMS

8.7.1 A coin is to be tested for bias (loaded). The test procedure is to toss it 10 times and reject the hypothesis $p = .5$ if 8 or more times heads show. What is the significance level of this test?

8.7.2 Repeat Prob. 8.3.8, using the method of this section.

8.7.3 Thirty students were paired according to equal ability (determined by a test) into two groups. Members in group I were taught their history lessons for one week by directed library assignments, whereas group II were taught the same topics by lecture method. The result of identical tests given to them yielded:

I: 65, 99, 80, 93, 69, 85, 90, 67, 81, 75, 84, 75, 65, 95, 84
II: 74, 99, 95, 90, 70, 91, 77, 62, 71, 81, 81, 91, 70, 93, 91

Do the two methods yield significantly different means? Apply the sign test.

8.7.4 Assuming $n = 15$ represents a large sample, use the normal approximation to the binomial to answer the question in Prob. 8.7.3.

9

Curve Fitting

9.0 INTRODUCTION

In the preceding chapters our considerations have been focused on finding a probability function of a random variable which describes closely the random phenomenon in question. In order to validate the correct choice of a probability function, hypotheses concerning the parameters of the selected probability function were tested. In testing hypotheses concerning the parameters of two or more populations, it was assumed that the random variables were stochastically independent. However, there are many practical applications in which the random variables are known to be *not* independent. If the degree of relationship between the two dependent variables could be measured—intuitively at least—one hopes to be able to predict the outcome of one variable from the other. This procedure in general will bring about a reduction in the experimental factors, which in turn brings about a lowering in cost for the experiment.

For instance, suppose x represents the heights of college students and y their weights. One may suspect that heights and weights are dependent variables and reason as follows: Joe Hoe is 74 inches tall, which represents an above-average height and hence his weight must be above average also—say 175 pounds. For another example, if x represents the IQ scores and y the grade points of high school students and Mary Smart has an IQ score of 150, one might expect her to have better-than-average grade points. Furthermore, if she seeks admission to a college, the admission officer might within certain tolerance limits (confidence interval) *predict* whether or not she will be successful in college—that is, whether her college grade points will be above a stated level. Naturally, a correct decision on the part of the admission officer cannot be expected on every occasion, but to maximize the number of correct decision would be highly desirable.

In this chapter two questions concerning dependent random variables will be discussed: (1) What is the most suitable form of mathematical equation (model) to predict average values of a random variable from another related variable? (2) What measure to use for the degree of association between the two related variables? The first is presented under the title of *regression*, the second under *correlation*.

9.1 LINEAR REGRESSION

Let us consider the following concrete example next. In a certain manufacturing process of an alloy it is desired to find if there exists any relationship between the percent of zinc content and the tensile strength in tons of the alloy. An experiment with a small sample size yields the following result:

Zinc Content of Alloy, percent	Tensile Strength of Alloy, tons
4.7	1.2
4.8	1.4
4.9	1.5
5.0	1.5
5.1	1.7

This information is plotted in Fig. 9.1.

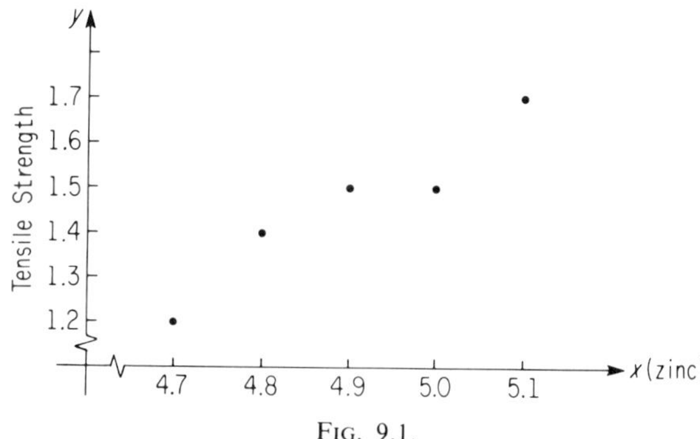

FIG. 9.1.

Figure 9.1 is called a *scatter diagram* or simply a *scattergram*. One might note that there exists a general trend of direct variation between the zinc content and the tensile strength of the alloy. However, the model or the mathematical equation that describes this relationship is not easy to determine and often an assumption has to be made about its nature. Examining the above scattergram someone might suggest that $f(x) = a + bx$, a line, describes the underlying relationship. Another person might suggest $g(x) = a + bx + cx^2 + dx^3$, a cubic, describes the relationship and still another person might suggest the use of two different lines, such as

$$f_1(x) = a_1 + b_1x \quad \text{if} \quad 4.7 < x < 4.9$$
$$f_2(x) = a_2 + b_2x \quad \text{if} \quad x > 4.9$$

to represent specific portions of the interval x (zinc content). Each suggestion might be valid under different situations but we shall consider the linear case in detail only. Clearly this notion can be extended to nonlinear expressions.

Assuming that the relationship between the tensile strength and the zinc content is linear, we confront the problem of finding the coefficients a and b in $f(x) = a + bx$ which describe the relationship with minimum error. However, what minimum error is, and what criteria should be included in the definition of the minimum error, are questions whose answers also must be stated as assumptions.

The notion of minimum distances from the line and the scatter points—points on the scattergram—is a criterion commonly included in the definition of minimum error. Another criterion usually included is the requirement of unbiased estimator. The method of least squares to be discussed next meets these conditions and will be utilized in the derivation of the formulas for the coefficients a and b.

Suppose Fig. 9.2 represents the linear relationship between the tensile strength and zinc content. We note that among several choices there are

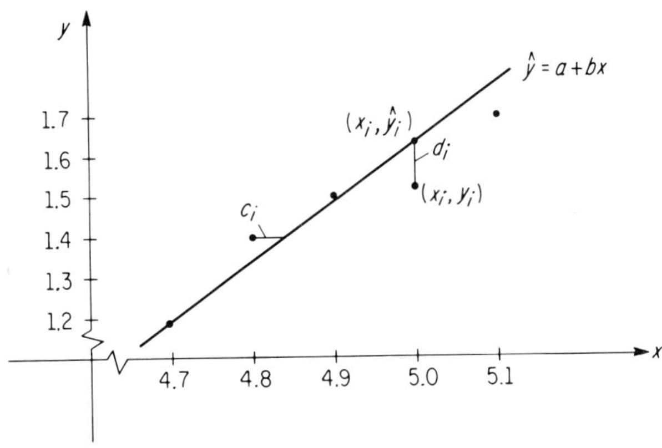

FIG. 9.2.

two different types of distances that can be considered for minimization criterion—distances measured perpendicular to the x axis and distances measured perpendicular to the y axis. In Fig. 9.2 these are labeled d_i and c_i respectively. If the sums of squares of the d_i's are minimized, $y = a + bx$ is called the *regression line of y on x*, where both a and b are unknown parameters of the regression line. However, if the sums of squares of the c_i's are minimized, $x = a_1 + b_1 y$ is called the *regression line of x on y*, where similarly a_1 and b_1 are unknown parameters. The two regression lines are related but in general are not the same. We shall discuss the regression line of y on x and obtain the expression for the regression line of x on y by interchanging variables.

The following assumptions will be made in our discussion of the regression line of y on x.

1. The true regression line is linear.
2. The values of y for any given x are independent and normally distributed with mean $\alpha + \beta x$ and variance σ_e^2.

These assumptions are graphically shown in Fig. 9.3.

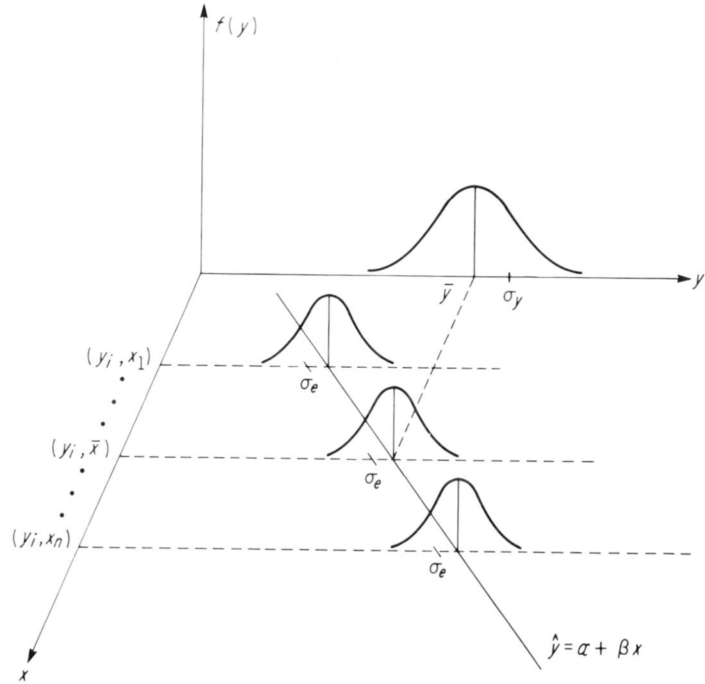

FIG. 9.3. Underlying assumptions for a regression line of y on x.

The method of least squares requires that the quantity $\sum_{i=1}^{n} d_i^2$ be a minimum, where n represents the total number of paired observations. When $D = \sum_{i=1}^{n} d_i^2$ is differentiated partially with respect to α and β and set equal to zero, one obtains a pair of linear equations whose unique solutions will be defined as "best" estimates of α and β. Now,

$$D = \sum_{i=1}^{n} d_i^2 = \sum_{i=1}^{n} (y_i - \hat{y}_i)^2 = \sum_{i=1}^{n} [y_i - (\alpha + \beta x_i)]^2 \qquad (9.1.1)$$

The partial derivatives of D with respect to α and β yield

$$\frac{\partial D}{\partial \alpha} = -2 \sum_{i=1}^{n} [y_i - \alpha - \beta x_i] \tag{9.1.2}$$

$$\frac{\partial D}{\partial \beta} = -2 \sum_{i=1}^{n} [y_i - \alpha - \beta x_i] x_i \tag{9.1.3}$$

which when set equal to zero can be expressed equivalently as

$$\sum_{i=1}^{n} y_i = \sum_{i=1}^{n} \alpha + \sum_{i=1}^{n} \beta x_i = n\alpha + \beta \sum_{i=1}^{n} x_i \tag{9.1.4}$$

$$\sum_{i=1}^{n} x_i y_i = \sum_{i=1}^{n} \alpha x_i + \sum_{i=1}^{n} \beta x_i^2 = \alpha \sum_{i=1}^{n} x_i + \beta \sum_{i=1}^{n} x_i^2 \tag{9.1.5}$$

Equations (9.1.4) and (9.1.5) are called the *normal equations* of linear regression.

The solution for β from the normal equation equals to

$$\beta = \frac{\begin{vmatrix} n & \sum_{i=1}^{n} y_i \\ \sum_{i=1}^{n} x_i & \sum_{i=1}^{n} x_i y_x \end{vmatrix}}{\begin{vmatrix} n & \sum_{i=1}^{n} x_i \\ \sum_{i=1}^{n} x_i & \sum_{i=1}^{n} x_i^2 \end{vmatrix}} = \frac{n \sum_{i=1}^{n} x_i y_i - \left(\sum_{i=1}^{n} x_i\right)\left(\sum_{i=1}^{n} y_i\right)}{n \sum_{i=1}^{n} x_i^2 - \left(\sum_{i=1}^{n} x_i\right)^2} \tag{9.1.6}$$

Now, for a given sample of k paired observations Eq. (9.1.6) will yield an estimate of β, say b, which will be unique unless all x_i's are equal. That is,

$$b = \frac{k \sum_{i=1}^{k} x_i y_i - \left(\sum_{i=1}^{k} x_i\right)\left(\sum_{i=1}^{k} y_i\right)}{k \sum_{i=1}^{k} x_i^2 - \left(\sum_{i=1}^{k} x_i\right)^2}$$

$$= \frac{\sum_{i=1}^{k} x_i y_i - k\bar{x}\bar{y}}{\sum_{i=1}^{k} x_i^2 - k\bar{x}^2} = \frac{\sum_{i=1}^{k} (x_i - \bar{x})(y_i - \bar{y})}{\sum_{i=1}^{k} (x_i - \bar{x})^2} \tag{9.1.7}$$

$$b = \frac{\dfrac{1}{(k-1)} \sum_{i=1}^{k} (x_i - \bar{x})(y_i - \bar{y})}{\dfrac{1}{(k-1)} \sum_{i=1}^{k} (x_i - \bar{x})^2} \tag{9.1.8}$$

Clearly, the denominator of Eq. (9.1.8) represents the sample variance for the r.v. x. The numerator of Eq. (9.1.8) is defined as the *sample covariance* of the random variables x and y and designated by the symbol s_{xy}. Therefore

$$b = \frac{s_{xy}}{s_x^2} \tag{9.1.9}$$

In words, b, the slope of the sample regression line, is equal to the ratio of sample covariance divided by the sample variance of x. It can be shown that b is an unbiased estimate of β.

Similarly, the solution for α from the normal equations will yield for a given sample

$$a = \frac{\begin{vmatrix} \sum_{i=1}^{k} y_i & \sum_{i=1}^{k} x_i \\ \sum_{i=1}^{k} x_i y_i & \sum_{i=1}^{k} x_i^2 \end{vmatrix}}{\begin{vmatrix} n & \sum_{i=1}^{k} x_i \\ \sum_{i=1}^{k} x_i & \sum_{i=1}^{k} x_i^2 \end{vmatrix}} \tag{9.1.10}$$

However, from Eq. (9.1.4) we deduce that

$$\bar{y} = a + b\bar{x}$$

and equivalently,

$$a = \bar{y} - b\bar{x} \tag{9.1.11}$$

Substituting the values of the estimates a and b into the regression equation, we obtain

$$\hat{y} = (\bar{y} - b\bar{x}) + bx = \bar{y} + b(x - \bar{x}) \tag{9.1.12}$$

which is equivalent to

$$\hat{y} = \bar{y} + \frac{s_{xy}}{s_x^2}(x - \bar{x}) \tag{9.1.13}$$

This expression is used for predicting y for a given x and sometimes referred to as the *line of prediction for y*.

Example 9.1.1 Find the equation of the regression line of y on x for the data of Fig. 9.1. Predict the tensile strength for $x = 5.05$.

Solution:

x_i	y_i	x_i^2	y_i^2	$x_i y_i$
4.7	1.2	22.09	1.44	5.64
4.8	1.4	23.04	1.96	6.72
4.9	1.5	24.01	2.25	7.35
5.0	1.5	25.00	2.25	7.50
5.1	1.7	26.01	2.89	8.67
24.5	7.3	120.15	10.79	35.88

From Eq. (9.1.6) we obtain

$$b = \frac{5(35.88) - (24.5)(7.3)}{5(120.15) - (24.5)^2} = 1.10$$

From Eq. (9.1.8),

$$a = \frac{7.3}{5} - 1.10\left(\frac{24.5}{5}\right) = -3.93$$

Hence $\hat{y} = -3.93 + 1.10x$ is the required line. If $x = 5.05$, $y_e = -3.93 + 1.10(5.05) = 1.62$ tons, where y_e is read "y estimated."

The regression line of x on y or the line of prediction for x is obtained by interchanging the variables x and y. That is,

$$\hat{x} = \bar{x} + b_1(y - \bar{y}) \tag{9.1.14}$$

where

$$b_1 = \frac{s_{yx}}{s_y^2} = \frac{s_{xy}}{s_y^2} \tag{9.1.15}$$

The intercept a_1 for the regression line of x on y is

$$a_1 = \bar{x} - b_1\bar{y} \tag{9.1.16}$$

PROBLEMS

9.1.1 Given the following data

x:	2	5	8	11	14
y:	3	6	10	11	15

(a) Plot the scattergram.

(b) Calculate $\bar{x}$ and $\bar{y}$ and plot this point on the scattergram also.

(c) Without calculating either the slope or the intercept of the regression line, draw the line of minimum error by sight.

(d) In part (c) what is the approximate value of the slope of the line? The intercept?

(e) Calculate the sample covariance for this data.

(f) Find the equation of the regression line of y on x and compare your result with part (d).

9.1.2 In Prob. 9.1.1 find (a) the equation of the regression line of x on y. (*Hint:* Interchange the random variables x and y.) (b) The geometric mean of the slopes of the two regression lines.

9.1.3 Repeat Prob. 9.1.1 for the data

x:	12	13	14	15	16	17	18	19	20
y:	9	14	12	13	18	16	15	17	24

9.1.4 Heights and weights of college students are assumed to be linearly related. In a random sample of 25 students the heights and weights were measured to the nearest inch and pound, respectively and the following data observed:

Height, in.	Weight, lb	Height, in.	Weight, lb
62	105	71	225
68	147	67	160
63	110	66	194
65	160	71	184
67	175	66	155
70	190	70	175
68	164	64	128
68	134	67	183
73	200	68	188
67	147	65	139
65	157	71	196
67	150	68	152
67	170		

(a) Find the regression line to predict weights from the heights.

(b) Predict the weight of a student who is 70 inches tall.

9.1.5 Suppose $\hat{y} = \alpha x^{\beta}$ where $\alpha > 0$ and β is an arbitrary constant. The transformation $\log \hat{y}$ yields

$$\log \hat{y} = \log \alpha + \beta \log x$$

which indicates that the logarithm of y and x are linearly related. Using this transformation and the method of least squares, find the estimates of α and β from the following data:

y:	.3	1.2	2.5	5.0	7.5
x:	1.1	1.9	3.0	4.1	4.9

9.1.6 Suppose $y = \alpha + \beta x + \gamma x^2$ and the y's corresponding to each x are independent and normally distributed. Using the least-squares approach, determine the estimates of α, β, and γ.

9.1.7 Suppose $y = \alpha + \beta_1 x_1 + \beta_2 x_2$, and the y's corresponding to different x_1's and x_2's are independent. Using the least-squares approach, determine the estimates of α, β_1, and β_2.

9.2 PEARSON PRODUCT MOMENT CORRELATION COEFFICIENT

In the previous section the use of the regression lines as predictors were discussed. A related topic describing the degree of association between the two variables will be considered here—namely, the concept of correlation coefficient.

The geometric mean of the slopes of the two regression lines is defined as the *correlation coefficient* between the two variables and is usually denoted by the Greek letter ρ. In symbols, $\rho = \sqrt{\beta\beta_1}$. Since b and b_1 are estimates of β and β_1,

$$r = \sqrt{b\,b_1} = \frac{s_{xy}}{s_x s_y} \tag{9.2.1}$$

represents an estimate of ρ. ρ and r defined in this fashion are referred to as *Pearson's product moment correlation coefficient* for bivariate populations and samples respectively. We shall refer to these merely by correlation coefficient.

For computational purposes Eq. (9.2.1) may be equivalently expressed as

$$r = \frac{k\sum_{i=1}^{k} x_i y_i - \sum_{i=1}^{k} x_i \sum_{i=1}^{k} y_i}{\sqrt{\left[k\sum_{i=1}^{k} x_i^2 - \left(\sum_{i=1}^{k} x_i\right)^2\right]\left[k\sum_{i=1}^{k} y_i^2 - \left(\sum_{i=1}^{k} y_i\right)^2\right]}} \tag{9.2.2a}$$

$$= \frac{\sum_{i=1}^{k} (x_i - \bar{x})(y_i - \bar{y})}{\sqrt{\sum_{i=1}^{k} (x_i - \bar{x})^2 \sum_{i=1}^{k} (y_i - \bar{y})^2}} \tag{9.2.2b}$$

The correlation coefficient is a pure number and is used as a measure of the degree of association or relationship between x and y and is always less than or equal to one in absolute value. This we show next.

If the expression in Eq. (9.1.1) is divided by $(k - 1)$, it yields

$$\frac{\sum_{i=1}^{k} d_i^2}{k - 1} = \frac{\sum_{i=1}^{k} (y_i - \hat{y}_i)^2}{k - 1} \tag{9.2.3}$$

Substituting the relation $\hat{y} = \bar{y} + b(x - \bar{x})$ in the last equation, one gets

$$\frac{\sum_{i=1}^{k} d_i^2}{k - 1} = \frac{\sum_{i=1}^{k} [(y_i - \bar{y}) - b(x - \bar{x})]^2}{k - 1} \tag{9.2.4a}$$

$$= s_y^2 - 2bs_{xy} + b^2 s_x^2 \tag{9.2.4b}$$

However, since $b = \dfrac{s_{xy}}{x_x^2}$, Eq. (9.2.4$b$) becomes

$$\frac{\sum_{i=1}^{k} d_i^2}{k - 1} = s_y^2 - \frac{2(s_{xy})^2}{s_x^2} + \left(\frac{s_{xy}}{s_x^2}\right)^2 s_x^2 \tag{9.2.5a}$$

$$= s_y^2 - \left(\frac{s_{xy}}{s_x}\right)^2 \tag{9.2.5b}$$

$$= s_y^2 \left[1 - \left(\frac{s_{xy}}{s_x s_y}\right)^2\right] \tag{9.2.5c}$$

$$= s_y^2 (1 - r^2) \tag{9.2.5d}$$

Because $\sum_{i=1}^{k} d_i^2$ and s_y^2 are nonnegative quantities, it follows that

$$1 - r^2 \geq 0 \tag{9.2.6}$$

or equivalently,

$$|r| \leq 1 \tag{9.2.7}$$

which is the desired relationship. Now, if we substitute for r in Eq. (9.2.7) the expression given by Eq. (9.2.1), the result becomes

$$\left|\frac{s_{xy}}{s_x s_y}\right| \leq 1 \tag{9.2.8a}$$

or equivalently,

$$|s_{xy}| \leq s_x s_y \tag{9.2.8b}$$

which states that the magnitude of the covariance is less than or equal to the product of the respective standard deviations.

We shall define the quantity $\dfrac{\sum_{i=1}^{k} d_i^2}{k-1} = s_e^2$ and hence Eq. (9.2.5d) becomes

$$s_e^2 = s_y^2(1 - r^2) \tag{9.2.9}$$

Taking the square root of both sides in the last equation, we obtain

$$s_e = s_y \sqrt{1 - r^2} \tag{9.2.10}$$

The value s_e is called the *standard error of estimate*, which is basically a measure of the degree of accuracy in predicting one variable from the other.

Equation (9.2.7) in conjunction with Eq. (9.2.10) suggests important interrelationships, and we shall consider the three cases where $r = \pm 1$ and $r = 0$.

In the case where $r = \pm 1$. one gets

$$s_e = s_y \sqrt{1 - 1} = 0$$

What does this indicate? Since

$$s_e^2 = \frac{\sum_{i=1}^{k} d_i^2}{k-1}$$

by definition, $\sum_{i=1}^{k} d_i^2$ must be equal to zero, which in turn implies that all d_i^2 are zero. That is, $(y_i - \hat{y}_i)^2 = 0$. In words, all values of y are located on the regression line $\hat{y} = \alpha + \beta x$ and σ_e in Fig. 9.3 is equal to zero. Figures 9.4a and 9.4b represent data for which $r = 1$ and $r = -1$ respectively.

In the case where $r = 0$, one gets $s_y^2 = s_e^2$. What are the implications of this relationship? Referring to the definitions, we note that

$$s_e^2 = \frac{\sum_{i=1}^{k} d_i^2}{k-1} = \frac{\sum_{i=1}^{k} (y_i - \bar{y})^2}{k-1} = s_y^2$$

However, since $d_i = y_i - \hat{y}_i$, we get

$$\sum_{i=1}^{k} (y_i - \hat{y}_i)^2 = \sum_{i=1}^{k} (y_i - \bar{y})^2 \tag{9.2.11}$$

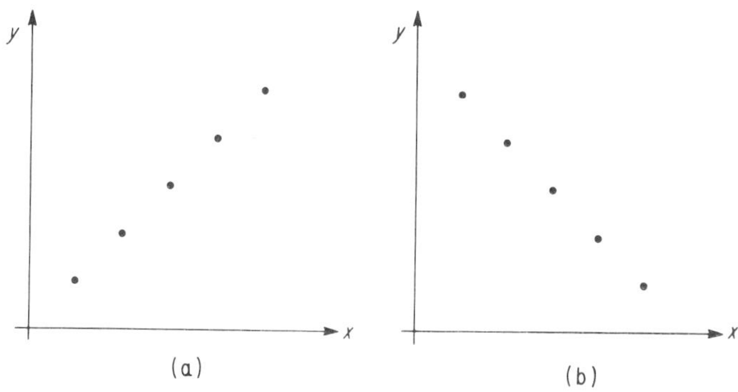

FIG. 9.4. (a) Data for $r = 1$. (b) Data for $r = -1$.

which implies that $\hat{y}_i = \bar{y}$. Now, referring to Eq. (9.1.12) we observe that the quantity $b(x - \bar{x})$ must be equal to zero in order to have $\hat{y}_i = \bar{y}$. However, if all x are not equal, $b(x - \bar{x}) = 0$ implies $b = 0$. That is, the slope of the regression line is zero and hence Fig. 9.3 can be represented as in Fig. 9.5. Clearly, the variance of the y's becomes identical with the variance of the errors. Figure 9.6 represents a scattergram for $r = 0$.

The choice between the regression lines of y on x and x on y is often a matter of convenience. For instance, one might just as well predict heights from weights instead of weights from heights. A question that

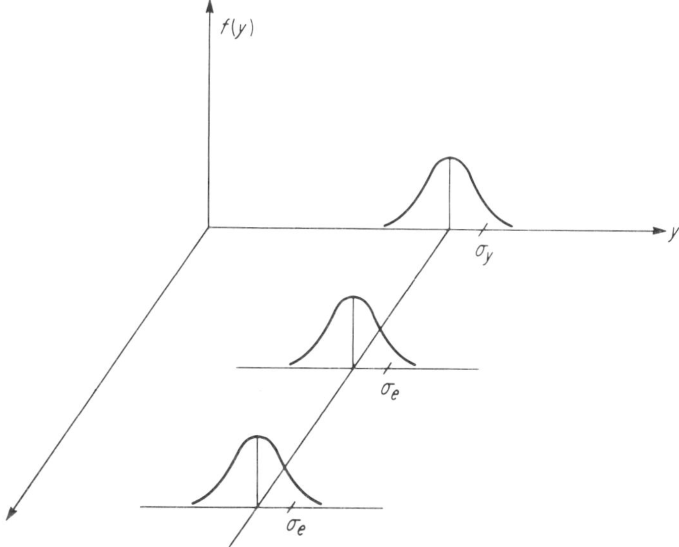

FIG. 9.5.

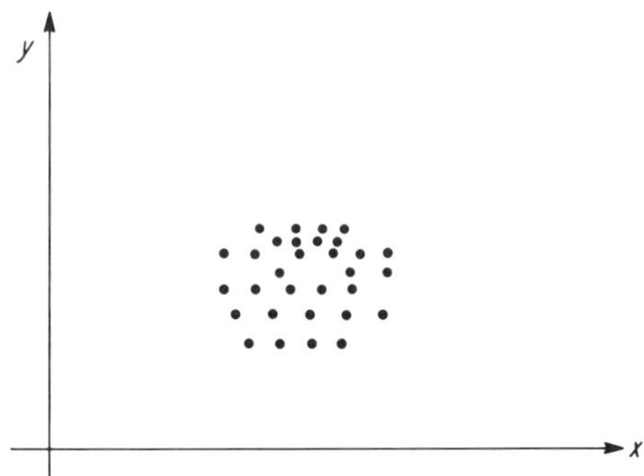

FIG. 9.6. Data for $r = 0$.

might be of interest to ask at this juncture is: Under what conditions are the two regression lines identical? That is, do the two regression lines ever coincide? In order for any two lines to coincide we recall from analytic geometry that the respective slopes and intercepts of the lines must be equal. If $y = \alpha + \beta x$ and $x = \alpha_1 + \beta_1 y$ are the two regression lines, and if we solve the first for x, we obtain $x = (y - \alpha)/\beta$. Setting the slopes equal yields $\beta_1 = 1/\beta$ or $\beta \cdot \beta_1 = 1$. Since $\beta \cdot \beta_1 = r^2$ by definition, $r^2 = 1$. This implies that the two regression lines are identical if all the points of a given data lie on a straight line. One obtains the same result when the intercepts are set equal.

In many of the topics presented previously the assumption of independence was made repeatedly. This assumption implied that the correlation coefficient between the variables was zero; i.e., if two variables are independent there is no correlation or no covariation. The converse of this assumption is not necessarily true. When $r = 0$, the variables are not necessarily independent, and in fact can be highly dependent. For example, suppose the random variables x and y are related by the formula $y = x^2$. The set of points on this particular parabola, for $|x| < 1$, have a correlation coefficient that is nearly zero but clearly the variables are related. $r = 0$ merely indicates that the random variables are not *linearly* related. Therefore if two variables are uncorrelated, $r = 0$, it is possible for the variables to be related and be dependent.

Example 9.2.1 Calculate the correlation coefficient for the data of Example 9.1.1.

Solution: Substituting in Eq. (9.2.2*a*) the values obtained in Example 9.1.1, we get

$$r = \frac{5(35.88) - (24.5)(7.3)}{\sqrt{[5(120.15) - (24.5)^2][5(10.79) - (7.3)^2]}}$$

$$= .96$$

This indicates a high degree of positive correlation.

A theorem concerning the correlation coefficient, which is of practical usefulness, will be presented next.

Theorem 9.2.1 The correlation coefficient is independent of origin and unit of measurements.

Proof: Consider the transformations

$$z = \frac{x - x_0}{k_0}$$

and

$$w = \frac{y - y_1}{k_1}$$

or equivalently

$$x = x_0 + zk_0$$

and

$$y = y_1 + k_1 w$$

for two random variables x and y where x_0, y_1, k_0 and k_1 are arbitrary constants. Recalling that

$$\bar{x} = x_0 + k_0\bar{z}$$
$$\bar{y} = y_1 + k_1\bar{w}$$

and

$$s_x = k_0 s_z,$$
$$s_y = k_1 s_w$$

one obtains

$$\frac{x - \bar{x}}{s_x} = \frac{k_0(z - \bar{z})}{k_0 s_z} = \frac{z - \bar{z}}{s_z} \tag{9.2.12}$$

and

$$\frac{y - \bar{y}}{s_y} = \frac{k_1(w - \bar{w})}{k_1 s_w} = \frac{w - \bar{w}}{s_w} \tag{9.2.13}$$

Therefore

$$\frac{(x_i - \bar{x})(y_i - \bar{y})}{s_x s_y} = \frac{(z_i - \bar{z})(w_i - \bar{w})}{s_z s_w} \tag{9.2.14}$$

which implies

$$\frac{s_{xy}}{s_x s_y} = \frac{s_{zw}}{s_z s_w} = r \tag{9.2.15}$$

Example 9.2.2 Calculate r for the data of Example 9.1.1 using the transformation

$$z = x - 4.7 \quad \text{and} \quad w = y - 1.2$$

Solution:

z_i	w_i	z_i^2	w_i^2	$z_i w_i$
0	0	0	0	0
.1	.2	.01	.04	.02
.2	.3	.04	.09	.06
.3	.3	.09	.09	.09
.4	.5	.16	.25	.20
1.0	1.3	.30	.47	.37

$$r = \frac{5\,(.37) - (1.0)(1.3)}{\sqrt{[5\,(.30) - (1.0)^2][5\,(.47) - (1.3)^2]}}$$

$$= \frac{1.85 - 1.30}{\sqrt{(.5)(.66)}} = \frac{.55}{\sqrt{.33}}$$

$$= \frac{.55}{.574} = .96$$

which agrees with the result of Example 9.2.1.

PROBLEMS

9.2.1 Find the correlation coefficient for the data of Prob. 9.1.3.

9.2.2 Find the standard error of estimate for the data in Prob. 9.2.1.

9.2.3 Review Prob. 9.1.2 and interpret the result of part (b) in the light of this section.

9.2.4 Using the transformation $H - 66 = h$ and $W - 150 = w$, find the correlation coefficient for the data of Prob. 9.1.4.

9.2.5 For the data of Prob. 9.1.4, find the regression line to predict the heights from the weights. (*Hint*: Use the results of Probs. 9.1.4 and 9.2.4.)

9.2.6 Show that for a bivariate population

$$\sigma_{x \pm y}^2 = \sigma_x^2 \pm 2\rho\sigma_x\sigma_y + \sigma_y^2$$

9.3 SPEARMAN'S RANK-ORDER CORRELATION COEFFICIENT

Suppose in the computation of r in the previous section the values of x were ranked according to magnitude without altering the pairs of observations. For instance, suppose (x_i, y_i) represents the height and weight of the ith individual in your statistics class. Now, ordering the heights in descending order will yield

$$(x_1, y_1), (x_2, y_2), \ldots, (x_k, y_k)$$

such that

$$x_1 \geq x_2 \geq \cdots \geq x_k$$

if there are ties, and

$$x_1 > x_2 > \cdots > x_k$$

if there are no ties. If we assume the latter situation and furthermore assume that all y_i's are different, we could express the ranked data as $R(x_1) = 1$, $R(x_2) = 2$, $\cdots$, $R(x_k) = k$. Clearly, there is associated with each $R(x_i)$ a corresponding $R(y_i)$. It is the correlation coefficient between $R(x_i)$ and $R(y_i)$ (heights and weights) that we wish to calculate. Specifically, we shall derive an expression for

$$r_s = \frac{S_{R(x)R(y)}}{S_{R(x)}S_{R(y)}} \tag{9.3.1}$$

and refer to it as *Spearman's rank-order correlation coefficient*, or simply Spearman's formula.

Letting $d_i = R(x_i) - R(y_i)$, we obtain for the sums of squares

$$\sum_{i=1}^{k} d_i^2 = \sum_{i=1}^{k} [R(x_i)]^2 + \sum_{i=1}^{k} [R(y_i)]^2 - 2\sum_{i=1}^{k} [R(x_i)][R(y_i)] \tag{9.3.2}$$

or equivalently

$$\sum_{i=1}^{k} [R(x_i)][R(y_i)] = \frac{1}{2}\left[\sum_{i=1}^{k} [R(x_i)]^2 + \sum_{i=1}^{k} [R(y_i)]^2 - \sum_{i=1}^{k} d_i^2\right] \tag{9.3.3}$$

Because of the assumptions on $R(x_i)$ and $R(y_i)$, we have

$$\sum_{i=1}^{k} R(x_i) = \sum_{i=1}^{k} R(y_i) = \sum_{i=1}^{k} i = \frac{k(k+1)}{2} \tag{9.3.4}$$

and

$$\sum_{i=1}^{k} [R(x_i)]^2 = \sum_{i=1}^{k} [R(y_i)]^2 = \sum_{i=1}^{k} i^2 = \frac{k(k+1)(2k+1)}{6} \tag{9.3.5}$$

Now the computational formula of Eq. (9.3.1) according to Eq. (9.2.2*a*) is

$$r = \frac{k\,\Sigma\, R(x_i)\, R(y_i) - \Sigma\, R(x_i)\,\Sigma\, R(y_i)}{\sqrt{[k\,\Sigma\, R(x_i)^2 - (\Sigma\, R(x_i))^2][k\,\Sigma\, R(y_i)^2 - (\Sigma\, R(y_i))^2]}} \tag{9.3.6}$$

Substituting Eqs. (9.3.3), (9.3.4), and (9.3.5) in the last expression yields after simplifications

$$r_s = 1 - \frac{6\sum_{i=1}^{k} d_i^2}{k(k^2 - 1)} \tag{9.3.7}$$

which is Spearman's formula.

Spearman's formula and Pearson's formula, Eq. (9.2.2*a*), yield equivalent results if there are no ties in the ranks. If however, as the case often is, there are ties in the ranks, the two formulas are not equivalent. This is expected because the underlying assumptions are not satisfied. If there are ties in the ranks and Spearman's formula is used, the average of the ranks is assigned to the tied ranks.

Example 9.3.1 Calculate r_s for the data of Example 9.1.1.
Solution:

$R(x_i)$	$R(y_i)$	d_i
1	1	0
2	2.5	−.5
3	2.5	.5
4	4	0
5	5	0

$$\sum_{i=1}^{5} d_i^2 = (.25) + (.25) = .50$$

$$r_s = 1 - \frac{6\,(.5)}{5\,(25 - 1)}$$

$$= 1 - \frac{1}{40} = \frac{39}{40} = .975$$

The closeness of the results in Example 9.2.1 and 9.3.1 should be noticed.

PROBLEMS

9.3.1 Verify Eq. (9.3.4). That is, show that

$$\sum_{i=1}^{k} i = \frac{k(k+1)}{2}$$

9.3.2 Verify Eq. (9.3.5). That is,

$$\sum_{i=1}^{k} i^2 = \frac{k(k+1)(2k+1)}{6}$$

9.3.3 Using the results of Probs. 9.3.1 and 9.3.2, verify Eq. (9.3.7).

9.3.4 Find the correlation coefficient for the data in Prob. 9.1.3, using Spearman's formula.

9.4 THE SAMPLING DISTRIBUTION OF THE COEFFICIENTS OF REGRESSION LINE

As in the case of sampling distributions of the mean and the variance, one needs to know the sampling distribution of the statistics a and b, the coefficients of the regression line $\hat{y} = a + bx$, in order to make inferences concerning the parameters α and β. If the distribution of a and b were known, one could test hypotheses concerning the population regression line and write confidence intervals around the true values of α and β. These distributions contain parameters which are functions of σ_e^2 and hence we need to obtain an estimate for σ_e^2.

Following similar steps as in the proof of $E[s_x^2] = \sigma_x^2$ (see Sec. 7.5), it can be shown that

$$E[s_e^2] = \frac{k-2}{k-1}\sigma_e^2 \tag{9.4.1}$$

where k represents the sample size of paired observations. Therefore, the quantity

$$\frac{k-1}{k-2}s_e^2 \tag{9.4.2}$$

represents an unbiased estimate of σ_e^2. Since $s_e^2 = s_y^2(1 - r^2)$, the expression

$$\frac{k-1}{k-2}s_y^2(1-r^2) \tag{9.4.3}$$

is also an unbiased estimate of σ_e^2.

Now it can be shown that the random variable b, the slope of the sample regression line, is normally distributed with mean β and variance $\sigma_e^2/(k-1)s_x^2$. Therefore, the random variable

$$z = \frac{b-\beta}{\sqrt{\dfrac{\sigma_e^2}{(k-1)s_x^2}}} = \frac{\sqrt{k-1}\,s_x}{\sigma_e}(b-\beta) \tag{9.4.4}$$

is unit normal. Substituting for σ_e^2 its unbiased estimate, $\dfrac{k-1}{k-2}s_e^2$, Eq. (9.4.4) reduces to

$$t = \frac{\sqrt{k-2}\,s_x}{s_e}(b-\beta) \tag{9.4.5}$$

which is the Student's t distributed with $k - 2$ d.f.

Substituting $s_e = s_y\sqrt{1-r^2}$ in Eq. (9.4.5) we get

$$t = \sqrt{\frac{k-2}{1-r^2}}\,\frac{s_x}{s_y}(b-\beta) \tag{9.4.6}$$

Recalling that $b = \dfrac{s_{xy}}{s_x^2}$ and $r = \dfrac{s_{xy}}{s_x s_y}$, we note that $\dfrac{s_x}{s_y} = \dfrac{r}{b}$ and therefore

$$t = \sqrt{\frac{k-2}{1-r^2}}\,\frac{r}{b}(b-\beta) \tag{9.4.7}$$

is equivalent to both Eqs. (9.4.5) and (9.4.6). Using either of the last three expressions, we can make a probability statement regarding β. That is, a confidence interval around the parameter β could be obtained. From

Eq. (9.4.5) we deduce

$$P\left[-t_{\alpha/2,k-2} < \frac{\sqrt{k-2}}{s_e} s_x(b-\beta) < t_{\alpha/2,k-2}\right] = 1-\alpha \qquad (9.4.8)$$

which after simplification reduces to

$$P\left[b - t_{\alpha/2,k-2}\frac{s_e}{s_x\sqrt{k-2}} < \beta < b + t_{\alpha/2,k-2}\frac{s_e}{s_x\sqrt{k-2}}\right] = 1-\alpha \qquad (9.4.9)$$

Similarly, from Eq. (9.4.7) we derive

$$P\left[b - t_{\alpha/2,k-2}\frac{b}{r}\sqrt{\frac{1-r^2}{k-2}} < \beta < b + t_{\alpha/2,k-2}\frac{b}{r}\sqrt{\frac{1-r^2}{k-2}}\right] = 1-\alpha \qquad (9.4.10)$$

which is equivalent to

$$P\left[b\left(1 - t_{\alpha/2,k-2}\frac{1}{r}\sqrt{\frac{1-r^2}{k-2}}\right) < \beta < b\left(1 + t_{\alpha/2,k-2}\frac{1}{r}\sqrt{\frac{1-r^2}{k-2}}\right)\right] = 1-\alpha \qquad (9.4.11)$$

Let us illustrate these concepts with specific examples.

Example 9.4.1 Using the results of Example 9.1.1 and 9.2.1, write a 95 percent confidence interval for the slope of the population regression line.

Solution: In this illustration $k = 5$, $b = 1.10$ and $r = .96$. From Table V, Appendix, we find $t_{.025,3} = 3.182$ and hence

$$\frac{3.182}{.96}\sqrt{\frac{1-(.96)^2}{3}} = \frac{3.182}{.96}\frac{\sqrt{.0782}}{\sqrt{3}}$$

$$= \frac{3.182}{.96}\frac{(.281)}{1.73} = \frac{.894}{1.66} = .54$$

Therefore, $[1.10(.46) < \beta < 1.10(1.54)] = [.51 < \beta < 1.69]$ is the 95 percent confidence interval.

Example 9.4.2 For a random sample of 11 paired observations b was computed to be equal to 0.16 and r equal to 0.6. Test the hypothesis that $\beta = .48$.

Solution:

H_0: $\beta = .48$.

H_1: $\beta \neq .48$.

$\alpha = .05$.

Test statistic: Eq. (9.4.7) with 9 d.f.
Critical region: $|t| > t_{.025,9} = 2.262$.

Calculations: $t = \sqrt{\dfrac{11 - 2}{1 - (.6)^2}}\,\dfrac{.6}{.16}\,(.16 - .48)$

$= \sqrt{\dfrac{9}{.64}}\,\dfrac{.6}{.16}\,(-.32) = \dfrac{3}{.8}\,(.6)\,2 = \dfrac{3.6}{.8} = -4.5.$

Reject H_0.

The sampling distribution of the random variable a, similar to that of b, can be shown to obey the normal p.f. with mean α and variance

$$\sigma_\alpha^2 = \left[\frac{1}{k} + \frac{\bar{x}^2}{(k-1)\,s_x^2}\right]\sigma_e^2 \qquad (9.4.12)$$

$$= \left[\frac{(k-1)\,s_x^2 + k\bar{x}^2}{k\,(k-1)\,s_x^2}\right]\sigma_e^2 \qquad (9.4.13)$$

That is, the random variable

$$z = \frac{a - \alpha}{\sigma_e\sqrt{\dfrac{1}{k} + \dfrac{\bar{x}^2}{(k-1)\,s_x^2}}} \qquad (9.4.14)$$

is unit normal. Now substituting the unbiased estimate of σ_e in Eq. (9.4.14) we obtain the random variable

$$t = \frac{a - \alpha}{s_e\sqrt{\dfrac{k-1}{k-2}\left[\dfrac{1}{k} + \dfrac{\bar{x}^2}{(k-1)\,s_x^2}\right]}} \qquad (9.4.15)$$

which is the Student's t distributed with $(k - 2)$ d.f. This last expression is equivalent to

$$t = \frac{a - \alpha}{s_y\sqrt{1 - r^2}\sqrt{\dfrac{k-1}{k-2}\left[\dfrac{1}{k} + \dfrac{\bar{x}^2}{(k-1)\,s_x^2}\right]}} \qquad (9.4.16)$$

Equations (9.4.15) and (9.4.16) could be used to test hypotheses concerning the parameter α, the intercept of the population regression line. We leave to the reader as an exercise to find the expression for $(1 - \alpha)$ 100 percent confidence interval around α. (See Prob. 9.4.1.) There should be no confusion between α, the significance level of a test, and the parameter α, the intercept of the population regression line.

PROBLEMS

9.4.1 Using Eq. (9.4.16) write an expression for $(1 - \alpha)$ 100 percent confidence interval around the intercept of the population regression line.

9.4.2 Using the result of Prob. 9.4.1, write a 95 percent confidence interval around the intercept of the population regression line of the data in Prob. 9.1.1.

9.4.3 Write a 95 percent confidence interval around the slope of the population regression line of the data in Prob. 9.1.1.

9.4.4 Repeat Prob. 9.4.2 for the data of Prob. 9.1.4.

9.4.5 Repeat Prob. 9.4.3 for the data of Prob. 9.1.4.

9.5 THE SAMPLING DISTRIBUTION OF THE CORRELATION COEFFICIENT

In some practical applications one might be particularly interested in the degree of association between two random variables and wish to make an inference concerning the population correlation coefficient ρ. Unlike the sampling distributions discussed in previous sections, the sampling distribution of the correlation coefficient r is relatively complex and therefore we shall consider a special case and present an approximate distribution.

In Sec. 9.2 the correlation coefficient was defined to be equal to the geometric mean of the slopes of the two regression lines (i.e., y on x and x on y). That is, $\rho = \sqrt{\beta\beta_1}$. Obviously ρ will be equal to zero if either β or β_1 is equal to zero. Since $\rho = 0$ implies that the two random variables are not correlated, one could reach the same conclusion if β is equal to zero. In other words, if β is equal to zero, we conclude that the two random variables are not correlated. Now substituting $\beta = 0$ in Eq. (9.4.7) we obtain

$$t = \sqrt{\frac{k-2}{1-r^2}}\, r \tag{9.5.1}$$

which obeys, as seen before, the Student's t distribution with $k - 2$ degrees of freedom.

Example 9.5.1 A random sample of size 18 from a bivariate normal distribution yields a correlation coefficient of .30, test the hypothesis that these samples come from uncorrelated populations.

Solution: We wish to test the hypothesis that $\rho = 0$.

H_0: $\rho = 0$.
H_1: $\rho \neq 0$.
$\alpha = .05$.

Test statistic: Eq. (9.5.1).
Critical region: $|t| > t_{.025,16}$.

Calculations:

$$t = \sqrt{\frac{18-2}{1-(.3)^2}}\,(.30) = \frac{4\,(.30)}{\sqrt{.91}} = \frac{1.2}{.95} = 1.26.$$

Therefore, we accept the null hypothesis.

Example 9.5.2 In Example 9.5.1 for what values of r will the hypothesis $\rho = 0$ be accepted?

Solution: The expression

$$t_{.025,16} = \sqrt{\frac{18 - 2}{1 - r^2}}\, r = 2.12$$

when solved for r yields

$$\frac{16r^2}{1 - r^2} = (2.12)^2$$

$$16r^2 = (2.12)^2 - (2.12)^2 r^2$$

$$[16 + (2.12)^2]\, r^2 = (2.12)^2$$

$$r^2 = \frac{2.12^2}{16 + 2.12^2}$$

$$r = \frac{2.12}{\sqrt{16 + 4.49}} = \frac{2.12}{\sqrt{20.49}}$$

$$= \frac{2.12}{4.52} = .468$$

Hence for $|r| < .468$ the hypothesis $\rho = 0$ will be accepted.

As mentioned priorly, the sampling distribution of r from correlated normal populations is complex and in general skewed. However, it has been shown that the random variable ζ, obtained by the transformation

$$\zeta = \frac{1}{2} \ln \frac{1 + r}{1 - r} \tag{9.5.2}$$

approximately obeys the normal p.f. with mean* $\mu_\zeta = \frac{1}{2} \ln \frac{1 + \rho}{1 - \rho}$ and variance $\sigma_\zeta^2 = \frac{1}{k - 3}$ where k is the number of paired observations in the sample. That is,

$$z = \frac{\zeta - \mu_\zeta}{\sigma_\zeta} \tag{9.5.3}$$

Table IX in the Appendix represents the tabulated values for ζ with increments of .01 of r. For instance, if $r = .05$, $\zeta = .0500$ and if $r = .78$, $\zeta = 1.0454$.

Table X in the Appendix represents the tabulated values of $r = \frac{e^{2\zeta} - 1}{e^{2\zeta} + 1}$ which is the inverse of the transformation given in Eq. (9.5.2).

*More accurately, $\mu_\zeta = \frac{1}{2} \ln \frac{1 + \rho}{1 - \rho} + \frac{\rho}{2(k - 1)}$.

For example, if $\zeta = 1.04$ $r = .77788$ and if $\zeta = 2.50$ $r = .98661$. In order to test hypotheses concerning ρ, the values obtained from Table IX suffice. However, if a confidence interval around the population correlation coefficient ρ is required, Table X is needed. Let us illustrate the concepts under discussion.

Example 9.5.3 If a random sample of size 28 from a bivariate normal population yields $r = .86$, (a) test the hypothesis that the sample is taken from correlated populations with $\rho = .95$. (b) write a 95 percent confidence interval around ρ.

Solution:

(a) H_0: $= .95$.
H_1: $\neq .95$.
$\alpha = .05$.

Test statistic: Eq. (9.5.3).

Critical region: $|z| > 1.96$.

Now from Table IX we get $\zeta = 1.2933$ and $\mu_\zeta = 1.8318$,

$$\sigma_\zeta = \frac{1}{\sqrt{28 - 3}} = \frac{1}{5}$$

which yield

$$z = \frac{1.2933 - 1.8318}{\frac{1}{5}} = -2.6925.$$

Therefore reject H_0.

(b) Recalling that

$$P[\zeta - z_{\alpha/2}\,\sigma_\zeta < \mu_\zeta < \zeta - z_{\alpha/2}\,\sigma_\zeta] = 1 - \alpha$$

we have for the 95 percent confidence interval

$$[1.2933 - 1.96\,\frac{1}{5} < \mu_\zeta < 1.2933 + 1.96\,\frac{1}{5}]$$
$$= [.9013 < \mu_\zeta < 1.6853]$$

From Table X we find

$$r = .71629 \text{ for } \zeta = .90$$

and

$$r = .93414 \text{ for } \zeta = 1.69$$

Therefore, the 95 percent confidence interval is

$$[.71629 < \rho < .93414]$$

However, since our computation of r is significant to two decimal places, we write

$$[.72 < \rho < .93]$$

PROBLEMS

9.5.1 Using the information in Prob. 9.2.1, write a 95 percent confidence interval around the population correlation coefficient.

9.5.2 Using the information in Prob. 9.2.4, test the hypothesis that the population correlation coefficient $\rho = .60$. Let $\alpha = 0.01$.

9.5.3 Using the information in Examples 9.1.1 and 9.2.1, test the hypothesis that the population correlation coefficient $\rho = .90$. Let $\alpha = .05$.

10

Design of Experiments and Analysis of Variance

10.0 INTRODUCTION

In Sec. 8.4 we discussed different methods to test the equality of the means of two independent populations. Suppose one has occasion to test hypotheses concerning the equality of means of three populations. For instance, if we have three types of light bulbs of equal wattage and wish to test whether or not the average life of each kind is the same, the only available method is to test each of the following hypotheses separately.

1. H_0: $\mu_1 = \mu_2$
   H_1: $\mu_1 \neq \mu_2$
2. H_0: $\mu_1 = \mu_3$
   H_1: $\mu_1 \neq \mu_3$
3. H_0: $\mu_2 = \mu_3$
   H_1: $\mu_2 \neq \mu_3$

If, however, instead of 3 types of light bulbs one has 10 different types and wishes to test the hypothesis of equality of means, there are $\binom{10}{2} = 45$ different hypotheses that need to be tested by the methods of Chapter 8. In general, if there are n populations, one needs to test $\binom{n}{2}$ separate hypotheses. A question that might be asked at this point is whether one can test these hypotheses simultaneously. That is, can the hypothesis

$$H_0. \mu_1 = \mu_2 = \cdots = \mu_n \tag{10.0.1}$$

against the alternative

$$H_1: \mu_i \neq \mu_j \qquad \text{for at least one } i \neq j$$

be tested with a single test? In this chapter we shall answer this question.

Usually, it is during an experiment that one needs to compare the means of more than two populations. For this reason the concepts of *design* of experiments are commonly presented in this context. The design or the planning of the experiment should always proceed the analysis.

The outline below represents the sequence of steps one must follow in planning an experiment.

A. The experiment
   1. Statement of the problem
   2. Choice of the dependent variable(s) to be studied
   3. Selection of the factors, quantitative or qualitative, to be varied in the experiment
   4. Choice of levels of each factor (values for quantitative factors, and states for qualitative factors) and whether these levels are set at fixed values or chosen at random or combinations of fixed and random, namely mixed

B. The design (mathematical model)
   1. Choosing the mathematical model which describes the experiment
   2. The number of observations to be taken
   3. The order in which the experiment is to be performed
   4. The method of randomization to be used

C. The analysis
   1. Analysis of variance
   2. Test of significance and components of variance
   3. Interpretation of the analysis

10.1 EXPERIMENTS

In Sec. 1.1 an experiment was defined as a repetitive process that results in anyone of a number of possible outcomes such that the particular outcome is determined by chance and is impossible to predict *a priori*. In this definition it was tacitly assumed that the conditions of the experiment did not change or alter from trial to trial. That is, a repetitive process was assumed to mean "repeated under identical conditions." However, in a more general sense the purpose of any experiment is to introduce changes in the experimental conditions to determine the effects of such changes upon the previously obtained results. For example, consider the experiment where the breakdown voltages at room temperature of certain types of transistors are being determined. If instead of room temperature (20 degrees centigrade) the experiment is carried out under a different temperature (say 75 degrees centigrade), a change is thereby introduced into the experimental conditions of the experiment, so that the experimenter must determine the effects of this change upon the previously obtained results.

The observations obtained under one set of experimental conditions are called the *experimental unit*. If temperature and the breakdown voltages were the only two criteria under consideration in the above example, room temperature and breakdown voltage represent one experimental unit, while 75 degrees centigrade and breakdown voltage represent another experimental unit.

Experimental conditions are usually grouped according to certain common criteria or *factors*. Referring to the aforementioned illustration again, temperature and humidity represent two among many possible factors of the experiment. The different subgroupings or categories within a given factor are called *levels*. 25, 50, and 75 degrees centrigrade and 30, 45, 60, and 95 percent relative humidities are examples of levels of temperature and relative humidity.

A *treatment* is defined as a combination of conditions in a particular experimental unit. For instance, 25 degrees centigrade, 80 percent humidity, with an applied voltage of 100 millivolts constitute a treatment, and under identical conditions an applied voltage of 200 millivolts represents another treatment.

We have indicated that the objective of an experiment is to introduce changes in the experimental conditions to discover effects such changes bring about. The effects themselves are seldom of importance but rather the *differences* between the effects. Hence, an experiment alternatively may be defined as a process which contrasts the effects of two or more treatments.

In order to attribute the differences between effects to something, more than one pair of observations are needed. Suppose that Don and Ron, two identical twins, are selected for an experiment in problem solving. Don is given a booklet to read on how to solve a certain type of mathematical problem and Ron is taught the same material by a tutor. The next day when identical questions are given to the boys Don solves his problems in 15 minutes while Ron finishes in only three minutes. Is the difference in time of solving problems attributable to the treatments (reading method and tutor method), or to the natural variability between the two boys, or partly to each? We cannot answer this question without further information or assumptions. If we could assume that the boys were really identical then the difference is attributable completely to the treatments, because there would be no natural variability between them. However, the assumption that the boys are really identical is not justifiable. Therefore at least two and preferably more pairs of observations or *replications* are needed.

Suppose the same treatments are applied to two other pairs of identical twins and the following data obtained.

Booklet Reading		Tutor Teaching	
Don	15 minutes	Ron	3 minutes
Mel	18 minutes	Hal	4 minutes
Jean	20 minutes	Joan	5 minutes

Now it should be intuitively evident that the difference in the effects could be attributed to the treatments and not to natural variability. That is 12, 14, and 15 minutes, the differences between the pairs of identical twins, represent data which we use to reject the hypothesis that the average difference in the treatment effects is zero. (See Sec. 8.3 and Eq. 8.3.17.) The greater the variability among the pairs of replications, the larger is the number of pairs required to average or balance out the random differences. Hence replication leaves in the clear the treatment effects which constitutes the basis for inference.

The failure of two identically treated experimental units to yield identical results is called *experimental error*. Experimental errors fall into two categories, systematic and random. Systematic errors are due to consistent bias. For instance, in the above illustration Don, Mel, and Jean who were assigned to the reading treatment might have been the firstborns of the three pairs of twins. This is a consistent bias and should be avoided. In order to accomplish this objective, one member of every pair of replications should be selected at random to receive the reading treatment and the other member the second treatment. This process is called *randomization*. Randomization makes the individual errors stochastically independent of one another, which is one of the required assumptions of the analysis.

Experimental errors usually tend to be correlated if the experimental units are adjacent in time or space. Randomization, without destroying the correlational pattern, gives any two treatments an equal chance of being adjacent or nonadjacent, and hence the correlation between any two treatments tends to cancel out with increased replication. Replications provide an estimate of experimental errors and could be used as a test statistic concerning the true difference between treatment effects. On the other hand, randomization makes the test statistic valid. Therefore both replication and randomization are necessary requirements in the design of an experiment.

Another concept needs to be defined at this point. Any process or method that reduces the experimental error is called *local control*. Grouping into homogenous units and refining the experimental techniques are examples of local control. Local control may be considered as the design proper of the experiment. For instance, suppose we wish to compare the results of two methods of teaching. If there are 28 students, several ways exist in which 14 students can be assigned to the two methods. One approach is to assign 14 students at random to method A and the remaining to method B. After instructing the students with the respective methods for a certain period of time, identical examinations are given to the 28 students and the result of the examinations represents the outcome of the experiment. This method of design is commonly referred to as the *completely randomized design*. Another approach to this illustration is to order the 28 students into 14 groups of 2 students each (14 pairs) according to some criterion, and assign each member in a pair randomly to method A or method B. This method of assignment is an example of *randomized block* design. These two designs will be discussed in subsequent sections of this chapter and other designs in Chapters 11 and 12.

PROBLEMS

10.1.1 Discuss the differences between an experimental unit and the treatment in a given experiment.

10.1.2 Suppose the effects of five different makes of drills are to be studied. Three holes of each are drilled on four different types of metals at 45- and 60-degree angles (inclinations) with 100, 200, 400, and 800 rpm. In this situation

(a) What is the experimental unit?
(b) What are the factors?
(c) What are the levels of each factor?
(d) What are the treatments?

10.1.3 Give several examples of systematic and random errors.

10.2 ONE-WAY CLASSIFICATION

(a) Completely Randomized Design—No Restrictions in Any Direction. Suppose an experiment is composed of J treatments with a total of N observations and each treatment being replicated n_j times. Hence

$$N = \sum_{j=1}^{J} n_j$$

Let x_{ij} denote the ith observation for the jth treatment where $i = 1, 2, \ldots, n_j$ and $j = 1, 2, \ldots, J$. Then Table 10.2.1, the one-way classification table, represents the data in an orderly fashion. The dot notation introduced in Sec. 1.5 is used for convenience. That is,

$$T_{\cdot j} = \sum_{i=1}^{n_j} x_{ij} = \text{total for the } j\text{th column}$$

$$\bar{x}_{\cdot j} = \frac{T_{\cdot j}}{n_j} = \text{mean for } j\text{th column}$$

$$T_{\cdot\cdot} = \sum_{j=1}^{J} T_{\cdot j} = \sum_{j=1}^{J} \sum_{i=1}^{n_j} x_{ij} = \text{grand total of the sample}$$

$$\bar{x}_{\cdot\cdot} = \frac{T_{\cdot\cdot}}{N} = \text{grand mean of the sample}$$

$$s_j^2 = \frac{\sum_{i=1}^{n_j} (x_{ij} - \bar{x}_{\cdot j})^2}{n_j - 1} = \text{variance for } j\text{th column}$$

$$S^2 = \frac{\sum_{j=1}^{J} \sum_{i=1}^{n_j} (x_{ij} - \bar{x}_{..})^2}{N - 1} = \text{variance of the total sample}$$

TABLE 10.2.1
ONE-WAY CLASSIFICATION

$i \backslash j$		Treatments							
		1	2	3	$\cdots$	j	$\cdots$	J	
Observations	1	x_{11}	x_{12}	x_{13}	$\cdots$	x_{1j}	$\cdots$	x_{1J}	
	2	x_{21}	x_{22}	x_{23}	$\cdots$	x_{2j}	$\cdots$	x_{2J}	
	3	x_{31}	x_{32}	x_{33}	$\cdots$	x_{3j}	$\cdots$	x_{3J}	
	$\vdots$				$\vdots$		$\vdots$		
	n_j	$x_{n_1 1}$	$x_{n_2 2}$	$x_{n_3 3}$	$\cdots$	$x_{n_j j}$	$\cdots$	$x_{n_J J}$	
Sample size..................		n_1	n_2	n_3	$\cdots$	n_j	$\cdots$	n_J	N
Treatment totals		$T_{.1}$	$T_{.2}$	$T_{.3}$	$\cdots$	$T_{.j}$	$\cdots$	$T_{.J}$	$T_{..}$
Treatment means		$\bar{x}_{.1}$	$\bar{x}_{.2}$	$\bar{x}_{.3}$	$\cdots$	$\bar{x}_{.j}$	$\cdots$	$\bar{x}_{.J}$	$\bar{x}_{..}$
Treatment variances		s_1^2	s_2^2	s_3^2	$\cdots$	s_j^2	$\cdots$	s_J^2	S^2
Population means..........		μ_1	μ_2	μ_3	$\cdots$	μ_j	$\cdots$	μ_J	μ
Population variances		σ_1^2	σ_2^2	σ_3^2	$\cdots$	σ_j^2	$\cdots$	σ_J^2	σ^2

Let us illustrate the notation of Table 10.2.1 with a numerical example. The observations represent the lives in hours of three types of light bulbs with differing filaments.

TABLE 10.2.2

	Treatment (Type of Filaments)			Composite
	Sample 1	Sample 2	Sample 3	
Observations in hours	8 6 10	2 4 5 1	9 7 7 8 5 6	
Sample size	3	4	6	$N = 13$
Treatment totals..............	24	12	42	$T_{..} = 78$
Treatment means.............	8	3	7	$\bar{x}_{..} = 6$
Treatment variances..........	$s_1^2 = 4$	$s_2^2 = \frac{10}{3}$	$s_3^2 = 2$	$S^2 = \frac{41}{6}$
Population means	μ_1	μ_2	μ_3	μ
Population variances.........	σ_1^2	σ_2^2	σ_3^2	σ^2

The hypothesis to be tested in this illustration is

$$H_0: \quad \mu_1 = \mu_2 = \mu_3$$

against the alternative

$$H_1: \quad \text{at least two means are unequal}$$

and in general, the hypothesis to be tested is as in Eq. (10.0.1).

In the derivation of a test statistic for testing hypotheses of aforementioned type, certain assumptions have to be made regarding choosing samples from a population, the parameters of the population and the distribution of the underlying random variable. The following assumptions are made in this case.

1. All observations are random samples from different populations.
2. Each population is normally distributed with equal variances and is stochastically independent of the others.

These assumptions are equivalent to the statement that the random variables x_j are independent and normally distributed with mean μ_j and variance σ^2.

Next we need to consider the mathematical model which describes the experiment. Let us note that

$$x_{ij} = \mu + (\mu_j - \mu) + (x_{ij} - \mu_j) \tag{10.2.1}$$

is an identity, where

$$\mu = \frac{\sum_{j=1}^{J} n_j \mu_j}{N}$$

and

$$N = \sum_{j=1}^{J} n_j$$

Setting

$$\mu_j - \mu = \tau_j$$

and

$$x_{ij} - \mu_j = \epsilon_{ij}$$

reduces Eq. (10.2.1) to

$$x_{ij} = \mu + \tau_j + \epsilon_{ij} \tag{10.2.2}$$

Now if the hypothesis

$$H_0: \quad \mu_1 = \mu_2 \cdots = \mu_J$$

is true then the hypothesis

$$H_0: \quad \mu_1 = \mu_2 \cdots = \mu_J = \mu$$

is also true and hence H_0 is equivalent to the hypothesis

$$H_0: \quad \tau_j = 0, \qquad j = 1, 2, \ldots, J$$

The τ_j's are called the *differential* effects of the treatments and correspond to the deviation of the jth population mean from the mean of all J populations means. The quantity ϵ_{ij} is considered to be a random error associated with the ith observation in the jth treatment. Sometimes μ is referred to as the *common effect* for the entire experiment, but more generally it is called the *grand mean*. To obtain an intuitive feeling and insight for Eq. (10.2.2) and what is meant by common and differential effects, let us consider the following illustration. Suppose the observations in Table 10.2.2 were

Treatment (Type of Filaments)		
Sample 1	Sample 2	Sample 3
6	6	6
6	6	6
6	6	6
	6	6
		6
		6

representing no variability or errors within all populations, then the model given by Eq. (10.2.2) reduces to

$$x_{ij} = \mu = 6$$

since $\tau_j = 0$ as well as $\epsilon_{ij} = 0$ for all i and j. Therefore, all observations are due to a common effect. Next suppose that the treatments are different, and there is a different treatment effect in each sample but no variability or errors within the populations (type of filaments). If we suppose that the treatment effects are $+2$, -3, and $+1$ in the samples respectively, then the above samples will reduce to

Treatment (Type of Filaments)		
Sample 1	Sample 2	Sample 3
$6 + 2 = 8$	$6 - 3 = 3$	$6 + 1 = 7$
$6 + 2 = 8$	$6 - 3 = 3$	$6 + 1 = 7$
$6 + 2 = 8$	$6 - 3 = 3$	$6 + 1 = 7$
	$6 - 3 = 3$	$6 + 1 = 7$
		$6 + 1 = 7$
		$6 + 1 = 7$

and our model can be expressed as $x_{ij} = \mu + \tau_j$ where $\tau_1 = 2$, $\tau_2 = -3$, $\tau_3 = 1$ and all $\epsilon_{ij} = 0$. Note that $\Sigma \tau_j = 0$. However, in general we expect to have some variability in any population and hence some error within each sample. Consequently, the following sample values represent a more realistic situation.

Treatment (Type of Filaments)		
Sample 1	Sample 2	Sample 3
$8 - 0 = 8$	$3 - 1 = 2$	$7 + 2 = 9$
$8 - 2 = 6$	$3 + 1 = 4$	$7 + 0 = 7$
$8 + 2 = 10$	$3 + 2 = 5$	$7 + 0 = 7$
	$3 - 2 = 1$	$7 + 1 = 8$
		$7 - 2 = 5$
		$7 - 1 = 6$

where the added or subtracted quantities correspond to the random error and our model becomes

$$x_{ij} = \mu + \tau_j + \epsilon_{ij}$$

which is Eq. (10.2.2) and the sample values are those given in Table 10.2.2. Therefore the model under consideration states that any observation x_{ij} is comprised of two constants (effects), μ the common effect and τ_j the differential or treatment effect plus ϵ_{ij}, a random error. Let us remark that the mean of the ϵ_{ij}'s is zero and the variance σ^2.

Equations (10.2.1) or (10.2.2) represents the assumed mathematical model for the completely randomized design. In this design it is further assumed that J is a fixed number, and any inference made should apply to these J populations only. This is called the *fixed-effects model.* Had we chosen these J populations from a larger set of populations, say P, then the model would be referred to as *random-effects model.*

In two-way, three-way and higher-order classifications combinations of fixed-effects and random-effects models are considered. Such models are referred to as *mixed-effects models*. Some authors refer to the fixed, random, and mixed-effects models as Model I, Model II, and Model III respectively.

Now that the model has been described, we are ready to consider the analysis of the observations. However, prior to doing that let us comment about the number of observations to be taken. As a general rule, whenever possible equal number of observations should be taken from each population under consideration. One advantage in taking equal sample sizes is that the power of the test is maximized. Another advantage is that if the null hypothesis is rejected, other test statistics are available which can be used only if the sample sizes are equal. Other theoretical considerations as well as ease in computational techniques make it desirable to follow this rule.

(b) The Analysis of the Design. The mathematical model given by Eq. (10.2.1) can also be expressed as

$$\begin{aligned} x_{ij} - \mu &= (\mu_j - \mu) + (x_{ij} - \mu_j) \qquad (10.2.3) \\ &= \tau_j + \epsilon_{ij} \end{aligned}$$

The corresponding sample identity for Eq. (10.2.3) is

$$(x_{ij} - \bar{x}_{..}) = (\bar{x}_{.j} - \bar{x}_{..}) + (x_{ij} - \bar{x}_{.j}) \qquad (10.2.4)$$

Considering the N observations as a single sample of size N, we get

$$s^2 = \frac{\sum_{j=1}^{J}\sum_{i=1}^{n_j}(x_{ij} - \bar{x}_{..})^2}{N - 1}$$

or equivalently

$$(N - 1)s^2 = \sum_{j=1}^{J}\sum_{i=1}^{n_j}(x_{ij} - \bar{x}_{..})^2 \qquad (10.2.5)$$

Substituting Eq. (10.2.4) in the last expression yields

$$\begin{aligned}(N - 1)s^2 &= \sum_{j=1}^{J}\sum_{i=1}^{n_j}[(\bar{x}_{.j} - \bar{x}_{..}) + (x_{ij} - \bar{x}_{.j})]^2 \\ &= \sum_{j=1}^{J}\sum_{i=1}^{n_j}(\bar{x}_{.j} - \bar{x}_{..})^2 + 2\sum_{j=1}^{J}\sum_{i=1}^{n_j}(\bar{x}_{.j} - \bar{x}_{..})(x_{ij} - \bar{x}_{.j}) \\ &\quad + \sum_{j=1}^{J}\sum_{i=1}^{n_j}(x_{ij} - \bar{x}_{.j})^2\end{aligned}$$

But since

$$\sum_{j=1}^{J}\sum_{i=1}^{n_j}(\bar{x}_{.j} - \bar{x}_{..})(x_{ij} - \bar{x}_{.j}) = 0$$

(see Prob. 10.2.1)

$$(N - 1)s^2 = \sum_{j=1}^{J}\sum_{i=1}^{n_j}(\bar{x}_{.j} - \bar{x}_{..})^2 + \sum_{j=1}^{J}\sum_{i=1}^{n_j}(x_{ij} - \bar{x}_{.j})^2 \qquad (10.2.6)$$

Equation (10.2.6) can be further simplified and reduce to

$$(N - 1)s^2 = \sum_{j=1}^{J} n_j(\bar{x}_{.j} - \bar{x}_{..})^2 + \sum_{j=1}^{J}\sum_{i=1}^{n_j}(x_{ij} - \bar{x}_{.j})^2 \qquad (10.2.7)$$

If we let

$$SS = \sum_{j=1}^{J}\sum_{i=1}^{n_j}(x_{ij} - \bar{x}_{..})^2 \qquad (10.2.8)$$

$$SS_j = \sum_{j=1}^{J}\sum_{i=1}^{n_j}(\bar{x}_{.j} - \bar{x}_{..})^2 \qquad (10.2.9)$$

$$SS_{i(j)} = \sum_{j=1}^{J}\sum_{i=1}^{n_j}(x_{ij} - \bar{x}_{.j})^2 \qquad (10.2.10)$$

Equation (10.2.7) becomes

$$SS = SS_j + SS_{i(j)} \tag{10.2.11}$$

The symbol SS is chosen primarily because of mnemonic considerations and is read "sums of squares."

Let us observe that SS is equal to $(N - 1)s^2$, which is a measure of overall variation of the N observations. A large value of SS indicates a large variance (range, spread, dispersion from the mean) in the total N observations and a small value of SS implies a small variance in the total N observations. Furthermore, Eq. (10.2.11) represents an expression for the SS being partitioned into two parts, SS_j and $SS_{i(j)}$. Let us note that except for the weighting factor n_j,

$$SS_j = \sum_{j=1}^{J} n_j (\bar{x}_{.j} - \bar{x}_{..})^2$$

corresponds to the numerator of the variance for the column means, which reflects the variability of the J sample means. If the J means are nearly alike, then SS_j is small and if the means are very much different, then SS_j is large. Also,

$$\begin{aligned} SS_{i(j)} &= \sum_{j=1}^{J} \sum_{i=1}^{n_j} (x_{ij} - \bar{x}_{.j})^2 \\ &= \sum_{j=1}^{J} (n_j - 1) s_j^2 \end{aligned} \tag{10.2.12}$$

which reflects a measure of variance within the individual samples. Therefore, SS_j is the weighted sum of squares *between* (among) treatment means and $SS_{i(j)}$ is the error (residual) sums of squares *within* treatments.

In Sec. 7.5 it was shown that the sample variance is an unbiased estimate of the population variance. That is, $E[s^2] = \sigma^2$. Now

$$\begin{aligned} E[SS_{i(j)}] &= E\left[\sum_{j=1}^{J} (n_j - 1) s_j^2\right] \\ &= \sum_{j=1}^{J} (n_j - 1) E[s_j^2] \\ &= \sum_{j=1}^{J} (n_j - 1) \sigma^2 \\ &= (N - J) \sigma^2 \end{aligned} \tag{10.2.13}$$

or equivalently

$$E\left[\frac{SS_{i(j)}}{N - J}\right] = \sigma^2 \tag{10.2.14}$$

Therefore, the quantity $\dfrac{SS_{i(j)}}{N - J}$ represents an unbiased estimate of σ^2, the population variance.

Next, we wish to show that

$$E\left[\frac{SS_j}{J - 1}\right] = \sigma^2 + \frac{\sum_{j=1}^{J} n_j \tau_j^2}{J - 1} \qquad (10.2.15)$$

Consider the expectation of SS_j:

$$E[SS_j] = E\left[\sum_{j=1}^{J} n_j(\bar{x}_{.j} - \bar{x}_{..})^2\right] = \sum_{j=1}^{J} n_j E[\bar{x}_{.j} - \bar{x}_{..}]^2 \qquad (10.2.16)$$

Letting $z = \bar{x}_{.j} - \bar{x}_{..}$, we get

$$\begin{aligned} E[z] &= E[\bar{x}_{.j}] - E[\bar{x}_{..}] \\ &= \mu_j - \frac{1}{N} E\left[\sum_{j=1}^{J} n_j \bar{x}_{.j}\right] \\ &= \mu_j - \frac{1}{N} \sum_{j=1}^{J} n_j \mu_j \end{aligned}$$

But $\mu_j = \mu + \tau_j$ by definition, and hence

$$\begin{aligned} E[z] &= \mu + \tau_j - \frac{1}{N} \sum_{j=1}^{J} n_j(\mu + \tau_j) \\ &= \mu + \tau_j - \mu - \frac{1}{N} \sum_{j=1}^{J} n_j \tau_j \\ &= \tau_j \end{aligned} \qquad (10.2.17)$$

since $\sum_{j=1}^{J} n_j \tau_j = 0$.

Recalling that $\sigma_z^2 = E[z^2] - (E[z])^2$ [see Eq. (3.2.14)], we get

$$\sigma_z^2 = E[z^2] - \tau_j^2 \qquad (10.2.18)$$

or equivalently

$$E[z^2] = \sigma_z^2 + \tau_j^2 \qquad (10.2.19)$$

However, by definition $\sigma_z^2 = E[(z - \tau_j)^2]$ and hence

$$\sigma_z^2 = E[(\bar{x}_{.j} - \bar{x}_{..} - \tau_j)^2] \qquad (10.2.20)$$

Now,

$$\begin{aligned} \bar{x}_{..} &= \frac{1}{N} \sum_{k=1}^{J} n_k \bar{x}_{.k} \\ &= \frac{1}{N}\left[\sum_{k=1}^{j-1} n_k \bar{x}_{.k} + n_j \bar{x}_{.j} + \sum_{k=j+1}^{J} n_k \bar{x}_{.k}\right] \end{aligned} \qquad (10.2.21)$$

and therefore

$$z = \bar{x}_{\cdot j} - \bar{x}_{\cdot\cdot} = \bar{x}_{\cdot j} - \frac{1}{N}\left[\sum_{k=1}^{j-1} n_k \bar{x}_{\cdot k} + n_j \bar{x}_{\cdot j} + \sum_{k=j+1}^{J} n_k \bar{x}_{\cdot k}\right]$$

$$= \left(1 - \frac{n_j}{N}\right)\bar{x}_{\cdot j} - \frac{1}{N}\left[\sum_{k=1}^{j-1} n_k \bar{x}_{\cdot k} + \sum_{k=j+1}^{J} n_k \bar{x}_{\cdot k}\right] \qquad (10.2.22)$$

Applying Theorem 7.1.1 to Eq. (10.2.22) and recalling from Theorem 7.1.2 that $\sigma^2_{\bar{x}\cdot j} = \frac{\sigma^2}{n_j}$, we obtain

$$\sigma_z^2 = \left(1 - \frac{n_j}{N}\right)^2 \frac{\sigma^2}{n_j} + \sum_{k=1}^{j-1}\left(\frac{n_k}{N}\right)^2 \frac{\sigma^2}{n_k} + \sum_{k=j+1}^{J}\left(\frac{n_k}{N}\right)^2 \frac{\sigma^2}{n_k}$$

$$= \frac{\sigma^2}{N^2}\left[\frac{(N - n_j)^2}{n_j} + \sum_{k=1}^{j-1} n_k + \sum_{k=j+1}^{J} n_k\right]$$

$$= \frac{\sigma^2}{N^2}\left[\frac{(N - n_j)^2}{n_j} + N - n_j\right]$$

$$= \sigma^2\left[\frac{1}{n_j} - \frac{1}{N}\right] \qquad (10.2.23)$$

Substituting Eq. (10.2.23) in Eq. (10.2.19) we get

$$E[z^2] = \sigma^2\left[\frac{1}{n_j} - \frac{1}{N}\right] + \tau_j^2 \qquad (10.2.24)$$

Next, substituting Eq. (10.2.24) in Eq. (10.2.16)

$$E[SS_j] = \sum_{j=1}^{J} n_j\left\{\sigma^2\left[\frac{1}{n_j} - \frac{1}{N}\right] + \tau_j^2\right\}$$

$$= \sigma^2 \sum_{j=1}^{J}\left[1 - \frac{n_j}{N}\right] + \sum_{j=1}^{J} n_j \tau_j^2$$

$$= \sigma^2(J - 1) + \sum_{j=1}^{J} n_j \tau_j^2 \qquad (10.2.25)$$

which when divided by $J - 1$ reduces to Eq. (10.2.15). Therefore, if $\mu_1 = \mu_2 = \cdots = \mu_j = \mu$; that is, the null hypothesis stated by Eq. (10.0.1) is true, then $\sum_{j=1}^{J} n_j \tau_j^2$ will be equal to zero and $\frac{SS_j}{J - 1}$ represents an unbiased estimate of σ^2. Applying Eq. (7.3.18) to the results of Eqs. (10.2.14) and (10.2.15) we observe that the random variable

$$F = \frac{\dfrac{SS_j}{J - 1}}{\dfrac{SS_{i(j)}}{N - J}} \qquad (10.2.26)$$

obeys the F p.f. with $(J - 1)$ and $(N - J)$ degrees of freedom associated with numerator and denominator respectively, and represents the statistic to test the null hypothesis given in Eq. (10.0.1). Clearly, a large value of F obtained from Eq. (10.2.26) will make us question the validity of H_0 and the hypothesis of equal means will be rejected if

$$F > F_{\alpha: J-1, N-J}$$

Commonly, the sums of squares when divided by their corresponding degrees of freedom are referred to as *mean squares*. Since SS_j represents the sums of squares between or among the J treatments, $\dfrac{SS_j}{J-1}$ is called *mean square between* and designated by MS_B. Similarly $\dfrac{SS_{i(j)}}{N-J}$ is called *mean square within* and denoted by MS_W.

The result of the analysis of the data is tabulated in Table 10.2.3 and will be referred to as the Anova table (abbreviation for analysis of variance).

TABLE 10.2.3
ANOVA TABLE FOR ONE-WAY EXPERIMENT

Source of Variation	Sums of Squares	Degrees of Freedom	Mean Square	Expected Mean Square	*F*-Ratio
Between treatments	$SS_j = \sum_{j=1}^{J} n_j(\bar{x}_{\cdot j} - \bar{x}_{\cdot\cdot})^2$	$J - 1$	$MS_B = \dfrac{SS_j}{J-1}$	$\sigma^2 + \dfrac{\sum_{j=1}^{J} n_j \tau_j^2}{J-1}$	$F = \dfrac{MS_B}{MS_W}$
Within treatments	$SS_{i(j)} = \sum_{j=1}^{J}\sum_{i=1}^{n_j} (x_{ij} - \bar{x}_{\cdot j})^2$	$N - J$	$MS_W = \dfrac{SS_{i(j)}}{N-J}$	σ^2	
Total	$SS = \sum_{j=1}^{J}\sum_{i=1}^{n_j} (x_{ij} - \bar{x}_{\cdot\cdot})^2$	$N - 1$			

Example 10.2.1 Test the hypothesis

H_0: $\mu_1 = \mu_2 = \mu_3$
H_1: At least two means are unequal
for the data of Table 10.2.2

Solution:

$$\begin{aligned} SS_j &= \sum_{j=1}^{3}\sum_{i=1}^{n_j} (\bar{x}_{\cdot j} - \bar{x}_{\cdot\cdot})^2 \\ &= 3(8 - 6)^2 + 4(3 - 6)^2 + 6(7 - 6)^2 \\ &= 3(4) + 4(9) + 6(1) = 12 + 36 + 6 = 54 \end{aligned}$$

$$
\begin{aligned}
SS_{i(j)} &= \sum_{j=1}^{3}\sum_{i=1}^{n_j}(x_{ij} - \bar{x}_{.j})^2 \\
&= (8 - 8)^2 + (6 - 8)^2 + (10 - 8)^2 + (2 - 3)^2 \\
&\quad + \cdots + (1 - 3)^2 + (9 - 7)^2 + \cdots + (6 - 7)^2 \\
&= 28 \\
J - 1 &= 3 - 1 = 2 \\
N - J &= 13 - 3 = 10 \\
MS_B &= \frac{54}{2} = 27 \\
MS_W &= \frac{28}{10} = 2.8
\end{aligned}
$$

Therefore,

$$F = \frac{27}{2.8} = 9.6^{**}$$

From Table VII in the Appendix we find

$$F_{.05;2,10} = 4.10$$
$$F_{.01;2,10} = 7.56$$

Since 9.6 is greater than both $F_{.05;2,10}$ and $F_{.01;2,10}$, the hypothesis of equality of means is rejected. There is statistical evidence to believe that the mean lives of these three types of electric bulbs are not equal.

It is a common practice to indicate the computed F value in the Anova table by an asterisk (*) if $F > F_{.05}$ and by a double asterisk if $F > F_{.01}$. Since in Example 10.2.1 $F > F_{.01}$, $F = 9.6^{**}$ is the manner in which the result is reported.

(c) Equivalent Formulas for Calculating Sums of Squares. In computing SS, SS_j, and $SS_{i(j)}$ in Example 10.2.1, the reader might have observed that all values for the data were integers, and futhermore the means were also integers. Such will not be the case in general, hence hand calculations will be rather cumbersome.

We shall next derive equivalent formulas for SS, SS_j, and $SS_{i(j)}$ which are simpler to use during computations with desk calculators. Now

$$
\begin{aligned}
SS &= \sum_{j=1}^{J}\sum_{i=1}^{n_j}(x_{ij} - \bar{x}_{..})^2 \\
&= \sum_{j=1}^{J}\sum_{i=1}^{n_j}(x_{ij}^2 - 2x_{ij}\bar{x}_{..} + \bar{x}_{..}^2) \\
&= \sum_{j=1}^{J}\sum_{i=1}^{n_j}x_{ij}^2 - 2\bar{x}_{..}\sum_{j=1}^{J}\sum_{i=1}^{n_j}x_{ij} + N\bar{x}_{..}^2
\end{aligned}
$$

But

$$\sum_{j=1}^{J}\sum_{i=1}^{n_j} x_{ij} = N\bar{x}_{..} = \frac{NT_{..}}{N} = T_{..}$$

Therefore

$$SS = \sum_{j=1}^{J}\sum_{i=1}^{n_j} x_{ij}^2 - N\bar{x}_{..}^2$$

$$= \sum_{j=1}^{J}\sum_{i=1}^{n_j} x_{ij}^2 - N\left(\frac{T_{..}}{N}\right)^2$$

$$= \sum_{j=1}^{J}\sum_{i=1}^{n_j} x_{ij}^2 - \frac{T_{..}^2}{N} \qquad (10.2.27)$$

Next let us consider SS_j:

$$SS_j = \sum_{j=1}^{J}\sum_{i=1}^{n_j} (\bar{x}_{.j} - \bar{x}_{..})^2$$

$$= \sum_{j=1}^{J} n_j(\bar{x}_{.j} - \bar{x}_{..})^2$$

$$= \sum_{j=1}^{J} n_j(\bar{x}_{.j}^2 - 2\bar{x}_{.j}\bar{x}_{..} + \bar{x}_{..}^2)$$

$$= \sum_{j=1}^{J} n_j\bar{x}_{.j}^2 - 2\bar{x}_{..}\sum_{j=1}^{J} n_j\bar{x}_{.j} + \sum_{j=1}^{J} n_j\bar{x}_{..}^2$$

$$= \sum_{j=1}^{J} \frac{n_j^2\bar{x}_{.j}^2}{n_j} - 2N\bar{x}_{..}^2 + N\bar{x}_{..}^2$$

$$= \sum_{j=1}^{J} \frac{T_{.j}^2}{n_j} - N\bar{x}_{..}^2$$

$$= \sum_{j=1}^{J} \frac{T_{.j}^2}{n_j} - \frac{T_{..}^2}{N} \qquad (10.2.28)$$

Since $SS = SS_j + SS_{i(j)}$,

$$SS_{i(j)} = \sum_{j=1}^{J}\sum_{i=1}^{n_j} x_{ij}^2 - \sum_{j=1}^{J}\frac{T_{.j}^2}{n_j} \qquad (10.2.29)$$

The quantity $\frac{T_{..}^2}{N}$ is referred to as the *correction for the mean.*

Example 10.2.2 Compute SS, SS_j and $SS_{i(j)}$ for the data of Example 10.2.1 by using Eqs. (10.2.27), (10.2.28), and (10.2.29).

Solution:

$$T_{..} = 78 \text{ and hence } \frac{T_{..}^2}{N} = \frac{6084}{13}$$

$$SS = 8^2 + 6^2 + \cdots + 6^2 - \frac{6084}{13}$$

$$= 550 - 468 = 82$$

$$SS_j = \frac{24^2}{3} + \frac{12^2}{4} + \frac{42^2}{6} - \frac{78^2}{13}$$

$$= 292 + 36 + 294 - 468 = 54$$

$$SS_{i(j)} = 82 - 54 = 28$$

which agree with the previous computations.

PROBLEMS

10.2.1 Show that $\sum_{j=1}^{j} \sum_{i=1}^{n_j} (\bar{x}_{\cdot j} - \bar{x}_{\cdot\cdot})(x_{ij} - \bar{x}_{\cdot j}) = 0$

10.2.2 Given the following measurements

A	B	C	D	E
13	26	17	20	5
12	28	18	22	10
14	27	16	21	6

(a) Calculate the mean of each sample of size three.

(b) Calculate the grand mean.

(c) Assuming $x_{ij} = \mu + \tau_j + \epsilon_{ij}$, represent each measurement as the sum of three components.

(d) Calculate the mean of ϵ_{ij}'s in part (c).

(e) Calculate the sums of squares of the given data considering the 15 observations as a single sample.

(f) Partition the total sums of squares in part (e) using Eq. (10.2.7).

(g) Find the mean squares and write out the complete Anova table for this data.

(h) Test the hypothesis of equal treatment effects using $\alpha = .05$.

10.2.3 A random sample of 6 Ph.D. candidates from 4 different universities had the following ages

A	B	C	D
31	26	31	32
33	28	30	28
29	22	29	33
27	25	30	29
46	23	27	30
32	26	30	31

(a) Calculate the sums of squares for this data, using the computational formulas.

(b) Write out the complete Anova table.

(c) Test the hypothesis that the mean ages of the candidates at these universities are equal.

10.3 HYPOTHESES CONCERNING LINEAR COMBINATIONS OF POPULATION MEANS

In the previous section we discussed testing hypotheses of equality of means for more than two populations. Although testing the equality of means represents an important class of hypotheses, there are other hypotheses that could be considered. For example, if we have reason to believe that all means are not equal, then we may wish to compare the mean of the first treatment with the mean of the second and third treatments. That is,

$$H_0\colon \mu_1 = \frac{\mu_2 + \mu_3}{2}$$

$$H_1\colon \mu_1 \neq \frac{\mu_2 + \mu_3}{2}$$

Or similarly, we may wish to compare the equality of the mean of the first and the second with the mean of the third, fourth, and fifth:

$$H_0\colon \frac{\mu_1 + \mu_2}{2} = \frac{\mu_3 + \mu_4 + \mu_5}{3}$$

$$H_1\colon \frac{\mu_1 + \mu_2}{2} \neq \frac{\mu_3 + \mu_4 + \mu_5}{3}$$

The above hypotheses may be rewritten as

$$H_0\colon 2\mu_1 - \mu_2 - \mu_3 = 0$$
$$H_1\colon 2\mu_1 - \mu_2 - \mu_3 \neq 0$$

and

$$H_0\colon 3\mu_1 + 3\mu_2 - 2\mu_3 - 2\mu_4 - 2\mu_5 = 0$$
$$H_1\colon 3\mu_1 + 3\mu_2 - 2\mu_3 - 2\mu_4 - 2\mu_5 \neq 0 \quad \text{respectively}$$

It may be readily generalized that H_0 could be written in the form

$$\sum_{j=1}^{J} \alpha_j \mu_j = 0 \tag{10.3.1}$$

That is, H_0 represents a linear combination of μ_j's, the treatment means.

If $\sum_{j=1}^{J} \alpha_j = 0$, then Eq. (10.3.1) is called a *contrast*. If there exists another contrast, say, $\sum_{j=1}^{J} \beta_j \mu_j = 0$, such that

$$\sum_{j=1}^{J} \alpha_j \beta_j = 0 \tag{10.3.2}$$

then $\sum_{j=1}^{J} \alpha_j \mu_j = 0$ and $\sum_{j=1}^{J} \beta_j \mu_j = 0$ are called *orthogonal contrasts*. An unbiased estimate for Eq. (10.3.1) is the statistic

$$\sum_{j=1}^{J} \alpha_j \bar{x}_{\cdot j} \tag{10.3.3}$$

If the assumption in Eq. (10.3.1) is true, then it can be shown that the quantity

$$L = \frac{n\left[\sum_{j=1}^{J} \alpha_j \bar{x}_{\cdot j}\right]^2}{\sum_{j=1}^{J} \alpha_j^2} \tag{10.3.4}$$

is an estimate of σ^2 with one degree of freedom, where n is the number of observations in each treatment. Multiplying numerator and denominator of Eq. (10.3.4) by n, yields

$$L = \frac{\left[n\sum_{j=1}^{J} \alpha_j \bar{x}_{\cdot j}\right]^2}{n\sum_{j=1}^{J} \alpha_j^2} = \frac{\left[\sum_{j=1}^{J} \alpha_j T_{\cdot j}\right]^2}{n\sum_{j=1}^{J} \alpha_j^2} \tag{10.3.5}$$

Recalling that MS_W is an estimate of σ^2 with $(N - J)$ degrees of freedom, we observe that the random variable

$$F = \frac{L}{MS_W} \tag{10.3.6}$$

is F distributed with 1 and $(N - J)$ numerator and denominator degrees of freedom. From Eq. (7.3.26) we note that the random variable

$$t = \sqrt{\frac{L}{MS_W}} = \frac{\sum_{j=1}^{J} \alpha_j T_{\cdot j}}{\sqrt{n\sum_{j=1}^{J} \alpha_j^2 MS_W}} = \frac{\sum_{j=1}^{J} \alpha_j \bar{x}_{\cdot j}}{\sqrt{\sum_{j=1}^{J} \alpha_j^2 MS_W\left(\frac{1}{n}\right)}} \tag{10.3.7}$$

obeys the Student's t p.f. with $(N - J)$ degrees of freedom.

Let us illustrate the notion under consideration by a numerical example.

Example 10.3.1 In the manufacturing of certain transformers there are five different methods of core winding, namely

1. All winding done by hand
2. All winding done by machine
3. First half by hand, second half by machine
4. First half by machine, second half by hand
5. First third by machine, second third by hand, and last third by machine.

Seven transformers of identical cores were wound by the above methods and tested for breakdown voltage. The result is tabulated in Table 10.3.1.

(a) Test the hypothesis of no difference in the means of the five methods of core winding.

(b) If H_0 is rejected, examine the comparisons:

1. The mean of 1st and 2nd treatment vs. 3rd.
2. The mean of the 4th vs. 5th treatment.
3. The mean of 1st and 2nd vs. the mean of 3rd, 4th, and 5th.
4. The mean of 1st vs. 2nd treatment.

(c) Find a set of mutually orthogonal contrasts and compute the corresponding value of L given by Eq. (10.3.5).

TABLE 10.3.1

BREAKDOWN VOLTAGES OF TRANSFORMERS

	Methods of Winding				
	1	2	3	4	5
1	250	400	350	450	375
2	290	390	375	400	275
3	300	430	325	395	300
4	330	450	340	405	330
5	350	390	360	275	320
6	250	420	370	300	310
7	310	370	430	375	320

Solution: (a) For computational purposes, we code the data of Table 10.3.1 by $x_i - 300$, which does not alter the respective variances. The coded data is shown in Table 10.3.2.

TABLE 10.3.2

	Treatments					
	1	2	3	4	5	
1	−50	100	50	150	75	
2	−10	90	75	100	−25	
3	0	130	25	95	0	
4	30	150	40	105	30	
5	50	90	60	−25	20	
6	−50	120	70	0	10	
7	10	70	130	75	20	
Totals....	−20	750	450	500	130	1,810

The computation of the sums of squares yield

$$SS = (-50)^2 + \cdots + (20)^2 - \frac{(1{,}810)^2}{35}$$

$$= 196{,}100 - \frac{3{,}276{,}100}{35} = 196{,}100 - 93{,}602.86$$

$$= 102{,}497.14$$

$$SS_j = \frac{1}{7}\,[(-20)^2 + \cdots + (130)^2] - \frac{(1{,}810)^2}{35}$$
$$= \frac{1{,}032{,}300}{7} - 93{,}602.86$$
$$= 147{,}471.42 - 93{,}602.86 = 53{,}868.56$$
$$SS_W = SS_{i(j)} = SS - SS_j$$
$$= 102{,}497.14 - 53{,}868.56 = 48{,}628.58$$

The Anova table for this example is given in Table 10.3.3.

TABLE 10.3.3

Source of Variation	Sums of Squares	D.F.	Mean Square	EMS	F-Ratio
Between treatments	$SS_j = 53{,}868.56$	4	$MS_B = 13{,}467.14$	$\sigma^2 + \frac{7\sum_{j=1}^{5}\tau_j^2}{4}$	$\frac{MS_B}{MS_W} = 8.3^{**}$
Within treatments	$SS_{i(j)} = 48{,}628.58$	30	$MS_W = 1{,}620.95$	σ^2	
Total	$SS = 102{,}497.14$	34			

Referring to Table VII in the Appendix we observe that

$$F_{.05;4,30} = 2.69$$

and

$$F_{.01;4,30} = 4.02$$

Therefore we reject the hypothesis of equal breakdown voltage for the various methods of transformer core winding.

(b) The four comparisons along with related computations are given in Table 10.3.4.

TABLE 10.3.4

Comparison	Coefficients α_i					$n\sum_{i=1}^{k}\alpha_i^2$
i:	1	2	3	4	5	
1. Between 1st and 2nd vs. 3rd treatment	1	1	−2	0	0	7(6) = 42
2. Between 4th vs. 5th treatment	0	0	0	1	−1	7(2) = 14
3. Between 1st and 2nd vs. 3rd, 4th, and 5th	3	3	−2	−2	−2	7(30) = 210
4. Between 1st vs. 2nd	1	−1	0	0	0	7(2) = 14

Now we apply Eq. (10.3.6) to each of the above comparisons, to test the corresponding hypotheses. For the first case we have:

$$\frac{[(+1)(-20) + (+1)750 + (-2)(450)]^2}{1{,}620.95(42)} = \frac{[-170]^2}{1{,}620.95[42]} = \frac{28{,}900}{68{,}159.7} = .42$$

Since $F_{.05;1,30} = 4.17$, the first comparison is not significant on the 5 percent level. For the 4th comparison above we have:

$$\frac{[1(-20) + (-1)(750)]^2}{1{,}620.95(14)} = 26$$

which is highly significant, since

$$F_{.01;1,30} = 7.56$$

The last comparison is important because it will test all hand-wound and all machine-wound effects, and is equivalent to the hypotheses

$$H_0: \mu_1 = \mu_2$$
$$H_1: \mu_1 \neq \mu_2$$

The interpretations of the other comparisons should be clear to the reader. The student is requested to examine the second and the third comparisons. (See Prob. 10.3.1.)

(c) A set of mutually orthogonal contrasts is given in Table 10.3.5. Note that the given set of mutually orthogonal contrasts is not unique. Many other such sets exist.

TABLE 10.3.5

Comparison	Coefficients α_i					$n \sum_{i=1}^{k} \alpha_i^2$
i:	1	2	3	4	5	
Mean 1 vs. 2	+1	−1	0	0	0	7(2) = 14
Mean 1, 2 vs. 3	+1	+1	−2	0	0	7(6) = 42
Mean 1, 2, 3 vs. 4	+1	+1	+1	−3	0	7(12) = 84
Mean 1, 2, 3, 4 vs. 5	+1	+1	+1	+1	−4	7(20) = 140

Calculating the values for the different L's, we obtain

$$L_1 = \frac{[(+1)(-20) + (-1)(750)]^2}{14} = \frac{770^2}{14} = 42{,}350.00$$

$$L_2 = \frac{[-20 + 750 - 2(450)]^2}{42} = \frac{170^2}{42} = 688.095$$

$$L_3 = \frac{[-20 + 750 + 450 - 1500]^2}{84} = \frac{320^2}{84} = 1{,}219.047$$

$$L_4 = \frac{[-20 + 750 + 450 + 500 - 520]^2}{140} = \frac{1{,}160^2}{140} = 9{,}611.420$$

Let us remark about the result of part (c) in Example 10.3.1. The sum of the 4 L's is 53,868.56 which is equal to the SS_j value computed in part (a) of the same example. This technique is sometimes used to check the calculated value of SS_j. In general, SS_j can be partitioned into $J - 1$ *components*, composed of mutually orthogonal contrasts, such that one degree of freedom is associated with each. The hypotheses corresponding to the contrast can be tested as shown in part (b) of Example 10.3.1.

Often it is desired to locate contrasts which are responsible for the rejection of the null hypothesis, *after* the experiment has been performed. This is important because it allows to extract more information from the analysis. A method due to Scheffé states that for *all possible* contrasts

$$P\left[\left|\sum_{i=1}^{k} (c_i\mu_i - c_i\bar{x}_{.j})\right| < \sqrt{K}\right] = 1 - \alpha \tag{10.3.8}$$

where α is the significance level associated with the null hypothesis, c_i are the coefficients of contrasts, and

$$K = (J - 1)\left(MS_W \sum_{i=1}^{k} \frac{c_i^2}{n_i}\right)[F_{\alpha;\, J-1,\, N-J}] \tag{10.3.9}$$

It can be shown that if the interval

$$\left(\sum_{i=1}^{k} c_i\bar{x}_{.i} \pm \sqrt{K}\right) \tag{10.3.10}$$

includes zero, the rejection of the hypothesis cannot be ascribed to the considered contrast. An example should suffice to illustrate the notion.

Example 10.3.2 Examine the 3rd comparison in Table 10.3.5 to determine whether or not it may have been responsible for the rejection of the hypothesis in part (a) of Example 10.3.1. Use 5 percent significance level.

Solution: For this case

$$J - 1 = 5 - 1 = 4$$

$$MS_W = 1620.95$$

$$F_{.05;4,30} = 2.69$$

$$\sum_{i=1}^{5} \frac{c_i^2}{n_i} = \frac{1}{7}[1 + 1 + 1 + 9] = \frac{12}{7}$$

Hence

$$K = 4(1{,}620.95)\left(\frac{12}{7}\right)(2.96) = 29{,}899.58$$

and

$$\sqrt{K} = 172.92$$

Also

$$\sum_{i=1}^{5} c_i \bar{x}_{.i} = \frac{1}{7}[-20 + 750 + 450 - 3(500)]$$
$$= -45.72$$

Since the interval -45.72 ± 172.92 includes zero, the rejection of H_0 cannot be attributed to this particular comparison. This result also implies that the hypothesis

$$H_0: \frac{\mu_1 + \mu_2 + \mu_3}{3} = \mu_4 \quad \text{against the alternative}$$

$$H_1: \frac{\mu_1 + \mu_2 + \mu_3}{3} \neq \mu_4$$

cannot be rejected on .05 significance level.

Several other methods besides Scheffé's have been given in literature to test hypotheses concerning linear combinations of population means. We shall not present in this introductory volume any other method than the one discussed.

PROBLEMS

10.3.1 In Example 10.3.1 part (b), examine the stated comparisons 2 and 3.

10.3.2 Refer to Prob. 10.2.2.

1. Examine the comparisons (a) the mean of the 1st and 2nd treatments vs. the 4th; (b) the mean of the 1st vs. the 3rd.
2. Find a set of mutually orthogonal contrasts and compute the corresponding L value given by Eq. (10.3.5).
3. Find the sum of the L's in part 2.
4. Using the Scheffé method, locate the possible combination of means which led to the rejection of the equal means hypothesis.

10.4 RANDOMIZED BLOCK DESIGN (RESTRICTION IN ONE DIRECTION)

Let us refer to Table 10.2.1 and consider the special case where all $n_j = n$. Suppose furthermore that the n observations in each treatment are subclassified according to another criterion. For instance, in the illustration of Table 10.2.2, the light bulbs could be classified according to high, medium, and low wattage where the definition of high, medium, and low is arbitrary. Then, the following might be one possible result of an experiment

TABLE 10.4.1
LIVES IN HOURS FOR DIFFERENT KINDS OF LIGHT BULBS

	Treatment (Type of Filaments)		
	Sample 1	Sample 2	Sample 3
High	8	3	7
Medium	6	4	8
Low	10	5	6

This is an example of *randomized block* design. Design of experiments is one branch of applied mathematics which had its foundation and start in agricultural research, and hence many terminologies are borrowed from that discipline. The term block is used to mean grouping or classifying of experimental units according to a certain common criterion. In the above example the light bulb of high wattage constitutes a block, and the medium and the low represent two other blocks. The primary difference between the completely randomized design and this approach lies in the manner by which the various experimental units are assigned to the treatments. In the former the experimental units are assigned with no restriction whatsoever. In the randomized blocks the experimental units are randomly assigned to the treatments *after* they have been classified into homogeneous groups.

Blocking is a method of local control and analogous to partitioning the within sums of squares into two parts; say, $SS_W = SS_i + SS_{ij}$. In so doing the likelihood of MS_{ij} being smaller than MS_W is increased and therefore it is more likely for MS_B/MS_{ij} to be larger than MS_B/MS_W. This implies that significant differences among the treatments will be detected with higher probability. (See Prob. 10. 4.4.)

TABLE 10.4.2
SUMMARY—RANDOMIZED BLOCK DESIGN

		Treatments				Block Total	Block Means
		1	2	⋯	j		
Blocks	1	x_{11}	x_{12}	⋯	x_{1J}	$T_{1\cdot}$	$\bar{x}_{1\cdot}$
	2	x_{21}	x_{22}	⋯	x_{2J}	$T_{2\cdot}$	$\bar{x}_{2\cdot}$
	3	x_{31}	x_{32}	⋯	x_{3J}	$T_{3\cdot}$	$\bar{x}_{3\cdot}$
	⋮	⋮	⋮	⋮	⋮	⋮	⋮
	I	x_{I1}	x_{I2}	⋯	x_{IJ}	$T_{I\cdot}$	$\bar{x}_{I\cdot}$
Treatment totals		$T_{\cdot 1}$	$T_{\cdot 2}$	⋯	$T_{\cdot J}$	$T_{\cdot\cdot}$	Grand total
Treatment means................		$\bar{x}_{\cdot 1}$	$\bar{x}_{\cdot 2}$	⋯	$\bar{x}_{\cdot J}$	Grand mean $x_{\cdot\cdot}$	

A summary for the randomized block designs is given in Table 10.4.2. Since these subclasses arrange themselves naturally into rows and columns, the blocks are referred to as *rows* and the treatments as *columns*.

As in the case of completely randomized design, we need to make certain assumptions in order to derive an expression for testing hypotheses. In the randomized block design it is assumed that:

1. All IJ populations are normally distributed with equal variances.
2. A random sample of size 1 is drawn from each of the IJ populations. (This condition can be extended to taking samples of size K and considering its mean as the single sample. The mean of a sample of size 1, is the sample value itself.)
3. The analysis falls into the fixed-effects case.
4. Blocks and treatment effects do not *interact*.

The last assumption needs to be discussed and illustrated prior to our proceeding with the analysis. The block treatment effects are said *not to interact* if

$$\mu_{ij} - \mu_{ij^*} = \mu_{i^*j} - \mu_{i^*j^*} \tag{10.4.1}$$

for all i, i^*, j and j^*. That is to say, the difference between the means for treatments j and j^* is equal in *every* block and the difference between the means for blocks i and i^* is equal in *every* treatment. If this condition is not satisfied the factors are said to interact.

An example of data with no interaction is given in Table 10.4.3, where $I = 5$ and $J = 3$.

TABLE 10.4.3

		Treatments		
		1	2	3
Blocks	1	1	5	3
	2	6	10	8
	3	4	8	6
	4	9	13	11
	5	12	16	14

Let us notice that the means for treatment 1 are 4 less than the means for treatment 2 in *every block*, and 2 less than means for treatment 3. Also, block differences remain constant from treatment to treatment. The no interaction assuption is rather strict and we shall modify it subsequently.

The expression (10.4.1) could be presented differently if the summation over i^* and j^* is considered. That is,

$$\sum_{i^*=1}^{I} \sum_{j^*=1}^{J} \mu_{ij} - \sum_{i^*=1}^{I} \sum_{j^*=1}^{J} \mu_{ij^*} = \sum_{i^*=1}^{I} \sum_{j^*=1}^{J} \mu_{i^*j} - \sum_{i^*=1}^{I} \sum_{j^*=1}^{J} \mu_{i^*j^*} \tag{10.4.2}$$

Rewriting Eq. (10.4.2), we get

$$IJ\mu_{ij} - I\sum_{j^*=1}^{J}\mu_{ij^*} = J\sum_{i^*=1}^{I}\mu_{i^*j} - IJ\frac{\sum_{i^*=1}^{I}\sum_{j^*=1}^{J}\mu_{i^*j^*}}{IJ} \tag{10.4.3}$$

$$IJ\mu_{ij} - IJ\frac{\sum_{j^*=1}^{J}\mu_{ij^*}}{J} = IJ\frac{\sum_{i^*=1}^{I}\mu_{i^*j}}{I} - IJ\mu_{..}$$

$$\mu_{ij} - \mu_{i.} = \mu_{.j} - \mu_{..} \tag{10.4.4}$$

which is equivalent to

$$\mu_{ij} - \mu_{i.} - \mu_{.j} + \mu_{..} = 0 \tag{10.4.5}$$

where $\mu_{..}$ is the grand mean. Therefore, no interaction assumption is equivalent to the hypothesis that Eq. (10.4.5) holds true.

Now let us consider the mathematical model for this case. An analogous identity to Eq. (10.2.3) is

$$\mu_{ij} - \mu_{..} = (\mu_{i.} - \mu_{..}) + (\mu_{.j} - \mu_{..}) + (\mu_{ij} - \mu_{i.} - \mu_{.j} + \mu_{..}) \tag{10.4.6}$$

where

$$\mu_{i.} = \frac{\sum_{j=1}^{J}\mu_{ij}}{J} \tag{10.4.7}$$

$$\mu_{.j} = \frac{\sum_{i=1}^{I}\mu_{ij}}{I} \tag{10.4.8}$$

and

$$\mu_{..} = \frac{\sum_{i=1}^{I}\sum_{j=1}^{J}\mu_{ij}}{IJ} \tag{10.4.9}$$

Letting

$$\alpha_i = \mu_{i.} - \mu_{..} \quad \text{(block effect)} \tag{10.4.10}$$

$$\beta_j = \mu_{.j} - \mu_{..} \quad \text{(treatment effect)} \tag{10.4.11}$$

and recalling that no interaction assumption is made, Eq. (10.4.6) reduces to

$$\mu_{ij} - \mu_{..} = \alpha_i + \beta_j \tag{10.4.12}$$

which is equivalent to

$$x_{ij} = \mu_{..} + \alpha_i + \beta_j + (x_{ij} - \mu_{ij}) \tag{10.4.13}$$

This last expression states that any observation is comprised of three different constants and an error term. That is, $\mu_{..}$ is the overall mean,

α_i is a constant due to block effect, β_j is a constant due to treatment effect and $x_{ij} - \mu_{ij} = \epsilon_{ij}$ is a random error.

Now the hypothesis

$$H_0:\quad \alpha_i = 0 \qquad i = 1, 2, \cdots, I$$

against the alternative

$$H_1:\quad \alpha_i \neq 0 \qquad \text{for at least one } i \neq i^*$$

can be tested to determine equal block effects and the hypothesis

$$H_0:\quad \beta_j = 0 \qquad j = 1, 2, \cdots, J$$

against the alternative

$$H_1:\quad \beta_j \neq 0 \qquad \text{for at least one } j \neq j^*$$

can be tested to determine equal treatment effects.

The corresponding sample identity for Eq. (10.4.6) may be written as

$$x_{ij} - \bar{x}_{..} = (\bar{x}_{i.} - \bar{x}_{..}) + (\bar{x}_{.j} - \bar{x}_{..}) + (x_{ij} - \bar{x}_{i.} - \bar{x}_{.j} + \bar{x}_{..}) \qquad (10.4.14)$$

Squaring both sides of the last equation yields

$$(x_{ij} - \bar{x}_{..})^2 = (\bar{x}_{i.} - \bar{x}_{..})^2 + (\bar{x}_{.j} - \bar{x}_{..})^2 + (x_{ij} - \bar{x}_{i.} - \bar{x}_{.j} + \bar{x}_{..})^2 + (3 \text{ cross-product terms}) \qquad (10.4.15)$$

Summing Eq. (10.4.15) with respect to i and j reduces the cross-product terms to zero and we get

$$\sum_{i=1}^{I}\sum_{j=1}^{J} (x_{ij} - \bar{x}_{..})^2 = \sum_{i=1}^{I}\sum_{j=1}^{J} (\bar{x}_{i.} - \bar{x}_{..})^2 + \sum_{i=1}^{I}\sum_{j=1}^{J} (\bar{x}_{.j} - \bar{x}_{..})^2 + \sum_{i=1}^{I}\sum_{j=1}^{J} (x_{ij} - \bar{x}_{i.} - \bar{x}_{.j} + \bar{x}_{..})^2 \qquad (10.4.16)$$

Let

$$SS = \sum_{i=1}^{I}\sum_{j=1}^{J} (x_{ij} - \bar{x}_{..})^2 \qquad (10.4.17)$$

$$SS_i = \sum_{i=1}^{I}\sum_{j=1}^{J} (\bar{x}_{i.} - \bar{x}_{..})^2 \qquad (10.4.18)$$

$$SS_j = \sum_{i=1}^{I}\sum_{j=1}^{J} (\bar{x}_{.j} - \bar{x}_{..})^2 \qquad (10.4.19)$$

$$SS_{ij} = \sum_{i=1}^{I}\sum_{j=1}^{J} (x_{ij} - \bar{x}_{i.} - \bar{x}_{.j} + \bar{x}_{..})^2 \qquad (10.4.20)$$

and hence

$$SS = SS_i + SS_j + SS_{ij} \qquad (10.4.21)$$

The corresponding computational formulas for the above sums of squares are, respectively,

$$SS = \sum_{i=1}^{I} \sum_{j=1}^{J} x_{ij}^2 - \frac{T_{..}^2}{IJ} \tag{10.4.22}$$

$$SS_i = \sum_{i=1}^{I} \frac{T_{j.}^2}{J} - \frac{T_{..}^2}{IJ} \tag{10.4.23}$$

$$SS_j = \sum_{j=1}^{J} \frac{T_{.j}^2}{I} - \frac{T_{..}^2}{IJ} \tag{10.4.24}$$

$$SS_{ij} = SS - SS_i - SS_j \tag{10.4.25}$$

In order to complete our analysis, we need to find the expected values for SS_i, SS_j, and SS_{ij}. Following similar steps as in the derivation of Eq. (10.2.14) and (10.2.15), it can be shown that

$$E[SS_{ij}] = (I - 1)(J - 1)\sigma^2 \tag{10.4.26}$$

$$E[SS_i] = (I - 1)\sigma^2 + J \sum_{i=1}^{I} \alpha_i^2 \tag{10.4.27}$$

$$E[SS_j] = (J - 1)\sigma^2 + I \sum_{j=1}^{J} \beta_j^2 \tag{10.4.28}$$

Consequently

$$\frac{E[SS_{ij}]}{(I - 1)(J - 1)} = \sigma^2 \tag{10.4.29}$$

$$\frac{E[SS_i]}{I - 1} = \sigma^2 + \frac{J \sum_{i=1}^{I} \alpha_i^2}{I - 1} \tag{10.4.30}$$

$$\frac{E[SS_j]}{J - 1} = \sigma^2 + \frac{I \sum_{j=1}^{J} \beta_j^2}{J - 1} \tag{10.4.31}$$

Now if $\sum_{i=1}^{I} \alpha_i^2 = 0$, which implies

$$\mu_{1.} = \mu_{2.} = \cdots = \mu_{I.} = \mu$$

then Eqs. (10.4.29) and (10.4.30) will have comparable values because each estimates σ^2. Also if $\sum_{j=1}^{J} \beta_j^2 = 0$, then Eqs. (10.4.29) and (10.4.31) will have comparable values. Therefore the validity of the hypotheses suggested above can be tested by the F-ratios. Table 10.4.4 is the summary of this section.

TABLE 10.4.4
ANOVA TABLE FOR RANDOMIZED BLOCK DESIGN

Source of Variation	Sums of Squares	Degrees of Freedom	Mean Square	EMS	F-Ratio
Between blocks	SS_i	$I - 1$	$MS_{Bl} = \dfrac{SS_i}{I-1}$	$\sigma^2 + \dfrac{J}{I-1}\sum_{i=1}^{I} \alpha_i^2$	$\dfrac{MS_{Bl}}{MS_E}$
Between treatments.......	SS_j	$J - 1$	$MS_{Tr} = \dfrac{SS_j}{J-1}$	$\sigma^2 + \dfrac{I}{J-1}\sum_{j=1}^{J} \beta_j^2$	$\dfrac{MS_{Tr}}{MS_E}$
Error.............	SS_{ij}	$(I-1)(J-1)$	$MS_E = \dfrac{SS_{ij}}{(I-1)(J-1)}$	σ^2	
Total.............	SS	$IJ - 1$			

Example 10.4.1 Test the hypotheses of equality of treatment effects and block effects for the data in Table 10.4.1.

Solution:

	Treatments			Totals
Blocks	8	3	7	18
	6	4	8	18
	10	5	6	21
Totals	24	12	21	57

$$SS_i = \frac{18^2 + 18^2 + 21^2}{3} - \frac{57^2}{9} = 2$$

$$SS_j = \frac{24^2 + 12^2 + 21^2}{3} - \frac{57^2}{9} = 26$$

$$SS = 8^2 + 6^2 + 10^2 + \cdots + 8^2 + 6^2 - \frac{57^2}{9} = 38$$

$$SS_{ij} = 38 - 2 - 26 = 10$$

TABLE 10.4.5
ANOVA TABLE FOR EXAMPLE 10.4.1

Source of Variation	Sums of Squares	D.F.	Mean Square	EMS	F-Ratio
Between blocks.......	$SS_i = 2$	2	$\dfrac{2}{2} = 1$	$\sigma^2 + \dfrac{3}{2}\Sigma\alpha_i^2$	$\dfrac{1}{2.5} = .4$
Between treatments ..	$SS_j = 26$	2	$\dfrac{26}{2} = 13$	$\sigma^2 + \dfrac{3}{2}\Sigma\beta_j^2$	$\dfrac{13}{2.5} = 5.2$
Error	$SS_{ij} = 10$	4	$\dfrac{10}{4} = 2.5$	σ^2	
Total..................	$SS = 38$	8			

Since $F_{.05;2,4} = 6.94$, both hypotheses are accepted.

Example 10.4.2 Students are classified into 6 ability groups and 4 methods of teaching employed. The following table represents the means in each group.

TABLE 10.4.6

I \ J		Teaching Methods A	B	C	D	Totals
	1	36	27	31	43	137
	2	34	23	29	37	123
Ability	3	30	30	35	41	136
groups	4	40	20	25	39	124
	5	37	22	28	64	131
	6	33	28	32	36	129
Totals		210	150	180	240	780

Test the hypothesis of equal effects among the teaching methods.

Solution: The computation of the sums of squares yield

$$SS_i = \frac{137^2}{4} + \frac{123^2}{4} + \cdots + \frac{129^2}{4} - \frac{780^2}{24}$$

$$= \frac{101{,}572}{4} - \frac{608{,}400}{24} = 25{,}393 - 25{,}350 = 43$$

$$SS_j = \frac{210^2}{6} + \frac{150^2}{6} + \cdots + \frac{240^2}{6} - \frac{780^2}{24}$$

$$= \frac{156{,}600}{6} - \frac{608{,}400}{24} = 26{,}100 - 25{,}350 = 750$$

$$SS = 36^2 + 27^2 + \cdots + 36^2 - \frac{780^2}{24}$$

$$= 26{,}358 - \frac{608{,}400}{24} = 26{,}348 - 25{,}350 = 998$$

$$SS_{ij} = 998 - 750 - 43 = 205$$

TABLE 10.4.7
ANOVA TABLE FOR EXAMPLE 10.4.2

Source of Variance	SS	D.F.	MS	F-Ratio
Between blocks........	43	$6 - 1 = 5$	$\frac{43}{5} = 8.6$	$\frac{8.6}{13.6} = .63$
Between treatments ...	750	$4 - 1 = 3$	$\frac{750}{3} = 250$	$\frac{250}{13.6} = 18.38$**
Error	205	15	$\frac{205}{15} = 13.6$	
Total..................	998	23		

Since

$$F_{.05;3,15} = 3.29$$

and

$$F_{.01;3,15} = 5.42$$

we conclude that the teaching methods are significantly different.

PROBLEMS

10.4.1 The effects of three different makes of drills are to be studied. Holes are drilled at 30-, 45-, and 60-degree angles with 100, 200, 400, and 800 rpm and the time in seconds of drilling observed. The following represents the result.

Drilling Angle	Speed, rpm			
	100	200	400	800
30°	43	33	24	20
45°	54	34	30	20
60°	62	53	37	27

(a) Analyze the result as a randomized block design.

(b) Set up orthogonal contrasts among the rpms.

(c) If the hypothesis of equal rpm effects is rejected in part (a), locate possible combination of means which led to the rejection of the hypothesis.

10.4.2 Refer to Prob. 10.2.3 and assume that the rows represent blocks corresponding to six different major fields of study. Test the hypothesis that the mean ages of the candidates within the major fields of study are equal.

10.4.3 Show that summing Eq. (10.4.15) with respect to i and j reduces the three cross-product terms to zero.

10.4.4 If the data of Example 10.4.2 is analyzed as completely randomized design, (a) show that $SS_W = SS_i + SS_{ij}$; (b) write the new Anova table.

11

Latin and Graeco-Latin Squares

11.0 INTRODUCTION

In Sec. 10.4 the randomized block design was contrasted with the completely randomized design and shown that its main advantage was in reducing the experimental error by removing a source of variation in which the investigator had little or no interest. Whenever two sources of variation are to be controlled, the Latin-square technique provides one method of accomplishing this purpose. If there are three sources of variation to be controlled an extension of Latin-squares, the Graeco-Latin-square technique, may sometimes be used. In this chapter these two designs and their analysis will be considered.

11.1 LATIN-SQUARES (RESTRICTION IN TWO DIRECTIONS)

Let us introduce the concept of a Latin-square by a numerical example. Suppose in an experiment the variables to be considered are gas mileage, air pressures in the tires, and types of roads on which the cars travel. The following configuration illustrates a possible use of Latin-square principle, where A, B, and C represent the three types of gasoline.

TABLE 11.1.1

Air Pressure in Tires	Road Types		
	1	2	3
28	B	A	C
30	A	C	B
32	C	B	A

We note that gasoline type B is used with road type 1 and tire pressure 28; road type 2 and tire pressure 32; and road type 3 and tire pressure 30. Similarly, gasoline types A and C are used with each type of road and each type of tire pressure. There are obvious limitations to this design, namely (1) each treatment occurs exactly once in each row and in each column, and (2) the number of levels must be equal in both variables considered. The limitations compensate in the fewer number of sample sizes required. That is, we need n^2 instead of n^3 samples for the Latin-square of dimension n.

The following list gives a few selected Latin-squares. Note that each letter appears only once in each row and in each column.

TABLE 11.1.2

3 × 3

(a)

A	B	C
C	A	B
B	C	A

(b)

A	B	C
B	C	A
C	A	B

4 × 4

(a)

A	B	C	D
B	A	D	C
D	C	A	B
C	D	B	A

(b)

A	B	C	D
B	A	D	C
D	C	B	A
C	D	A	B

(c)

A	B	C	D
C	D	A	B
D	C	B	A
B	A	D	C

5 × 5

(a)

A	B	C	D	E
E	A	B	C	D
D	E	A	B	C
C	D	E	A	B
B	C	D	E	A

(b)

A	B	C	D	E
B	C	D	E	A
E	A	B	C	D
D	E	A	B	C
C	D	E	A	B

6 × 6

A	B	C	D	E	F
F	A	B	C	D	E
E	F	A	B	C	D
D	E	F	A	B	C
C	D	E	F	A	B
B	C	D	E	F	A

7 × 7

A	B	C	D	E	F	G
G	A	B	C	D	E	F
F	G	A	B	C	D	E
E	F	G	A	B	C	D
D	E	F	G	A	B	C
C	D	E	F	G	A	B
B	C	D	E	F	G	A

Now suppose we wish to use a 5 × 5 Latin-square and decide to select the model (b) in Table 11.1.2. We choose a random order for 5 numbers, say 4, 2, 5, 3, 1, and then arrange the *rows* of model (b) according to this random order. This yields

TABLE 11.1.3

D	E	A	B	C
B	C	D	E	A
C	D	E	A	B
E	A	B	C	D
A	B	C	D	E

If instead of arranging the rows we arrange the *columns* in the random order 4, 1, 5, 3, 2, we obtain

TABLE 11.1.4

D	A	E	C	B
E	B	A	D	C
C	E	D	B	A
B	D	C	A	E
A	C	B	E	D

Once a Latin-square is selected, the observations are classified according to row, column, Latin-square configuration.

Referring to the illustration of gasoline mileage, road types, and tire pressure, we see the need to use three subscripts to represent each cell observation. Let x_{ijk} be the kth treatment in cell ij, then Table 11.1.1 is represented as follows:

TABLE 11.1.5

x_{112}	x_{121}	x_{133}
x_{211}	x_{223}	x_{232}
x_{313}	x_{322}	x_{331}

where for instance x_{112} corresponds to the observation of row 1, column 1 and gasoline type B and x_{313} corresponds to the observation of row 3, column 1 and gasoline type C.

In the analysis of Latin-square design one needs the sums of the rows, columns and Latin-square classification. In general, for a Latin-square of order n, let

$T_{i\cdot\cdot}$ = total for ith row; $i = 1, 2, \ldots, n$

$T_{\cdot j\cdot}$ = total for jth column; $j = 1, 2, \ldots, n$

$T_{\cdot\cdot k}$ = total for all observations receiving kth treatment; $k = 1, 2, \ldots, n$

$T_{\cdots}$ = total for all observations in the Latin-square

We wish to point out that the dot notation for Latin-squares is *not* the same as previously used. It is slightly modified. When we sum up the ith-row observations, we are summing over both j and k but only in the combinations that j and k appear in that particular row of the selected Latin-square. Likewise, when we add up the jth column observations, we sum over both i and k but only in the combinations that i and k appear in that particular column. The double summation indices represent a way to describe the totals for Latin-squares. That is,

$$T_{i\cdot\cdot} = \sum_{j,k=1}^{n} x_{ijk} \tag{11.1.1}$$

$$T_{\cdot j \cdot} = \sum_{i,k=1}^{n} x_{ijk} \tag{11.1.2}$$

$$T_{\cdot\cdot k} = \sum_{i,j=1}^{n} x_{ijk} \tag{11.1.3}$$

For the above example the following are three totals out of the nine possible.

$$T_{1\cdot\cdot} = x_{112} + x_{121} + x_{133}$$
$$T_{\cdot 2 \cdot} = x_{121} + x_{223} + x_{322}$$
$$T_{\cdot\cdot 3} = x_{313} + x_{223} + x_{133}$$

The definition of the mean follows the same pattern as before.

$$\bar{x}_{i\cdot\cdot} = \frac{T_{i\cdot\cdot}}{n} \tag{11.1.4}$$

$$\bar{x}_{\cdot j\cdot} = \frac{T_{\cdot j\cdot}}{n} \tag{11.1.5}$$

$$\bar{x}_{\cdot\cdot k} = \frac{T_{\cdot\cdot k}}{n} \tag{11.1.6}$$

The underlying assumptions for the mathematical model are (1) random samples of size one are drawn from each of n^2 normally distributed populations with variance σ^2, (2) no interaction between row, column, and treatment effects, and (3) analysis falls into the fixed-effects case.

An identity analogous to Eq. (10.4.6) is

$$\mu_{ijk} - \mu = (\mu_{i\cdot\cdot} - \mu) + (\mu_{\cdot j\cdot} - \mu) + (\mu_{\cdot\cdot k} - \mu) + (\mu_{ijk} - \mu_{i\cdot\cdot} - \mu_{\cdot j\cdot} - \mu_{\cdot\cdot k} + 2\mu) \tag{11.1.7}$$

where

$$\mu_{i\cdot\cdot} = \frac{\sum_{j,k=1}^{n} \mu_{ijk}}{n}$$

$$\mu_{\cdot j\cdot} = \frac{\sum_{i,k=1}^{n} \mu_{ijk}}{n}$$

$$\mu_{\cdot\cdot k} = \frac{\sum_{i,j=1}^{n} \mu_{ijk}}{n}$$

Letting

$$\alpha_i = \mu_{i\cdot\cdot} - \mu \quad \text{(row effects)}$$
$$\beta_j = \mu_{\cdot j\cdot} - \mu \quad \text{(column effects)}$$
$$\gamma_k = \mu_{\cdot\cdot k} - \mu \quad \text{(treatment effects)}$$

and if the assumption of no interaction holds,

$$\mu_{ijk} - \mu_{i\cdot\cdot} - \mu_{\cdot j\cdot} - \mu_{\cdot\cdot k} + 2\mu = 0$$

Hence Eq. (11.1.7) can be expressed as

$$\mu_{ijk} = \mu + \alpha_i + \beta_j + \gamma_k \tag{11.1.8}$$

where the interpretation of α_i, β_j, and γ_k are analogous to those described in Sec. 10.2*a*. However, the assumption of no interaction effects is more complex. No interaction in this model implies that for all possible combinations of effects, (row and column), (row and treatment), (column and treatment), and (row and column and treatment) do not produce new effects acting together. Therefore, a random observation x_{ijk} is comprised of four separate effects plus a random error. That is,

$$x_{ijk} = \mu + \alpha_i + \beta_j + \gamma_k + \epsilon_{ijk} \tag{11.1.9}$$

is the desired model, where $\epsilon_{ijk} = x_{ijk} - \mu_{ijk}$ and

$$\mu = \frac{\sum_{i,j,k=1}^{n} \mu_{ijk}}{n^2}$$

The hypothesis of equality of treatment effects [or all γ_k's are equal to zero] is expressed as

$$H_0: \quad \gamma_1 = \gamma_2 = \cdots = \gamma_k = 0; k = 1, \cdots, n$$
$$H_1: \quad \gamma_k \neq \gamma_{k^*} \quad \text{for at least one } k \neq k^*$$

The corresponding sample identity for Eq. (11.1.7) is

$$(x_{ijk} - \bar{x}_{\ldots}) = (\bar{x}_{i\cdot\cdot} - \bar{x}_{\ldots}) + (\bar{x}_{\cdot j\cdot} - \bar{x}_{\ldots}) + (\bar{x}_{\cdot\cdot k} - \bar{x}_{\ldots}) + (x_{ijk} - \bar{x}_{i\cdot\cdot} - \bar{x}_{\cdot j\cdot} - \bar{x}_{\cdot\cdot k} + 2\bar{x}_{\ldots}) \tag{11.1.10}$$

Squaring both sides of the last equation and summing with respect to i, j, and k we obtain

$$\begin{aligned}\sum_{i,j,k=1}^{n} (x_{ijk} - \bar{x}_{\ldots})^2 = & \sum_{i,j,k=1}^{n} (\bar{x}_{i\cdot\cdot} - \bar{x}_{\ldots})^2 \\ & + \sum_{i,j,k=1}^{n} (\bar{x}_{\cdot j\cdot} - \bar{x}_{\ldots})^2 \\ & + \sum_{i,j,k=1}^{n} (\bar{x}_{\cdot\cdot k} - \bar{x}_{\ldots})^2 \\ & + \sum_{i,j,k=1}^{n} (x_{ijk} - \bar{x}_{i\cdot\cdot} - \bar{x}_{\cdot j\cdot} - \bar{x}_{\cdot\cdot k} + 2\bar{x}_{\ldots})^2 \\ & + \sum_{i,j,k=1}^{n} \left\{\binom{4}{2} \text{cross-product terms}\right\}\end{aligned} \tag{11.1.11}$$

The six cross-product terms in the last expression add up to zero, since the sum of the differences from the respective means adds to zero. Following a similar pattern of notation for the sums of squares as in the previous chapter, we let

$$SS = \sum_{i,j,k=1}^{n} (x_{ijk} - \bar{x}_{...})^2 \tag{11.1.12}$$

$$SS_i = \sum_{i,j,k=1}^{n} (\bar{x}_{i..} - \bar{x}_{...})^2 \tag{11.1.13}$$

$$SS_j = \sum_{i,j,k=1}^{n} (\bar{x}_{.j.} - \bar{x}_{...})^2 \tag{11.1.14}$$

$$SS_k = \sum_{i,j,k=1}^{n} (\bar{x}_{..k} - \bar{x}_{...})^2 \tag{11.1.15}$$

$$SS_{ijk} = \sum_{i,j,k=1}^{n} (x_{ijk} - \bar{x}_{i..} - \bar{x}_{.j.} - \bar{x}_{..k} + 2\bar{x}_{...})^2 \tag{11.1.16}$$

and hence Eq. (11.1.11) can be expressed as

$$SS = SS_i + SS_j + SS_k + SS_{ijk} \tag{11.1.17}$$

The computational formulas for the sums of squares can be shown to be equal to

$$SS = \sum_{i,j,k=1}^{n} x_{ijk}^2 - \frac{T_{...}^2}{n^2} \tag{11.1.18}$$

$$SS_i = \sum_{i=1}^{n} \frac{T_{i..}^2}{n} - \frac{T_{...}^2}{n^2} \tag{11.1.19}$$

$$SS_j = \sum_{j=1}^{n} \frac{T_{.j.}^2}{n} - \frac{T_{...}^2}{n^2} \tag{11.1.20}$$

$$SS_k = \sum_{k=1}^{n} \frac{T_{..k}^2}{n} - \frac{T_{...}^2}{n^2} \tag{11.1.21}$$

and

$$SS_{ijk} = SS - SS_i - SS_j - SS_k \tag{11.1.22}$$

In order to test the hypothesis of equality of treatment effects, we need to compute the expected values for the quantities SS_i, SS_j, SS_k. For the fixed-effects model that we are considering, following similar steps as in the proof of Eq. (10.2.15), it can be shown that

$$E[SS_i] = (n - 1)\sigma^2 + n\sum_{i=1}^{n} \alpha_i^2 \tag{11.1.23}$$

$$E[SS_j] = (n - 1)\sigma^2 + n\sum_{i=1}^{n} \beta_j^2 \tag{11.1.24}$$

$$E[SS_k] = (n - 1)\sigma^2 + n \sum_{i=1}^{n} \gamma_k^2 \tag{11.1.25}$$

Also, following similar steps as in the proof of Eq. (10.2.14), it can be shown that

$$E[SS_{ijk}] = (n - 1)(n - 2)\sigma^2 \tag{11.1.26}$$

Now if the treatment effects are equal, the summation in Eq. (11.1.25) will be zero and $\frac{E[SS_k]}{n - 1}$ will represent an unbiased estimate of σ^2. Similarly, the expression $\frac{E[SS_{ijk}]}{(n - 1)(n - 2)}$ will be an unbiased estimate of σ^2. Therefore, the F-ratio, $F = \frac{MS_k}{MS_E}$ is the test statistic to test the suggested null hypothesis.

The Anova table summarizing the discussion of this section is given in Table 11.1.6.

TABLE 11.1.6
ANOVA TABLE FOR LATIN-SQUARE

Source of Variation	SS	D.F.	Mean Square	EMS	F-Ratio
Row variable.	SS_i	$n - 1$	$\frac{SS_i}{n-1} = MS_i$	$\sigma^2 + \frac{n}{n-1}\sum_{i=1}^{n} \alpha_i^2$	$\frac{MS_i}{MS_E}$
Column variable.	SS_j	$n - 1$	$\frac{SS_j}{n-1} = MS_j$	$\sigma^2 + \frac{n}{n-1}\sum_{j=1}^{n} \beta_j^2$	$\frac{MS_j}{MS_E}$
Treatment variable.	SS_k	$n - 1$	$\frac{SS_k}{n-1} = MS_k$	$\sigma^2 + \frac{n}{n-1}\sum_{k=1}^{n} \gamma_k^2$	$\frac{MS_k}{MS_E}$
Error.	SS_{ijk}	$(n - 1)(n - 2)$	$\frac{SS_{ijk}}{(n-1)(n-2)} = MS_E$	σ^2	
Total.	SS	$n^2 - 1$			

We conclude this section by a numerical example.

Example 11.1.1 Referring to Table 11.1.5, suppose the mileages yielded were as follows.

TABLE 11.1.7

Air Pressures	Road Types		
	1	2	3
1	15B	18A	20C
2	17A	20C	16B
3	23C	14B	19A

Test the hypothesis of equal-treatment effects.

Solution:

Air Pressure in Tire	Road Types 1	2	3	
1	15B	18A	20C	$T_{1..} = 53$
2	17A	20C	16B	$T_{2..} = 53$
3	23C	14B	19A	$T_{3..} = 56$
	$T_{.1.} = 55$	$T_{.2.} = 52$	$T_{.3.} = 55$	$T_{...} = 162$

$$T_{..1} = 17 + 18 + 19 = 54$$
$$T_{..2} = 15 + 14 + 16 = 45$$
$$T_{..3} = 23 + 20 + 20 = 63$$
$$T_{...} = 162$$

$$SS = 15^2 + 17^2 + \cdots + 19^2 - \frac{162^2}{9}$$
$$= 2{,}980 - \frac{26{,}244}{9} = \frac{576}{9} = 64$$
$$SS_i = \frac{53^2 + 53^2 + 56^2}{3} - \frac{26{,}244}{9} = \frac{18}{9} = 2$$
$$SS_j = \frac{55^2 + 52^2 + 55^2}{3} - \frac{26{,}244}{9} = \frac{18}{9} = 2$$
$$SS_k = \frac{54^2 + 45^2 + 63^2}{3} - \frac{26{,}244}{9} = \frac{486}{9} = 54$$
$$SS_{ijk} = 64 - 2 - 2 - 54 = 6$$

Hence the Anova table becomes

Source	SS	D.F.	MS	EMS	*F*-Ratio
Air pressure.......	2	2	1		
Road type.........	2	2	1		
Mileage	54	2	27	$\sigma^2 + \frac{3}{2}\sum_{k=1}^{3} \gamma_k^2$	$\frac{27}{3} = 9$
Error..............	6	2	3		
Total..............	64	8			

Referring to Table VII we note that $F_{.05;2,2} = 19.0$. Therefore the hypothesis of equal-treatment effects is accepted.

PROBLEMS

11.1.1 Show that any one of the six possible cross product terms in Eq. (11.1.11) has zero value.

11.1.2 Verify Eqs. (11.1.18) and (11.1.19).

11.1.3 Analyze the Latin-square

	1		2		3	
1	A	17.5	B	23.5	C	28.5
2	C	24.5	A	17.0	B	18.5
3	B	19.0	C	25.5	A	17.5

11.1.4 The following Latin-square represents the number of electrodes used per welding machine per day and A, B, C, D, and E correspond to the operators of the machines. Analyze the given data.

		Machines									
		1		2		3		4		5	
Days	M	C	125	A	118	D	115	E	122	B	114
	T	B	132	E	137	C	131	D	122	A	118
	W	E	116	C	135	B	134	A	127	D	121
	T	A	130	D	115	E	126	B	128	C	122
	F	D	112	B	130	A	128	C	133	E	121

11.2 GRAECO-LATIN SQUARES (RESTRICTION IN THREE DIRECTION)

The Graeco-Latin squares are a natural extension of Latin-square designs discussed in Sec. 11.1. Here also, let us introduce the underlying notions by an illustration. In an experiment concerning transformer core winding, an engineer considers the variables to be wire size, insulation thickness, and insulation material. There are 5 different insulation materials and 5 insulation thicknesses to consider. This problem lends itself to a Latin-square analysis.

Suppose the hypothesis to consider is that insulation thicknesses have equal effects. Then we choose a random sample of 5 × 5 Latin-square similar to that of Table 11.2.1.

TABLE 11.2.1

Insulation Material	Wire Size				
	I	II	III	IV	V
1	C	E	A	D	B
2	B	D	E	C	A
3	A	C	D	B	E
4	E	B	C	A	D
5	D	A	B	E	C

A, B, C, D, and E correspond to the five insulation thicknesses. Now in the design of the experiment we assign C to type 1 insulation material and type I wire size, etc. We fill the matrix according to this particular pattern of Latin-square. After the readings have been taken, the analysis of variance is carried out, and conclusions drawn accordingly.

Suppose now that there were a fourth variable, say 5 wire manufacturers, to be included in the analysis. The only addition to the given Latin-square is the assigning of 5 manufacturers at random to each insulation thickness. For example, referring to the first row we see that the insulation thickness is given in the order $C\,E\,A\,D\,B$. Now we assign at random the manufacturers' numbers to these, say, $C_4 E_1 A_3 D_5 B_2$. We repeat the random assigning of manufacturers to insulation thickness of the remaining rows such that Latin-square principle is preserved. Completion of the rows and columns may look something like Table 11.2.2.

TABLE 11.2.2

Insulation Material	Wire Size				
	I	II	III	IV	V
1	C_4	E_1	A_3	D_5	B_2
2	B_2	D_4	E_1	C_3	A_5
3	A_5	C_2	D_4	B_1	E_3
4	E_1	B_3	C_5	A_2	D_4
5	D_3	A_5	B_2	E_4	C_1

In Table 11.2.2 the subscripts are the manufacturers' code number. This type of configuration is referred to as a *Graeco-Latin square*. The representation of Graeco-Latin squares is not any more complex than the Latin-squares. A very slight extension of the notions of Latin-squares will yield the desired results. Here we need four subscripts to represent each cell observation. Let x_{ijkm} be the kth treatment in cell ij (as was in Latin-squares) with the mth variable associated with it. The notion is very simple but the "bookkeeping" can become tedious. The above illustration is summarized in Table 11.2.3.

TABLE 11.2.3
(Representation of Table 11.2.2)

Insulation Material	Wire Size				
	I	II	III	IV	V
1	x_{1134}	x_{1251}	x_{1313}	x_{1445}	x_{1522}
2	x_{2122}	x_{2244}	x_{2351}	x_{2433}	x_{2515}
3	x_{3115}	x_{3232}	x_{3344}	x_{3421}	x_{3553}
4	x_{4151}	x_{4223}	x_{4335}	x_{4412}	x_{4544}
5	x_{5143}	x_{5215}	x_{5322}	x_{5454}	x_{5531}

Let

$T_{i\cdots}$ = total for the ith row; $i = 1, 2, \ldots, n$

$T_{\cdot j\cdot\cdot}$ = total for the jth column: $j = 1, 2, \ldots, n$

$T_{\cdot\cdot k\cdot}$ = total for each level of the third source of variation

$T_{\cdots m}$ = total for each level of the fourth source of variation

$T_{\cdots\cdot}$ = total for all observations in the Graeco-Latin square

The dot notation here is analogous to the one described in Latin-square section, except that the summation is over 3 variables, in the combinations that they appear and in the particular Graeco-Latin square. Again, the "bookkeeping" is the difficult part and not the notion. For example, in our example above

$$T_{..\underline{3}.} = x_{11\underline{3}4} + x_{32\underline{3}2} + x_{43\underline{3}5} + x_{24\underline{3}3} + x_{55\underline{3}1}$$

which is the total corresponding to the 3rd insulation thickness. The underlying assumptions for the mathematical model for the Graeco-Latin squares are identical to the Latin-square with one addition to the no interaction assumption. Here the no interaction implies that all possible combinations of 4 variables taken 2, 3, and 4 at a time, do not produce new effects acting together. For instance, it is assumed that the column and the third variable acting together do not introduce a new effect. With these assumptions we proceed with the mathematical model. The identity

$$\mu_{ijkm} = \mu + (\mu_{i...} - \mu) + (\mu_{.j..} - \mu) + (\mu_{..k.} - \mu) + (\mu_{...m} - \mu) + (\mu_{ijkm} - \mu_{i...} - \mu_{.j..} - \mu_{..k.} - \mu_{...m} + 3\mu) \quad (11.2.1)$$

is analogous to Eq. (11.1.7).

Let

$$\alpha_i = \mu_{i...} - \mu \quad (11.2.2)$$

$$\beta_j = \mu_{.j..} - \mu \quad (11.2.3)$$

$$\gamma_k = \mu_{..k.} - \mu \quad (11.2.4)$$

$$\delta_m = \mu_{...m} - \mu \quad (11.2.5)$$

If the no interaction assumption holds, then

$$\mu_{ijkm} - \mu_{i...} - \mu_{.j..} - \mu_{..k.} - \mu_{...m} + 3\mu = 0 \quad (11.2.6)$$

Also, it may be shown that

$$\sum_{i=1}^{n} \alpha_i = 0, \quad \sum_{j=1}^{n} \beta_j = 0, \quad \sum_{k=1}^{n} \gamma_k = 0, \quad \sum_{m=1}^{n} \delta_m = 0 \quad (11.2.7)$$

Equation (11.2.1) may now be written as

$$x_{ijkm} = \mu + x_{ijkm} - \mu_{ijkm} + \alpha_i + \beta_j + \gamma_k + \delta_m \quad (11.2.8)$$

Letting

$$x_{ijkm} - \mu_{ijkm} = \epsilon_{ijkm}$$

Eq. (11.2.8) becomes

$$x_{ijkm} = \mu + \alpha_i + \beta_j + \gamma_k + \delta_m + \epsilon_{ijkm} \quad (11.2.9)$$

Now if we were to test the equality of the 3rd variable effects, that is, all γ's are equal to zero; or if we were to test the equality of the 4th variable, that is all δ's are equal to zero; we follow the same procedure as was used previously. The only difference is that the degrees of freedom for the error become $(n - 1)(n - 3)$, as we shall indicate in the new Anova table.

The computational formulas for the partitioned sums of squares can easily be shown to be:

$$SS = \sum_{i,j,k,m=1}^{n} x_{ijkm}^2 - \frac{T_{....}^2}{n^2} \qquad (11.2.10)$$

$$SS_i = \frac{\sum_{i=1}^{n} T_{i...}^2}{n} - \frac{T_{....}^2}{n^2} \qquad (11.2.11)$$

$$SS_j = \frac{\sum_{j=1}^{n} T_{.j..}^2}{n} - \frac{T_{....}^2}{n^2} \qquad (11.2.12)$$

$$SS_k = \frac{\sum_{k=1}^{n} T_{..k.}^2}{n} - \frac{T_{....}^2}{n^2} \qquad (11.2.13)$$

$$SS_m = \frac{\sum_{m=1}^{n} T_{...m}^2}{n} - \frac{T_{....}^2}{n^2} \qquad (11.2.14)$$

$$SS_E = SS - SS_i - SS_j - SS_k - SS_m \qquad (11.2.15)$$

For the fixed-effects model it can be shown that the expected values for the sums of squares are as follows:

$$E[SS_i] = (n - 1)\sigma^2 + n\sum_{i=1}^{n} \alpha_i^2 \qquad (11.2.16)$$

$$E[SS_j] = (n - 1)\sigma^2 + n\sum_{j=1}^{n} \beta_j^2 \qquad (11.2.17)$$

$$E[SS_k] = (n - 1)\sigma^2 + n\sum_{k=1}^{n} \gamma_k^2 \qquad (11.2.18)$$

$$E[SS_m] = (n - 1)\sigma^2 + n\sum_{m=1}^{n} \delta_m^2 \qquad (11.2.19)$$

$$E[SS_E] = (n - 1)(n - 3)\sigma^2 \qquad (11.2.20)$$

Now if the 3rd and 4th variable effects were equal, the summation portion of Eqs. (10.2.18) and (11.2.19) will be equal to zero, and

$$\frac{E[SS_k]}{n - 1}, \quad \frac{E[SS_m]}{n - 1} \quad \text{and} \quad \frac{E[SS_E]}{(n - 1)(n - 3)}$$

will be unbiased estimates of σ^2. Then the Anova table becomes

TABLE 11.2.4
ANOVA TABLE FOR GRAECO-LATIN SQUARE

Source of Variation	SS	D.F.	Mean Square	EMS	F-Ratio
Row variable.	SS_i	$n-1$	$\dfrac{SS_i}{n-1} = MS_i$	$\sigma^2 + \dfrac{n}{n-1}\sum_{i=1}^{n} \alpha_i^2$	
Column variable.	SS_j	$n-1$	$\dfrac{SS_j}{n-1} = MS_j$	$\sigma^2 + \dfrac{n}{n-1}\sum_{j=1}^{n} \beta_j^2$	
3rd variable	SS_k	$n-1$	$\dfrac{SS_k}{n-1} = MS_k$	$\sigma^2 + \dfrac{n}{n-1}\sum_{k=1}^{n} \gamma_k^2$	$\dfrac{MS_k}{MS_E}$
4th variable	SS_m	$n-1$	$\dfrac{SS_m}{n-1} = MS_m$	$\sigma^2 + \dfrac{n}{n-1}\sum_{m=1}^{n} \delta_m^2$	$\dfrac{MS_m}{MS_E}$
Error.	SS_E	$(n-1)(n-3)$	$\dfrac{SS_E}{(n-1)(n-3)} = MS_E$	σ^2	
Total	SS	n^2-1			

We shall conclude this section with a numerical example.

Example 11.2.1 Suppose the data for Table 11.2.2 were as follows.

TABLE 11.2.5

1	$C_4 = 9$	$E_1 = 8$	$A_3 = 7$	$D_5 = 6$	$B_2 = 5$	35
2	$B_2 = 9$	$D_4 = 9$	$E_1 = 4$	$C_3 = 6$	$A_5 = 5$	33
3	$A_5 = 8$	$C_2 = 9$	$D_4 = 5$	$B_1 = 5$	$E_3 = 5$	32
4	$E_1 = 8$	$B_3 = 7$	$C_5 = 3$	$A_2 = 5$	$D_4 = 5$	28
5	$D_3 = 10$	$A_5 = 6$	$B_2 = 3$	$E_4 = 3$	$C_1 = 5$	27
	44	39	22	25	25	155

Test the hypotheses of equal-insulation-thickness effects and equal-manufacturer effects.

Solution: The row and column totals are shown in the margins of Table 11.2.5 and the other totals are

$$T_{..1.} = 8 + 6 + 7 + 5 + 5 = 31$$

$$T_{..2.} = 9 + 7 + 3 + 5 + 5 = 29$$

$$T_{..3.} = 9 + 9 + 3 + 6 + 5 = 32$$

$$T_{..4.} = 10 + 9 + 5 + 6 + 5 = 35$$

$$T_{..5.} = 8 + 8 + 4 + 3 + 5 = 28$$

$$T_{...1} = 8 + 8 + 4 + 5 + 5 = 30$$

$$T_{...2} = 9 + 9 + 3 + 5 + 5 = 31$$

$$T_{...3} = 10 + 7 + 7 + 6 + 5 = 35$$

$$T_{...4} = 9 + 9 + 5 + 3 + 5 = 31$$

$$T_{...5} = 8 + 6 + 3 + 6 + 5 = 28$$

Therefore, the computations for the sums of squares yield

$$SS = 9^2 + 9^2 + \cdots + 5^2 - \frac{155^2}{25}$$

$$= 2{,}175 - 961 = 1214$$

$$SS_i = \frac{1}{5}[35^2 + 33^2 + \cdots + 27^2] - 961$$

$$= 970.2 - 961.0 = 9.2$$

$$SS_j = \frac{1}{5}[44^2 + 39^2 + \cdots + 25^2] - 961$$

$$= 1{,}038.2 - 961 = 77.2$$

$$SS_k = \frac{1}{5}[31^2 + 29^2 + \cdots + 28^2] - 961$$

$$= 967 - 961 = 6$$

$$SS_m = \frac{1}{5}[30^2 + 31^2 + \cdots + 28^2] - 961$$

$$= 966.2 - 961 = 5.2$$

$$SS_E = 1{,}214 - 9.2 - 77.2 - 6 - 5.2 = 1{,}116.4$$

The Anova table for the analysis is shown in Table 11.2.6.

TABLE 11.2.6

Source of Variation	Sums of Squares	D.F.	Mean Square	EMS	*F*-Ratio
Insulation material	$SS_i = 9.2$	4	$\frac{9.2}{4} = 2.3$	(See Table 11.2.4)	
Wire size............	$SS_j = 77.2$	4	$\frac{77.2}{4} = 19.3$		
Insulation thickness............	$SS_k = 6$	4	$\frac{6}{4} = 1.5$		$\frac{1.5}{139.5} = .01$
Manufacturer........	$SS_m = 5.2$	4	$\frac{5.2}{4} = 1.3$		$\frac{1.3}{139.5} = .009$
Error...............	$SS_E = 1116.4$	8	$\frac{1116.4}{8} = 139.5$		
Total...............	1214	24			

Since $F_{.05;4,8} = 3.84$, the hypotheses of equal insulation thickness and equal manufacturer effects cannot be rejected.

PROBLEMS

11.2.1 Refer to Prob. 11.1.3 and suppose that the data represents a Graeco-Latin-square design. Provide the subscript numbers for such a design. Could this data be analyzed by the method of this section? Why?

11.2.2 In Prob. 11.1.4 assume that there are 5 brands of electrodes and the given data is applied to the following Graeco-Latin square.

Days	Machines 1	2	3	4	5
M	C_2	A_4	D_3	E_1	B_5
T	B_4	E_2	C_1	D_5	A_3
W	E_3	C_1	B_5	A_2	D_4
T	A_5	D_3	E_2	B_4	C_1
F	D_1	B_5	A_4	C_3	E_2

Analyze the data under these conditions.

12

Factorial Designs

12.0 INTRODUCTION

So far we have considered the cases where the major emphasis was on one specific factor. In the Graeco-Latin squares we introduced the case where two factors were considered in a limited manner. Now a more general discussion will be presented for the two-way classification, and later the three-way classification will be discussed. The new concept or rather the *extension* over the previous material is that more than one independent variable or factor is examined simultaneously. This extension is significant because the restriction of no interaction is not necessary.

The interaction of factors can be examined and hypotheses concerning them can be tested. Also, an economic advantage enters the picture here. In general, two separate experiments with one variable in each is more expensive than one experiment with two variables or factors. For example, we could test the breakdown voltages of transformers, and at the same time test core temperatures. Therefore the statistical inferences drawn from more than one factor experiment are usually more conclusive and broader in their applications. Experiments where two or more factors are examined simultaneously are called *factorial experiments*.

12.1 TWO-WAY CLASSIFICATION—FIXED-EFFECTS MODEL

The most elementary case of factorial experiments is the two-way classification and may be regarded as a natural extension of the randomized block designs. The relationships between these two designs will be more apparent as we state the underlying assumptions for the two-way classification. Before stating the assumptions, however, we will write a summary table for the data. Let us refer to Table 10.4.2. Suppose, instead of block and treatment we consider factor A and factor B respectively. Furthermore, instead of one observation per cell, we have K observations. Then the new table will be something like that shown in Table 12.1.1.

In Table 10.4.2 we had I blocks and J treatments. In Table 12.1.1 we have I *levels* for factor A and J *levels* for factor B. The dot notation

TABLE 12.1.1

A \ B	1	2	...	J	Totals	Means
1	x_{111} x_{112} $\vdots$ x_{11K}	x_{121} x_{122} $\vdots$ x_{12K}	... ...	x_{1J1} x_{1J2} $\vdots$ x_{1JK}	$T_{1..}$	$\bar{x}_{1..}$
2	x_{211} x_{212} $\vdots$ x_{21K}	x_{221} x_{222} $\vdots$ x_{22K}	... ...	x_{2J1} x_{2J2} $\vdots$ x_{2JK}	$T_{2..}$	$\bar{x}_{2..}$
3	x_{311} x_{312} $\vdots$ x_{31K}	x_{321} x_{322} $\vdots$ x_{32K}	... ...	x_{3J1} x_{3J2} $\vdots$ x_{3JK}	$T_{3..}$	$\bar{x}_{3..}$
$\vdots$	$\vdots$	$\vdots$		$\vdots$	$\vdots$	$\vdots$
I	x_{I11} x_{I12} $\vdots$ x_{I1K}	x_{I21} x_{I22} $\vdots$ x_{I2K}	... ...	x_{IJ1} x_{IJ2} $\vdots$ x_{IJK}	$T_{I..}$	$\bar{x}_{I..}$
Totals	$T_{.1.}$	$T_{.2.}$	...	$T_{.J.}$	$T_{...}$	
Means	$\bar{x}_{.1.}$	$\bar{x}_{.2.}$	...	$\bar{x}_{.J.}$		$\bar{x}_{...}$

is the same as was in randomized blocks. For example:

$$\sum_{k=1}^{K} x_{ijk} = T_{ij.}$$

$$\sum_{j=1}^{J} \sum_{k=1}^{K} x_{ijk} = T_{i..}$$

and

$$\sum_{i=1}^{I} \sum_{j=1}^{J} \sum_{k=1}^{K} x_{ijk} = T_{...}$$

Now we state the underlying assumptions for the fixed-effects model. The assumptions are:

A random sample of size K is drawn from each of the IJ populations all of which are normally distributed with equal variance, σ^2.

Notice that in the present case we do not include the assumption of no interaction. Furthermore, in the randomized block design the randomization is by blocks, whereas in the two-way classification of factorial design, the randomization is over all cells.

In order to carry out the analysis we follow an analogous pattern as in randomized block design, and use the identity

$$\mu_{ij} = \mu_{..} + (\mu_{i.} - \mu_{..}) + (\mu_{.j} - \mu_{..}) + (\mu_{ij} - \mu_{i.} - \mu_{.j} + \mu_{..}) \quad (12.1.1)$$

which is the same as Eq. (10.4.6). Again letting

$$\alpha_i = \mu_{i.} - \mu_{..} \quad (12.1.2)$$

$$\beta_j = \mu_{.j} - \mu_{..} \quad (12.1.3)$$

and defining

$$(\alpha\beta)_{ij} = \mu_{ij} - \mu_{i.} - \mu_{.j} + \mu_{..} \quad (12.1.4)$$

Eq. (12.1.1) may be written as

$$\mu_{ij} = \mu_{..} + \alpha_i + \beta_j + (\alpha\beta)_{ij} \quad (12.1.5)$$

Adding x_{ijk} to both sides of Eq. (12.1.5) and rewriting we obtain

$$x_{ijk} = x_{ijk} - \mu_{ij} + \mu_{..} + \alpha_i + \beta_j + (\alpha\beta)_{ij} \quad (12.1.6)$$

Let

$$x_{ijk} - \mu_{ij} = \epsilon_{(ij)k} \quad (12.1.7)$$

so that

$$x_{ijk} = \mu_{..} + \alpha_i + \beta_j + (\alpha\beta)_{ij} + \epsilon_{(ij)k} \quad (12.1.8)$$

It should be evident from the assumptions that $\epsilon_{(ij)k}$ is normally distributed with mean zero and variance σ^2.

Notice that the random-error term, $\epsilon_{(ij)k}$, has been written with parentheses around the subscript letters ij. This notion was not discussed prior to now. Thus far in the assumed mathematical models the random error was over each observation but here the error is assumed to be due to the K observations in each of the IJ cells. This implies that the K observations are taken within each of the IJ cells, and referred to as error being *nested* in each cell. *Nesting* of the replications means that the kth observation in the ijth cell has no correspondence to the kth observation in the i^*j^*th cell for all i, j, i^*, j^*. For this reason the random error ϵ_{ijk} is represented by $\epsilon_{(ij)k}$ which indicates that the replications are nested in the factors A and B. More will be said about nestedness subsequently; in particular when factors are considered to be nested within each other. In Eq. (12.1.8), α_i represents the row effects, β_j represents the column effects, and $(\alpha\beta)_{ij}$ represents interaction effect. The interaction effect is that effect which the combination of factors A and B produce, and cannot be attributed to either factor *alone*.

Equation (12.1.8) states that each observation from the IJ populations can be accounted for by five expressions, namely:

1. Overall mean of means, $\mu_{..}$
2. A constant arising from factor A, α_i
3. A constant arising from factor B, β_j
4. A constant arising from interaction of A and B, $(\alpha\beta)_{ij}$
5. A random error, $\epsilon_{(ij)k}$

Similar hypotheses to the ones discussed before can be considered here.

1. H_0: $\alpha_i = 0 \qquad i = 1, 2, \ldots, I$
   H_1: $\alpha_i \neq 0 \qquad$ for some i

With this the hypothesis of equal-treatment effect for factor A can be tested.

2. H_0: $\beta_j = 0 \qquad j = 1, 2, \ldots, J$
   H_1: $\beta_j \neq 0 \qquad$ for some j

With this the hypothesis of equal-treatment effect for factor B can be tested.

3. H_0: $(\alpha\beta)_{ij} = 0 \qquad$ For all $i = 1, 2, \ldots, I$; $j = 1, 2, \ldots, J$
   H_1: $(\alpha\beta)_{ij} \neq 0 \qquad$ for some i and j

And with this the hypothesis of no interaction for factors A and B can be tested.

The corresponding sample model will be patterned after Eq. (10.1.8). Consider the identity

$$x_{ijk} - \bar{x}_{...} = (\bar{x}_{i..} - \bar{x}_{...}) + (\bar{x}_{.j.} - \bar{x}_{...}) + (\bar{x}_{ij.} - \bar{x}_{i..} - \bar{x}_{.j.} + \bar{x}_{...}) + (x_{ijk} - \bar{x}_{ij.}) \qquad (12.1.9)$$

Squaring both sides of Eq. (12.1.9) and summing over i, j, and k, we obtain

$$\begin{aligned}\sum_{i=1}^{I}\sum_{j=1}^{J}\sum_{k=1}^{K}(x_{ijk} - \bar{x}_{...})^2 &= \sum_{i=1}^{I}\sum_{j=1}^{J}\sum_{k=1}^{K}(\bar{x}_{i..} - \bar{x}_{...})^2 \\ &+ \sum_{i=1}^{I}\sum_{j=1}^{J}\sum_{k=1}^{K}(\bar{x}_{.j.} - \bar{x}_{...})^2 \\ &+ \sum_{i=1}^{I}\sum_{j=1}^{J}\sum_{k=1}^{K}(\bar{x}_{ij.} - \bar{x}_{i..} - \bar{x}_{.j.} + \bar{x}_{...})^2 \\ &+ \sum_{i=1}^{I}\sum_{j=1}^{J}\sum_{k=1}^{K}(x_{ijk} - \bar{x}_{ij.})^2 \\ &+ \sum_{i=1}^{I}\sum_{j=1}^{J}\sum_{k=1}^{K}\text{(6 cross-product terms)} \qquad (12.1.10)\end{aligned}$$

When the 6 cross-product terms are summed over i, j, and k, all will yield a value of zero.

Let

$$SS = \sum_{i=1}^{I}\sum_{j=1}^{J}\sum_{k=1}^{K} (x_{ijk} - \bar{x}_{...})^2 \tag{12.1.11}$$

$$SS_i = \sum_{i=1}^{I}\sum_{j=1}^{J}\sum_{k=1}^{K} (\bar{x}_{i..} - \bar{x}_{...})^2 \tag{12.1.12}$$

$$SS_j = \sum_{i=1}^{I}\sum_{j=1}^{J}\sum_{k=1}^{K} (\bar{x}_{.j.} - \bar{x}_{...})^2 \tag{12.1.13}$$

$$SS_{ij} = \sum_{i=1}^{I}\sum_{j=1}^{J}\sum_{k=1}^{K} (\bar{x}_{ij.} - \bar{x}_{i..} - \bar{x}_{.j.} + \bar{x}_{...})^2 \tag{12.1.14}$$

$$SS_{(ij)k} = \sum_{i=1}^{I}\sum_{j=1}^{J}\sum_{k=1}^{K} (x_{ijk} - \bar{x}_{ij.})^2 \tag{12.1.15}$$

So that Eq. (12.1.10) can be written as

$$SS = SS_i + SS_j + SS_{ij} + SS_{(ij)k} \tag{12.1.16}$$

As in the previous cases, computational formulas can be obtained for each term of Eq. (12.1.16):

$$SS = \sum_{i=1}^{I}\sum_{j=1}^{J}\sum_{k=1}^{K} x_{ijk}^2 - \frac{T_{...}^2}{IJK} \tag{12.1.17}$$

$$SS_i = \frac{\sum_{i=1}^{I} T_{i..}^2}{JK} - \frac{T_{...}^2}{IJK} \tag{12.1.18}$$

$$SS_j = \frac{\sum_{j=1}^{J} T_{.j.}^2}{IK} - \frac{T_{...}^2}{IJK} \tag{12.1.19}$$

$$SS_{(ij)k} = \sum_{i=1}^{I}\sum_{j=1}^{J}\sum_{k=1}^{K} x_{ijk}^2 - \frac{\sum_{i=1}^{I}\sum_{j=1}^{J} T_{ij.}^2}{K} \tag{12.1.20}$$

$$SS_{ij} = SS - SS_i - SS_j - SS_{(ij)k} \tag{12.1.21}$$

Now we need to obtain the expected values for SS_i, SS_j, SS_{ij}, and $SS_{(ij)k}$. The pattern for derivation of the expected values was shown in Sec. 10.2. Following the same type of reasoning it can be shown that

$$E[SS_i] = (I - 1)\sigma^2 + JK \sum_{i=1}^{I} \alpha_i^2 \tag{12.1.22}$$

$$E[SS_j] = (J - 1)\sigma^2 + IK \sum_{j=1}^{J} \beta_j^2 \tag{12.1.23}$$

$$E[SS_{ij}] = (I - 1)(J - 1)\sigma^2 + K\sum_{i=1}^{I}\sum_{j=1}^{J}(\alpha\beta)_{ij}^2 \quad (12.1.24)$$

$$E[SS_{(ij)k}] = IJ(K - 1)\sigma^2 \quad (12.1.25)$$

These last four relations can be equivalently written and denoted as

$$MS_A = \frac{E[SS_i]}{I - 1} = \sigma^2 + JK\frac{\sum_{i=1}^{I}\alpha_i^2}{I - 1} \quad (12.1.26)$$

$$MS_B = \frac{E[SS_j]}{J - 1} = \sigma^2 + IK\frac{\sum_{j=1}^{J}\beta_j^2}{J - 1} \quad (12.1.27)$$

$$MS_{AB} = \frac{E[SS_{ij}]}{(I - 1)(J - 1)} = \sigma^2 + K\frac{\sum_{i=1}^{I}\sum_{j=1}^{J}(\alpha\beta)_{ij}^2}{(I - 1)(J - 1)} \quad (12.1.28)$$

$$MS_E = \frac{E[SS_{(ij)k}]}{IJ(K - 1)} = \sigma^2 \quad (12.1.29)$$

Now we can test the three proposed hypotheses. The correct F-ratios to use are

$$F[I - 1, IJ(K - 1)] = \frac{MS_A}{MS_E}$$

$$F[J - 1, IJ(K - 1)] = \frac{MS_B}{MS_E}$$

$$F[(I - 1)(J - 1), IJ(K - 1)] = \frac{MS_{AB}}{MS_E}$$

respectively. Table 12.1.2 is the summary of the fixed-effects model for the two-way classification.

TABLE 12.1.2
ANOVA FOR TWO-WAY CLASSIFICATION
(Fixed-Effects)

Source of Variation	SS	D.F.	Mean Square	EMS	F-Ratio
Factor A	SS_i	$I - 1$	$MS_A = \frac{SS_i}{I - 1}$	Eq. (12.1.26)	$\frac{MS_A}{MS_E}$
Factor B	SS_j	$J - 1$	$MS_B = \frac{SS_j}{J - 1}$	Eq. (12.1.27)	$\frac{MS_B}{MS_E}$
Interaction of A and B...........	SS_{ij}	$(I - 1)(J - 1)$	$MS_{AB} = \frac{SS_{ij}}{(I - 1)(J - 1)}$	Eq. (12.1.28)	$\frac{MS_{AB}}{MS_E}$
Error...............	$SS_{(ij)k}$	$IJ(K - 1)$	$MS_E = \frac{SS_{(ij)k}}{IJ(K - 1)}$	σ^2	
Total...............	SS	$IJK - 1$			

Example 12.1.1 Suppose an experiment is performed in transformer-core winding. There are 4 methods of winding the cores and there are 3 types of wires. Fifteen cores will be wound by each method, and the 3 types of wires are randomly assigned, such that there will be 5 replications of each wire. The coded results were as follows:

TABLE 12.1.3

Factor A	Factor B (Methods) 1	2	3	4	
Types of Wire 1	9 4 9 4 9	3 10 7 6 4	6 3 5 2 4	4 8 10 6 12	
	$T_{11\cdot} = 35$	$T_{12\cdot} = 30$	$T_{13\cdot} = 20$	$T_{14\cdot} = 40$	$T_{1\cdot\cdot} = 125$
2	6 8 10 2 14	6 7 10 9 8	5 4 6 3 2	16 14 20 15 25	
	$T_{21\cdot} = 40$	$T_{22\cdot} = 40$	$T_{23\cdot} = 20$	$T_{24\cdot} = 90$	$T_{2\cdot\cdot} = 190$
3	2 7 4 6 1	2 4 1 2 1	2 3 2 3 5	10 7 12 10 1	
	$T_{31\cdot} = 20$	$T_{32\cdot} = 10$	$T_{33\cdot} = 15$	$T_{34\cdot} = 40$	$T_{3\cdot\cdot} = 85$
	$T_{\cdot 1\cdot} = 95$	$T_{\cdot 2\cdot} = 80$	$T_{\cdot 3\cdot} = 55$	$T_{\cdot 4\cdot} = 170$	$T_{\cdot\cdot\cdot} = 400$

(a) Test the equality hypothesis of equal-treatment effect for factor A and factor B.

(b) Test the hypothesis of no interaction.

(c) If the results in part (a) prove significant, find a comparison to which the rejection can be attributed.

Solution: The computation of the sums of squares yields

$$SS = 9^2 + 4^2 + \cdots + 1^2 - \frac{(400)^2}{60}$$

$$= 4{,}034 - \frac{160{,}000}{60} = 4{,}034 - 2{,}666.67 = 1{,}367.33$$

$$SS_i = \frac{125^2 + 190^2 + 85^2}{20} - \frac{(400)^2}{60}$$

$$= \frac{58{,}950}{20} - \frac{160{,}000}{60} = 2{,}947.50 - 2{,}666.67 = 280.83$$

$$SS_j = \frac{95^2 + 80^2 + 55^2 + 170^2}{15} - \frac{(400)^2}{60}$$

$$= \frac{47{,}350}{15} - \frac{160{,}000}{60} = 3{,}156.67 - 2{,}666.67$$

$$= 490$$

$$SS_{(ij)k} = 9^2 + 4^2 + \cdots + 1^2 - \frac{35^2 + 30^2 + \cdots + 40^2}{5}$$

$$= 4{,}034 - \frac{18{,}150}{5}$$

$$= 4{,}034 - 3{,}630 = 404$$

$$SS_{ij} = SS - SS_i - SS_j - SS_{(ij)k} = 192.50$$

The following Anova table gives the summary for this experiment.

TABLE 12.1.4

Source of Variation	SS	D.F.	Mean Squares	EMS	F-Ratio
Between Factor A	280.83	2	140.41	$\sigma^2 + 10\Sigma\alpha_i^2$	16.68**
Between Factor B	490.00	3	163.33	$\sigma^2 + 5\Sigma\beta_j^2$	19.41**
Interaction between A and B .	192.50	6	32.08	$\sigma^2 + \frac{5}{6}\Sigma\Sigma(\alpha\beta)_{ij}^2$	3.81**
Error	404.00	48	8.42	σ^2	
Total	1,367.33	59			

(a) Since $F_{.01;2,48} = 5.10$ and $F_{.01;3,48} = 4.20$, the hypotheses of equal treatment effects for factor A and for factor B are rejected.

(b) Since $F_{.01;6,48} = 3.21$, the hypothesis of no interaction is also rejected.

(c) Since the results show significantly different effects, we can use Scheffé's method to test other hypotheses. For example, we can test the hypothesis that method 1 and method 4 of winding are the same, that is, $\mu_{.1} = \mu_{.4}$; or the hypothesis that type 1 wire and type 3 wire have the same effect, $\mu_{1.} - \mu_{3..}$ We shall calculate these two hypotheses on the .01 level of significance. The first:

$$\bar{x}_{.1} - \bar{x}_{.4} = \frac{1}{15}[95 - 170] = -5$$

$$K = (4 - 1)(8.42)\left[\frac{1}{15} + \frac{1}{15}\right] F_{.01;3,48}$$

$$= (3)(8.42)\left(\frac{2}{15}\right)(4.20)$$

$$= 14.11$$

$$\sqrt{K} = 3.76$$

Since the interval (-5 ± 3.76) does not include zero, the hypothesis $\mu_{.1} = \mu_{.4}$ must be rejected, and we have located one possible source responsible for the rejection of the equal methods effects of the original hypothesis.

The second:

$$\bar{x}_{1.} - \bar{x}_{3.} = \frac{1}{20}[125 - 85] = 2$$

$$K = (3 - 1)(8.42)\left[\frac{1}{20} + \frac{1}{20}\right] F_{.01;2,48}$$

$$= (2)(8.42)(.1)(5.10) = 8.588$$

$$\sqrt{K} = 2.93$$

Since the interval (2 ± 2.93) does include zero, the hypothesis $\mu_{1.} = \mu_{3.}$ cannot be rejected and this comparison is not regarded responsible for the rejection of equal wire effects of the original hypothesis.

In Example 12.1.1, interaction of factors A and B was shown to be highly significant, and hence sources for interaction effect should be considered. There are complex analytic as well as simple graphic methods to locate sources of interaction. In order to make use of the graphical method, a physical interpretation of interaction might be helpful. Interaction is the change in response for factor A due to the simultaneous combination of different levels of factor B with factor A, (or vice versa). To illustrate this, let us consider the following example.

Example 12.1.2 Using a graphic method interpret the interactions of factors A and B, in the following data.

TABLE 12.1.5

Factor A	Factor B 1	2	3	4
1	$\bar{x}_{11.} = 7$	$\bar{x}_{12.} = 6$	$\bar{x}_{13.} = 4$	$\bar{x}_{14.} = 7$
2	$\bar{x}_{21.} = 4$	$\bar{x}_{22.} = 2$	$\bar{x}_{23.} = 3$	$\bar{x}_{24.} = 8$
3	$\bar{x}_{31.} = 9$	$\bar{x}_{32.} = 8$	$\bar{x}_{33.} = 2$	$\bar{x}_{34.} = 4$

Solution: Graphing levels 1 and 2 of factor A, we get Fig. 12.1. The graph shows the source for interaction for these two factors to be in the simultaneous combination of levels 1 and 2 of factor A and levels 3 and 4 of factor B, that is, where the graph shows intersection. Therefore, we ascribe the interaction effect partly to the mentioned combinations of levels.

Before we discuss other mathematical models, let us consider the special case where $K = 1$. That is, the situation where there are single

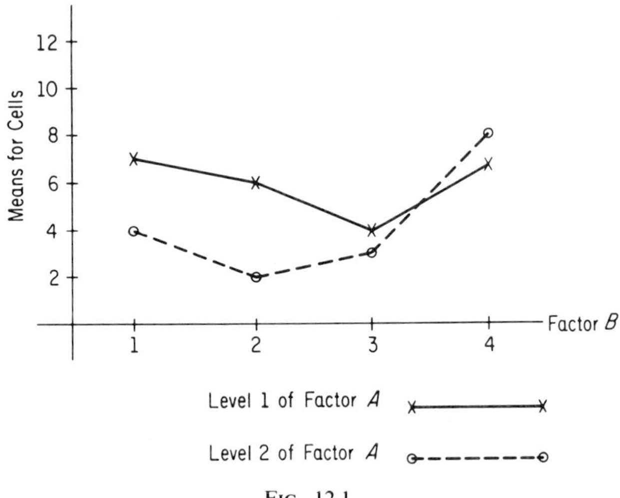

FIG. 12.1.

observations per cell. Referring to Eq. (12.1.29) we note that Error degrees of freedom becomes zero and MS_E is undefined. Also, the expected mean squares for factor A and factor B reduce to

$$\sigma^2 + \frac{J\Sigma\alpha_i^2}{I-1} \quad \text{and} \quad \sigma^2 + \frac{I\Sigma\beta_j^2}{J-1}$$

Next, assuming no interaction and referring to the table of the randomized block design, Table 10.4.4, we note that the EMS values are identical to the present case. Therefore when $K = 1$ the correct F-ratios for testing hypotheses of equality of means for factor A and factor B are

$$\frac{MS_A}{MS_{AB}} \quad \text{and} \quad \frac{MS_B}{MS_{AB}} \quad \text{respectively}$$

Notice the importance of determining which F-ratio to use in testing a specific hypothesis. This notion will be discussed and elaborated on in subsequent sections.

PROBLEMS

12.1.1 Show that any one of the six possible cross-product terms in Eq. (12.1.10) has zero value.

12.1.2 The following data represents the result of a completely randomized 3×2 factorial experiment with 3 replications. The observations are yields of 3 types of wheat and 2 kinds of fertilizers.

		Types of Wheat 1	2	3
Fertilizers	A	128	121	117
		136	117	122
		132	120	125
	B	126	130	115
		123	127	121
		121	121	119

Analyze this experiment at the .05 significance level for the row and column effects.

12.1.3 Refer to Example 12.1.2. Graph levels (a) 1 and 3 (b) 2 and 3 to locate other possible sources of interaction.

12.1.4 In Prob. 12.1.2 test the hypothesis of no interaction. If the hypothesis is rejected, use graphic method to locate sources of interactions.

12.1.5 To test the effectiveness of 3 different teaching methods, 4 teachers were randomly assigned 9 students each. These students were then randomly assigned to the different teaching methods and taught the same topics. At the conclusion of the experiment, identical examinations were given to the students with the result as follows:

		Teachers A	B	C	D
Methods	1	88	60	95	84
		74	90	70	82
		72	71	81	75
	2	90	76	89	81
		80	91	82	85
		92	72	90	70
	3	80	86	99	83
		85	80	81	72
		70	73	82	70

Analyze this data using $\alpha = .05$.

12.1.6 Show that

$$\sum_{i=1}^{I} \alpha_i = \sum_{j=1}^{J} \beta_j = \sum_{i=1}^{I} (\alpha\beta)_{ij} = \sum_{j=1}^{J} (\alpha\beta)_{ij} = 0$$

12.2 TWO-WAY CLASSIFICATION—RANDOM-EFFECTS MODEL

The primary difference between the fixed-effects model and this case lies in the levels of factors A and B. The implication in the random-effects model is that the I levels of Factor A and J levels of Factor B do represent a random sample from a *larger* set of populations, such that inferences could be made to the totality of the levels involved. The calculations for the various sums of squares and mean squares of the random-effects model are *identical* with the fixed-effects model. The only difference is the underlying assumption for the mathematical model, which *alters* the expected mean square values.

In Eq. (12.1.8) the assumed model for the fixed-effects case was

$$x_{ijk} = \mu_{..} + \alpha_i + \beta_j + (\alpha\beta)_{ij} + \epsilon_{(ij)k} \tag{12.2.1}$$

where the $\epsilon_{(ij)k}$ was assumed to be normally distributed with mean zero and variance σ^2. In the random-effects model the same model will be used with the further assumptions that α_i is normally distributed with mean zero and variance σ_α^2, β_j is normally distributed with mean zero and variance σ_β^2, and $(\alpha\beta)_{ij}$ is normally distributed with mean zero and variance $\sigma_{\alpha\beta}^2$.

With these assumptions it can be shown that

$$MS_A = E\left[\frac{SS_i}{I-1}\right] = \sigma^2 + JK\sigma_\alpha^2 + K\sigma_{\alpha\beta}^2 \tag{12.2.2}$$

$$MS_B = E\left[\frac{SS_j}{J-1}\right] = \sigma^2 + IK\sigma_\beta^2 + K\sigma_{\alpha\beta}^2 \tag{12.2.3}$$

$$MS_{AB} = E\left[\frac{SS_{ij}}{(I-1)(J-1)}\right] = \sigma^2 + K\sigma_{\alpha\beta}^2 \tag{12.2.4}$$

$$MS_E = E\left[\frac{SS_{(ij)k}}{IJ(K-1)}\right] = \sigma^2 \tag{12.2.5}$$

Comparing the expected mean squares of this case with those of the fixed-effects model, we may note the close similarities between the two.

The hypotheses of equal factor A effects, equal factor B effects, and no interactions can be expressed as

1. H_0: $\sigma_\alpha^2 = 0$
   H_1: $\sigma_\alpha^2 \neq 0$
2. H_0: $\sigma_\beta^2 = 0$
   H_1: $\sigma_\beta^2 \neq 0$
3. H_0: $\sigma_{\alpha\beta}^2 = 0$
   H_1: $\sigma_{\alpha\beta}^2 \neq 0$

respectively.

In order to test these hypotheses, the proper F-ratios must be selected. If $\sigma_\alpha^2 = 0$, we note that Eqs. (12.2.2) and (12.2.4) have comparable values. Therefore to test the hypothesis $\sigma_\alpha^2 = 0$, the proper F-ratio is

$$F[(I-1), (I-1)(J-1)] = \frac{MS_A}{MS_{AB}} \tag{12.2.6}$$

Similarly, to test the hypothesis $\sigma_\beta^2 = 0$, the proper F-ratio is

$$F[(J-1), (I-1)(J-1)] = \frac{MS_B}{MS_{AB}} \tag{12.2.7}$$

Finally, to test the hypothesis $\sigma^2_{\alpha\beta} = 0$ we note that Eqs. (12.2.4) and (12.2.5) have comparable values, hence

$$F[(I - 1)(J - 1), IJ(K - 1)] = \frac{MS_{AB}}{MS_E} \tag{12.2.8}$$

is the proper F-ratio. Therefore, the changes that appear in a corresponding Anova table is as shown in the EMS and F-ratio columns in Table 12.2.1.

TABLE 12.2.1

Source of Variation	EMS	F-Ratio
Factor A	$\sigma^2 + JK\sigma^2_\alpha + K\sigma^2_{\alpha\beta}$	$\frac{MS_A}{MS_{AB}}$
Factor B	$\sigma^2 + IK\sigma^2_\alpha + K\sigma^2_{\alpha\beta}$	$\frac{MS_B}{MS_{AB}}$
Interaction of A and B	$\sigma^2 + K\sigma^2_{\alpha\beta}$	$\frac{MS_{AB}}{MS_E}$
Error	σ^2	

Example 12.2.1 Repeat Example 12.1.1 (a) if factors A and B are considered to be random.

Solution: To test the hypothesis $\sigma^2_\alpha = 0$, we get from Table 12.1.4

$$F = \frac{140.41}{32.08} = 4.37$$

Since $F_{.01;2,6} = 10.92$, factor A does not have a significantly different effect on the 1 percent level. That is, the 3 types of wires randomly chosen from a larger group have equal effects. To test the hypothesis $\sigma^2_\beta = 0$, we get

$$F = \frac{163.33}{32.08} = 5.09$$

Because $F_{.01;3,6} = 9.76$ factor B does not have a significantly different effect on the 1 percent level either.

PROBLEMS

12.2.1 Repeat Prob. 12.1.2, assuming the types of wheat and the kinds of fertilizers are selected at random from many other available types of wheat and kinds of fertilizers.

12.2.2 Repeat Prob. 12.1.5, assuming that the four teachers and the three methods of teaching are selected at random from a larger group of teachers and teaching methods.

12.3 TWO-WAY CLASSIFICATION—MIXED-EFFECTS MODEL

Sometimes in a two-way classification occasions arise when one of the factors, say A, is fixed while B is random, or conversely B is fixed while A is random. These two possible situations are called *mixed-effects* model and will be referred to as case I and case II of mixed-effects models, respectively.

The computations for the various sums of squares and expected mean squares are similar to the fixed-effects model. As in the case of random-effects model, here also, the only changes are in the expected mean squares which alters the F-ratios. In the consideration of the hypotheses regarding the factor A and factor B, the assumptions depend on whether factors A and B are fixed and random or random and fixed respectively. The factor that is assumed fixed carries with itself the assumptions of the fixed-effects model and the factor that is random carries with itself the assumptions of random-effects model.

Therefore the assumptions of mixed-effects case I are

1. $\alpha_i = 0 \qquad i = 0, 1, 2, \ldots, I$
2. β_j's are normally distributed with mean zero and variance σ_β^2
3. $(\alpha\beta)_{ij}$'s are normally distributed with mean zero and variance $\sigma_{\alpha\beta}^2$

In case II the roles of α and β are simply interchanged.

For case I, the various EMS are as follows:

$$MS_A = E\left[\frac{SS_i}{I-1}\right] = \sigma^2 + \frac{JK\Sigma\alpha_i^2}{I-1} + K\sigma_{\alpha\beta}^2 \tag{12.3.1}$$

$$MS_B = E\left[\frac{SS_j}{J-1}\right] = \sigma^2 + IK\sigma_\beta^2 \tag{12.3.2}$$

$$MS_{AB} = E\left[\frac{SS_{ij}}{(I-1)(J-1)}\right] = \sigma^2 + K\sigma_{\alpha\beta}^2 \tag{12.3.3}$$

$$MS_E = E\left[\frac{SS_{(ij)k}}{IJ(K-1)}\right] = \sigma^2 \tag{12.3.4}$$

The hypothesis of equality of means for factor A is tested by the F ratio

$$F[(I-1), (I-1)(J-1)] = \frac{MS_A}{MS_{AB}} \tag{12.3.5}$$

because MS_A and MS_{AB} have comparable values if H_0 is true.

The hypothesis of equality of means for factor B is tested by

$$F[(J-1), IJ(K-1)] = \frac{MS_B}{MS_E} \tag{12.3.6}$$

Lastly, to test the interaction effects the correct F-ratio to use is,

$$F[(I-1)(J-1), IJ(K-1)] = \frac{MS_{AB}}{MS_E} \tag{12.3.7}$$

For case II, the various EMS are as follows:

$$MS_A = E\left[\frac{SS_i}{I-1}\right] = \sigma^2 + JK\sigma_\alpha^2 \tag{12.3.8}$$

$$MS_B = E\left[\frac{SS_j}{J-1}\right] = \sigma^2 + \frac{IK\Sigma\beta_j^2}{J-1} + K\sigma_{\alpha\beta}^2 \tag{12.3.9}$$

$$MS_{AB} = E\left[\frac{SS_{ij}}{(I-1)(J-1)}\right] = \sigma^2 + K\sigma_{\alpha\beta}^2 \tag{12.3.10}$$

$$MS_E = E\left[\frac{SS_{(ij)k}}{IJ(K-1)}\right] = \sigma^2 \tag{12.3.11}$$

The hypothesis of equality of means for factor A is tested by

$$F[(I-1), IJ(K-1)] = \frac{MS_A}{MS_E} \tag{12.3.12}$$

The hypothesis of equality of means for factor B is tested by

$$F[(J-1), (I-1)(J-1)] = \frac{MS_B}{MS_{AB}} \tag{12.3.13}$$

and the hypothesis of no interaction effect is tested by

$$F[(I-1)(J-1), IJ(K-1)] = \frac{MS_{AB}}{MS_E} \tag{12.3.14}$$

In each of the mathematical models discussed thus far, the reader may have recognized the basic reason for stipulating the particular mathematical model. Based upon the assumed model, the expected mean squares were determined which in turn were used to find comparable F-ratios in testing hypotheses concerning various effects. Therefore the gist of the analysis is the determining of the proper EMS values for a given experiment from the assumed model. If for every experiment the EMS values were to be computed as derived in Sec. 10.2, the difficulty of the task need not be elaborated.

There are several methods described in literature whereby one can determine the EMS values. C. A. Bennett and N. L. Franklin in their text *Statistical Analysis in Chemistry and the Chemical Industry* (New York: Wiley, 1954) have presented a general outline by which the various EMS values can be determined. Professor Hicks of Purdue University has presented another method for determining the EMS values. However, a very quick and simple method is due to Schultz, which we shall present at this point, deferring the Hicks method until the next section.

SCHULTZ'S METHOD TO DETERMINE EMS VALUES

Step 1. Write out all the variances corresponding to each source of variation. For instance, in the case of two factors they will be

$$\sigma_\alpha^2 + \sigma_\beta^2 + \sigma_{\alpha\beta}^2 + \sigma^2$$

where σ^2 is the error variance.

Step 2. Each term of Step 1 except the error variance σ^2, should have for coefficient the letters corresponding to the missing subscript(s) in that term. For the illustration given is Step 1, we will get:

$$JK\sigma_\alpha^2 + IK\sigma_\beta^2 + K\sigma_{\alpha\beta}^2 + \sigma^2$$

Step 3. To obtain the desired EMS value, maintain only those terms in Step 2 which contain the desired subscript(s) and the error variance. For the example under consideration it yields

$$\begin{aligned} MS_A &= JK\sigma_\alpha^2 + K\sigma_{\alpha\beta}^2 + \sigma^2 \\ MS_B &= IK\sigma_\beta^2 + K\sigma_{\alpha\beta}^2 + \sigma^2 \\ MS_{AB} &= K\sigma_{\alpha\beta}^2 + \sigma^2 \end{aligned}$$

Step 4. *Any of the terms, other than the factor(s) under consideration*, obtained in Step 3 which contain subscript letter(s) corresponding to fixed effects, DELETE.

Step 5. Any variance which has a subscript corresponding to a fixed-effects should be replaced by the appropriate summation.

For the mixed-effects case I these rules will yield

1. $MS_A = \dfrac{JK\Sigma\alpha_i^2}{I - 1} + K\sigma_{\alpha\beta}^2 + \sigma^2$

Because of Step 3, $IK\sigma_\beta^2$ is not included and because of Step 4 the term $K\sigma_{\alpha\beta}^2$ is not deleted. Note that β corresponds to the random factor B. Also, because of Step 5, σ_α^2 is replaced by $\dfrac{\Sigma\alpha_i^2}{I - 1}$.

2. $MS_B - IK\sigma_\beta^2 + \sigma^2$

because from Step 3 we get

$$MS_B = IK\sigma_\beta^2 + K\sigma_{\alpha\beta}^2 + \sigma^2$$

and by Step 4, $K\sigma_{\alpha\beta}^2$ is deleted since α corresponds to the fixed factor A.

3. $MS_{AB} = K\sigma_{\alpha\beta}^2 + \sigma^2$

because the subscripts $\alpha\beta$ are jointly under consideration. Lastly,

4. $MS_E = \sigma^2$, which holds true under all situations.

In Table 12.3.1 we summarize the various models, and the corresponding expected mean squares and the F-ratios.

TABLE 12.3.1
SUMMARY TABLE FOR EXPECTED MEAN SQUARES

Model	Source of Variation	EMS	Significant Tests
Fixed-effects	A (fixed)	Eq. (12.1.26)	*All* effects are tested against error mean square.
	B (fixed)	Eq. (12.1.27)	
	AB	Eq. (12.1.28)	
	Error	σ^2	
Random-effects	A (random)	$\sigma^2 + JK\sigma_\alpha^2 + K\sigma_{\alpha\beta}^2$	A and B effects are tested against interaction mean square. Interaction effect tested against error mean square.
	B (random)	$\sigma^2 + IK\sigma_\beta^2 + K\sigma_{\alpha\beta}^2$	
	AB	$\sigma^2 + K\sigma_{\alpha\beta}^2$	
	Error	σ^2	
Mixed-effects, case I	A (fixed)	$\sigma^2 + \dfrac{JK\Sigma\alpha_i^2}{I-1} + K\sigma_{\alpha\beta}^2$	A effect is tested against interaction mean square. B and interaction effects are tested against error mean square.
	B (random)	$\sigma^2 + IK\sigma_\beta^2$	
	AB	$\sigma^2 + K\sigma_{\alpha\beta}^2$	
	Error	σ^2	
Mixed-effects, case II	A (random)	$\sigma^2 + JK\sigma^2$	B effect is tested against interaction mean square. A and interaction effects are tested against error mean square.
	B (fixed)	$\sigma^2 + \dfrac{IK\Sigma\beta_j^2}{J-1} + K\sigma_{\alpha\beta}^2$	
	AB	$\sigma^2 + K\sigma_{\alpha\beta}^2$	
	Error	σ^2	

PROBLEMS

12.3.1 Repeat Prob. 12.1.2, assuming the types of wheat are selected at random from a larger set of varieties of wheat.

12.3.2 Repeat Prob. 12.1.5, assuming that the three methods of teaching are selected at random from an infinite population of methods of teaching.

12.3.3 Write out the expected mean squares for the analysis of variance, if in Prob. 12.1.2 the kinds of fertilizers are assumed to be selected at random from a larger set of available fertilizers.

12.3.4 Write out the expected mean squares for the analysis of variance, if in Prob. 12.1.5 the teachers are assumed to be selected at random from an infinite population of teachers.

12.4 NESTED CLASSIFICATION

In Chapter 10, when randomized block designs were considered, an example concerning light bulbs was given in which the three blocks were of high, medium, and low wattage. It was tacitly assumed and implied there that high, medium, and low wattage had the same definition for the various treatments. For example, high wattage may have been defined 100 to 149 watts, medium wattage 50 to 99 watts, and low wattage 1 to 49 watts.

Earlier in this chapter when two-way classification was discussed it was again tacitly implied that the ith level of factor A for the *first* level of factor B is identical with ith level of factor A and *any other* level of factor B. This type of classification is called *cross-classification.* In contrast, suppose in Chapter 10 high wattage were defined 110–135 watts for treatment 1, 100–125 watts for treatment 2, and 120–140 watts for treatment 3. Such classifications are called *nested classification.* Some authors refer to this as *hierarchical classification.* To illustrate how nested classifications arise, consider the following situation. Three makes of automobiles are being examined, say, Chevrolet, Dodge, and Ford, to take names at random. Suppose we consider 6- and 8-cylinder cars. Then Table 12.4.1 may be constructed.

TABLE 12.4.1

	Chevrolet	Dodge	Ford
6-Cylinder	C_1	D_1	F_1
8-Cylinder	C_2	D_2	F_2

Let factor A be number of cylinders and factor B the types of cars. Obviously C_1 is not the same as D_1, i.e., the first level of factor A for the first level of factor *B is not identical* with the second level of factor B. (A 6-cylinder Chevrolet is not the same as a 6-cylinder Dodge!)

Another example of nested classification is the following. Four manufacturers, I, II, III, and IV have supplied three different types of transformers with two replications of each type. The following is a tabular representation of it.

TABLE 12.4.2

Manufacturers I			II			III			IV		
Types 1	2	3	Types 1	2	3	Types 1	2	3	Types 1	2	3
A_{11}	A_{21}	A_{31}	B_{11}	B_{21}	B_{31}	C_{11}	C_{21}	C_{31}	D_{11}	D_{21}	D_{31}
A_{12}	A_{22}	A_{32}	B_{12}	B_{22}	B_{32}	C_{12}	C_{22}	C_{32}	D_{12}	D_{22}	D_{32}

Now it should be clear that I's type 1 transformers are not necessarily the same kind as II's type 1 transformer. Similarly, III's type 2 transformers and IV's type 2 transformers may or may not have anything in common. Therefore this too is an example of nested classification.

In order to avoid confusion, it may be desirable to represent in the above data all types to *be different* hence $4 \times 3 = 12$ types of transformers are referred to. (See Table 12.4.3.)

TABLE 12.4.3

Manufacturers											
I			II			III			IV		
Types			Types			Types			Types		
1	2	3	4	5	6	7	8	9	10	11	12
A_1	A_3	A_5	B_1	B_3	B_5	C_1	C_3	C_5	D_1	D_3	D_5
A_2	A_4	A_6	B_2	B_4	B_6	C_2	C_4	C_6	D_2	D_4	D_6

It is apparent from this representation what sources of variation should be considered. They are:

1. Between manufacturers
2. Between types *within* manufacturers
3. Between samples *within* types

Let factor A represent manufacturers with levels $i = 1, 2, \ldots, I$ and let B represent types with levels $j = 1, 2, \ldots, J$. Therefore, there are IJ cells to consider. Let K represent the replications per cell. In the above example $IJ = 4 \times 3 = 12$, $K = 2$, and total observations are $12 \times 2 = 24$.

The underlying assumptions for the nested classification are the same as in the cross-classification. That is, a random sample of size K is drawn from each of the IJ populations, all of which are normally distributed with equal variances, σ^2. Next let us consider the following identity which constitutes the basis for analysis of nested classifications.

$$\mu_{ij} = \mu_{..} + (\mu_{i.} - \mu_{..}) + (\mu_{ij} - \mu_{i.}) \tag{12.4.1}$$

Let

$$\mu_{i.} - \mu_{..} = \alpha_i \tag{12.4.2}$$

$$\mu_{ij} - \mu_{i.} = \beta_{(i)j} \tag{12.4.3}$$

then Eq. (12.4.1) becomes

$$\mu_{ij} = \mu_{..} + \alpha_i + \beta_{(i)j} \tag{12.4.4}$$

and note that

$$\sum_{i=1}^{I} \alpha_i = 0 \qquad \sum_{j=1}^{J} \beta_{(i)j} = 0$$

Let us remark that the symbolism $\beta_{(i)j}$ indicates the factor associated with i is nested within the factor corresponding to j. Adding x_{ijk} to both sides of Eq. (12.4.4) and rewriting, we obtain

$$x_{ijk} = \mu_{..} + \alpha_i + \beta_{(i)j} + x_{ijk} - \mu_{ij} \tag{12.4.5}$$

Letting

$$x_{ijk} - \mu_{ij} = \epsilon_{(i)j}$$

Eq. (12.4.5) becomes

$$x_{ijk} = \mu_{..} + \alpha_i + \beta_{(i)j} + \epsilon_{(ij)k} \tag{12.4.6}$$

This equation states that each observation can be accounted for by four expressions, namely:

1. An overall mean of means, $\mu_{..}$
2. A constant arising from factor A, α_i
3. A constant arising from jth level of factor B, *within* each ith level of factor A, $\beta_{(i)j}$
4. A random error, $\epsilon_{(ij)k}$

Before going on with the analysis, let us refer to Eqs. (12.1.3) and (12.1.4). Notice that

$$\beta_j + (\alpha\beta)_{ij} = \mu_{ij} - \mu_{i\cdot} = \beta_{(i)j} \tag{12.4.7}$$

which indicates that no interaction effect $(\alpha\beta)_{ij}$ exists because β_j is nested within α_i. We mention this to point out similarities between the crossed and nested classifications.

Now the corresponding sample model of Eq. (12.4.6) can be written as

$$x_{ijk} - \bar{x}_{\ldots} = (\bar{x}_{i\cdot\cdot} - \bar{x}_{\ldots}) + (\bar{x}_{ij\cdot} - \bar{x}_{i\cdot\cdot}) + (x_{ijk} - \bar{x}_{ij\cdot}) \tag{12.4.8}$$

Squaring both sides of Eq. (12.4.8) and summing over i, j, and k, we obtain

$$\begin{aligned}\sum_{i=1}^{I}\sum_{j=1}^{J}\sum_{k=1}^{K}(x_{ijk} - \bar{x}_{\ldots})^2 &= JK\sum_{i=1}^{I}(\bar{x}_{i\cdot\cdot} - \bar{x}_{\ldots})^2 \\ &\quad + K\sum_{i=1}^{I}\sum_{j=1}^{J}(\bar{x}_{ij\cdot} - \bar{x}_{i\cdot\cdot})^2 \\ &\quad + \sum_{i=1}^{I}\sum_{j=1}^{J}\sum_{k=1}^{K}(x_{ijk} - \bar{x}_{ij\cdot})^2 \\ &\quad + \sum_{i=1}^{I}\sum_{j=1}^{J}\sum_{k=1}^{K}\text{(3 cross-product terms)}\end{aligned} \tag{12.4.9}$$

As in the previous cases, the cross-product terms vanish when summed over i, j, and k. Let

$$SS = \sum_{i=1}^{I}\sum_{j=1}^{J}\sum_{k=1}^{K}(x_{ijk} - \bar{x}_{\ldots})^2 \tag{12.4.10}$$

$$SS_i = JK\sum_{i=1}^{I}(\bar{x}_{i\cdot\cdot} - \bar{x}_{\ldots})^2 \tag{12.4.11}$$

$$SS_{(i)j} = K\sum_{i=1}^{I}\sum_{j=1}^{J}(\bar{x}_{ij\cdot} - \bar{x}_{i\cdot\cdot})^2 \tag{12.4.12}$$

$$SS_{(ij)k} = \sum_{i=1}^{I}\sum_{j=1}^{J}\sum_{k=1}^{K}(x_{ijk} - \bar{x}_{ij\cdot})^2 \tag{12.4.13}$$

In the last expression $(ij)k$ implies that the variable associated with k is nested within the (ij)th cell. That is, the error variation is nested within factors A and B. Now note that

$$SS = SS_i + SS_{(i)j} + SS_{(ij)k} \tag{12.4.14}$$

The computational formulas for the above sums of squares are as follows

$$SS = \sum_{i=1}^{I}\sum_{j=1}^{J}\sum_{k=1}^{K} x_{ijk}^2 - \frac{T_{...}^2}{IJK} \tag{12.4.15}$$

$$SS_i = \frac{\sum_{i=1}^{I} T_{i..}^2}{JK} - \frac{T_{...}^2}{IJK} \tag{12.4.16}$$

$$SS_{(i)j} = \frac{\sum_{i=1}^{I}\sum_{j=1}^{J} T_{ij.}^2}{K} - \frac{\sum_{i=1}^{I} T_{i..}^2}{JK} \tag{12.4.17}$$

$$SS_{(ij)k} = SS - SS_i - SS_{(i)j} \tag{12.4.18}$$

The degrees of freedom associated with the sums of squares are as follows:

S.S.	Degrees of Freedom
SS_i	$I - 1$
$SS_{(i)j}$	$I(J - 1)$
$SS_{(ij)k}$	$IJ(K - 1)$

In order to complete the analysis the expected mean squares have to be derived for determining the proper F-ratios. Schultz's method described in the previous section could be used to obtain the EMS values but we will present next another procedure due to Hicks which is particularly useful for nested classifications.

HICKS' METHOD OF DETERMINING EMS VALUES

Step 1. Write out all the variables along with the associated subscript(s) in the mathematical model as row headings in a two-way table, and the subscripts used with the variables as column headings. For instance, in a two-way classification with K observations per cell we get

	i	j	k
α_i			
β_j			
$(\alpha\beta)_{ij}$			
$\epsilon_{(ij)k}$			

Step 2. Indicate by F or R which subscripts correspond to fixed levels of a factor and which to random levels, remembering that *errors are always considered random*. In the illustration of Step 1, if we assume factor A fixed and B random then, we get mixed-effects model case I.

F	R	R
i	j	k

Step 3. Examine each row heading subscript(s) and whichever subscript(s) is (are) missing, enter in the table the number of observations under the corresponding column heading.

In α_i the missing subscripts are j and k. Hence J and K are entered in the j and k columns. To continue with β_j, $(\alpha\beta)_{ij}$, and $\epsilon_{(ij)k}$ we obtain

	F	R	R
	i	j	k
α_i		J	K
β_j	I		K
$(\alpha\beta)_{ij}$			K
$\epsilon_{(ij)k}$			

Step 4. For any subscript letter(s) in the model which is (are) bracketed (nested within), place a 1 under the corresponding column heading(s). In the example under consideration we have $\epsilon_{(ij)k}$ and hence enter 1 in columns i and j.

	F	R	R
	i	j	k
α_i		J	K
β_j	I		K
$(\alpha\beta)_{ij}$			K
$\epsilon_{(ij)k}$	1	1	

Step 5. Fill in the remaining cells with 0 or 1, depending whether the subscript represents a fixed or random factor respectively.

The completed table for the example being discussed is

	F	R	R
	i	j	k
α_i	0	J	K
β_j	I	1	K
$(\alpha\beta)_{ij}$	0	1	K
$\epsilon_{(ij)k}$	1	1	1

To find the EMS values for the various factors, proceed as follows:

1. Cover the entries of the column(s) which contain nonbracketed subscript letter(s) in the corresponding term of the mathematical model. For example, in the computation of the EMS value for factor A, the term under question is α_i, hence cover the column containing the letter i.

2. Cover the row(s) which *does* (*do*) *not* contain the subscript letter(s) under question. In the above example it would be the row corresponding

to β_j. The entries remaining are

	F	R	R
	i	j	k
α_i	0	J	K
β_j	I	1	K
$(\alpha\beta)_{ij}$	0	1	K
$\epsilon_{(ij)k}$	1	1	1

3. Then the products of the entries of the remaining rows give the coefficients of the corresponding variances. That is,

$$MS_A = JK\sigma_\alpha^2 + K\sigma_{\alpha\beta}^2 + \sigma^2$$

Let us complete the remaining EMS values. For MS_B, we have

	F	R	R
	i	j	k
α_i	0	J	K
β_j	I	1	K
$(\alpha\beta)_{ij}$	0	1	K
$\epsilon_{(ij)k}$	1	1	1

$$MS_B = IK\sigma_\beta^2 + 0\sigma_{\alpha\beta}^2 + 1\sigma^2$$
$$= IK\sigma_\beta^2 + \sigma^2$$

For MS_{AB}, we have

	F	R	R
	i	j	k
α_i	0	J	K
β_j	I	1	K
$(\alpha\beta)_{ij}$	0	1	K
$\epsilon_{(ij)k}$	1	1	1

$$MS_{AB} = K\sigma_{\alpha\beta}^2 + \sigma^2$$

Lastly,

$$MS_E = \sigma^2$$

Technically, in this approach as in the Schultz's method, all variances whose subscript represent a fixed-effects should be replaced by the appropriate summation. However, as seen in all previous considerations, the primary purpose for computing the EMS values is to determine comparable expressions for each F-ratio. Since the EMS results do not affect the numerical values of the F-ratios, we may forego the substitutions of the appropriate summations. That is, we leave the EMS values in terms of the population variances.

Example 12.4.1 Suppose the data for Table 12.4.3 is as shown in Table 12.4.4.

TABLE 12.4.4

Manufacturers I			II			III			IV		
Types			Types			Types			Types		
1	2	3	4	5	6	7	8	9	10	11	12
25	30	40	15	30	50	20	30	40	30	40	50
30	35	35	25	40	30	25	35	45	35	45	55
55	65	75	40	70	80	45	65	85	65	85	105
195			190			195			255		

Test the equality of means for transformers *within* manufacturers.

Solution: Using Hicks method, we obtain the following scheme.

	F	R	R	
	i	j	k	
α_i	0	J	K	Manufacturers
$\beta_{(i)j}$	1	1	K	Types *within* manufacturers
$\epsilon_{(ij)k}$	1	1	1	Error

Therefore

$$MS_A = JK\sigma_\alpha^2 + K\sigma_\beta^2 + \sigma^2$$
$$MS_{B(i)j} = K\sigma_\beta^2 + \sigma^2$$
$$MS_E = \sigma^2$$

The complete Anova table becomes:

TABLE 12.4.5

Source of Variation	Degrees of Freedom	Sums of Squares	Mean Square	EMS	F-Ratio
(1) A	$I - 1$	SS_i	$\frac{SS_i}{I - 1}$	$JK\sigma_\alpha^2 + K\sigma_\beta^2 + \sigma^2$	(1) ÷ (2)
(2) $B_{(i)j}$	$I(J - 1)$	$SS_{(i)j}$	$\frac{SS_{(i)j}}{I(J - 1)}$	$K\sigma_\beta^2 + \sigma^2$	(2) ÷ (3)
(3) $\epsilon_{(ij)k}$	$IJ(K - 1)$	$SS_{(ij)k}$	$\frac{SS_{(ij)k}}{IJ(K - 1)}$	σ^2	
Total	$IJK - 1$	SS			

We compute the sums of squares values and obtain

$$SS = 25^2 + 30^2 + \cdots + 55^2 - \frac{835^2}{24} = 2{,}223.96$$

$$SS_i = \frac{1}{6}[195^2 + 190^2 + 195^2 + 255^2] - \frac{835^2}{24} = 478.12$$

$$SS_{(i)j} = \frac{1}{2}[55^2 + 65^2 + \cdots + 105^2] - \frac{1}{6}[195^2 + \cdots + 255^2] = 1{,}333.33$$

$$SS_E = 2{,}223.96 - 478.12 - 1{,}333.33 = 412.51$$

The Anova table is as follows:

TABLE 12.4.6

Source	D.F.	SS	MS	EMS (Mixed Model)	F-Ratio
Manufacturer	$4 - 1 = 3$	478.12	159.37	(See Table 12.4.5)	.956
Types within manufacturer............	$4(2) = 8$	1,333.33	166.66		4.85**
Error	$4(3)(1) = 12$	412.51	34.37		
Total	23	2,223.96			

Since $F_{.01;8,12} = 4.50$, we conclude that types of transformer effects *within* manufacturers are unequal.

Example 12.4.2 Partition the $SS_{(i)j}$ in Example 12.4.1 into 4 parts and examine the types *within* each manufacturer.

Solution: Let us note that

$$\begin{aligned} SS_{(i)j} &= \frac{1}{2}[55^2 + 65^2 + 75^2] - \frac{1}{6}[195^2] \\ &\quad + \frac{1}{2}[40^2 + 70^2 + 80^2] - \frac{1}{6}[190^2] \\ &\quad + \frac{1}{2}[45^2 + 65^2 + 85^2] - \frac{1}{6}[195^2] \\ &\quad + \frac{1}{2}[65^2 + 85^2 + 105^2] - \frac{1}{6}[255^2] \\ &= 100 + 433.33 + 400 + 400 = 1{,}333.33 \end{aligned}$$

That is,

$$SS_{(i)j} = SS_{(1)j} + SS_{(2)j} + SS_{(3)j} + SS_{(4)j}$$

The Anova table for this analysis is shown in Table 12.4.7.

Since $F_{.05;2,12} = 3.89$ and $F_{.01;2,12} = 6.93$, we note that only types within Manufacturer 1 are not significant. Furthermore, notice that in this analysis we are only testing types within manufacturers, and as in the previous case the proper F-ratios are

$$\frac{MS_{(i)j}}{MS_E}, \qquad i = 1, 2, 3, 4$$

TABLE 12.4.7

Source	D.F.	SS	MS	F-Ratio
Manufacturers	3	478.12	159.37	
Types within Manufacturer 1	2	100.00	50.00	$\frac{50.00}{34.37} = 1.45$
Types within Manufacturer 2	2	433.33	216.66	$\frac{216.66}{34.37} = 6.30^*$
Types within Manufacturer 3	2	400.00	200.00	$\frac{200.00}{34.37} = 5.82^*$
Types within Manufacturer 4	2	400.00	200.00	$\frac{200.00}{34.37} = 5.82^*$
Error	12	412.51	34.37	
Total	23	2,223.96		

PROBLEMS

12.4.1 Verify the entries of Table 12.3.1 by using the Hicks method.

12.4.2 To test the effectiveness of teaching methods in calculus classes, four universities are selected at random. Two sections of beginning calculus classes are chosen from each of these universities with 25 students in each class. One of the sections in each university is taught by the conventional method (lecture-textbook) while the other by an experimental approach. Write out a complete Anova table to analyze the result of the experiment, if final examination scores are the criterion of measurements.

12.4.3 To test whether morning and afternoon sections of the same classes have students of equal ability, three teachers were chosen who had a morning and an afternoon section of the same class at a college. Six students were chosen from each group and the following is their midterm scores.

T_1		T_2		T_3	
M	*A*	*M*	*A*	*M*	*A*
75	78	61	92	62	68
91	93	65	78	73	81
84	90	66	86	84	98
83	87	83	85	68	97
78	82	87	86	93	88
86	91	92	77	83	93

Analyze this data, making the necessary assumptions.

12.5 THREE-WAY CLASSIFICATION

In the discussion of the two-way classification of factorial designs, the fundamental principles of analysis of variance were presented. The extension of those principles to the next step up the ladder, the three-way classification, is rather simple and straightforward. All the assumptions made there hold true here. The mathematical model assumed is analogous. The calculations for the sums of squares are identical for the fixed-, random-, and mixed-effects models. The calculations of the EMS can be carried out by the Hicks or Schultz method.

Let us refer to Table 12.1.1. Suppose that we were to include a third variable, say C, to the experiment, with K levels associated with it. Now, the K observations per cell in Table 12.1.1 may be considered as K layers with one observation in each cell, that is, each observation represents an entry in the IJK cells of a three-dimensional configuration.

Next suppose that the experiment is replicated M times so that there are M observations in each cell. Then x_{ijkm} represents the mth observation in the ijkth cell. The table below represents the kth layer of the three-dimensional configuration.

TABLE 12.5.1
kTH LEVEL (LAYER) OF VARIABLE C

A \ B	1	2	...	J
1	x_{11k1} x_{11k2} ⋮ x_{11kM}	x_{12k1} x_{12k2} ⋮ x_{12kM}	... ...	x_{1Jk1} x_{1Jk2} ⋮ x_{1JkM}
2	x_{21k1} x_{21k2} ⋮ x_{21kM}	x_{22k1} x_{22k2} ⋮ x_{22kM}	... ...	x_{2Jk1} x_{2Jk2} ⋮ x_{2JkM}
⋮	⋮	⋮		⋮
I	x_{I1k1} x_{I1k2} ⋮ x_{I1kM}	x_{I2k1} x_{I2k2} ⋮ x_{I2kM}	... ...	x_{IJk1} x_{Ijk2} ⋮ x_{IJkM}

In the three-way classification the dot notation is carried over four variables. For example,

$$\sum_{i=1}^{I} x_{ijkm} = T_{\cdot jkm} \tag{12.5.1}$$

$$\sum_{i=1}^{I}\sum_{k=1}^{K} x_{ijkm} = T_{\cdot j \cdot m} \tag{12.5.2}$$

$$\sum_{i=1}^{I}\sum_{k=1}^{K}\sum_{m=1}^{M} x_{ijkm} = T_{\cdot j \cdot \cdot} \tag{12.5.3}$$

$$\sum_{i=1}^{I}\sum_{j=1}^{J}\sum_{k=1}^{K}\sum_{m=1}^{M} x_{ijkm} = T_{\cdot\cdot\cdot\cdot} \tag{12.5.4}$$

Next let us consider the underlying assumptions for the three-way classification.

1. A random sample of size M is drawn from each of the IJK populations.
2. All IJK populations are normally distributed.
3. The variances of each of the IJK populations are equal.

As in the two-way classification, here also the randomization is over all cells.

Next let us present the mathematical model that describes the underlying assumptions. Since there are 3 factors or variables in this discussion, there will be associated with it $\binom{3}{2} = 3$ second-order interaction effects, and $\binom{3}{3} = 1$ third-order interaction effect. Consequently there will be $3 + 3 + 1 = 7$ terms in the model due to the various effects. The following is the identity representing these effects.

$$\begin{aligned}
\mu_{ijk} = \mu_{\cdot\cdot\cdot} &+ && (12.5.5)\\
&(\mu_{i\cdot\cdot} - \mu_{\cdot\cdot\cdot}) + && \text{Factor } A \text{ effect}\\
&(\mu_{\cdot j\cdot} - \mu_{\cdot\cdot\cdot}) + && \text{Factor } B \text{ effect}\\
&(\mu_{\cdot\cdot k} - \mu_{\cdot\cdot\cdot}) + && \text{Factor } C \text{ effect}\\
&(\mu_{ij\cdot} - \mu_{i\cdot\cdot} - \mu_{\cdot j\cdot} + \mu_{\cdot\cdot\cdot}) + && AB \text{ interaction effect}\\
&(\mu_{i\cdot k} - \mu_{i\cdot\cdot} - \mu_{\cdot\cdot k} + \mu_{\cdot\cdot\cdot}) + && AC \text{ interaction effect}\\
&(\mu_{\cdot jk} - \mu_{\cdot j\cdot} - \mu_{\cdot\cdot k} + \mu_{\cdot\cdot\cdot}) + && BC \text{ interaction effect}\\
&(\mu_{ijk} - \mu_{ij\cdot} - \mu_{i\cdot k} - \mu_{\cdot jk}\\
&\quad + \mu_{i\cdot\cdot} + \mu_{\cdot j\cdot} + \mu_{\cdot\cdot k} - \mu_{\cdot\cdot\cdot}) && ABC \text{ interaction effect}
\end{aligned}$$

Factors A, B, and C are sometimes referred to as *main effects*. In Eq. (12.5.5) let

$$\mu_{i\cdot\cdot} - \mu_{\cdot\cdot\cdot} = \alpha_i \tag{12.5.6}$$

$$\mu_{\cdot j\cdot} - \mu_{\cdot\cdot\cdot} = \beta_j \tag{12.5.7}$$

$$\mu_{\cdot\cdot k} - \mu_{\cdot\cdot\cdot} = \gamma_k \tag{12.5.8}$$

$$\mu_{ij\cdot} - \mu_{i\cdot\cdot} - \mu_{\cdot j\cdot} + \mu_{\cdot\cdot\cdot} = (\alpha\beta)_{ij} \tag{12.5.9}$$

$$\mu_{i\cdot k} - \mu_{i\cdot\cdot} - \mu_{\cdot\cdot k} + \mu_{\cdot\cdot\cdot} = (\alpha\gamma)_{ik} \tag{12.5.10}$$

$$\mu_{\cdot jk} - \mu_{\cdot j\cdot} - \mu_{\cdot\cdot k} + \mu_{\cdots} = (\beta\gamma)_{jk} \tag{12.5.11}$$

$$\mu_{ijk} - \mu_{ij\cdot} - \mu_{i\cdot k} - \mu_{\cdot jk} + \mu_{i\cdot\cdot} + \mu_{\cdot j\cdot} + \mu_{\cdot\cdot k} - \mu_{\cdots} = (\alpha\beta\gamma)_{ijk} \tag{12.5.12}$$

Adding x_{ijkm} to both sides of Eq. (12.5.5) and rewriting we obtain

$$x_{ijkm} = x_{ijkm} - \mu_{ijk} + \mu_{\cdots} + \alpha_i + \beta_j + \gamma_k + (\alpha\beta)_{ij} + (\alpha\gamma)_{ik} + (\beta\gamma)_{jk} + (\alpha\beta\gamma)_{ijk} \tag{12.5.13}$$

Let

$$x_{ijkm} - \mu_{ijk} = \epsilon_{(ijk)m} \tag{12.5.14}$$

where $\epsilon_{(ijk)m}$ represents the random error which is independently normal distributed with mean zero and variance σ^2.

Let us mention next that in this model also error is assumed to be nested in the cells, and parentheses around ijk subscripts of ϵ refer to this. Therefore, Eq. (12.5.14) is written

$$x_{ijkm} = \mu_{\cdots} + \alpha_i + \beta_j + \gamma_k + (\alpha\beta)_{ij} + (\alpha\gamma)_{ik} + (\beta\gamma)_{jk} + (\alpha\beta\gamma)_{ijk} + \epsilon_{(ijk)m} \tag{12.5.15}$$

where each term of the last expression refers to the various effects mentioned in connection with Eq. (12.5.5) and the random error is denoted by the last term.

As in the two-way factorial design, here too the computational formulas for the sums of squares and mean squares are identical for the fixed, random, and mixed models. Therefore, we derive these formulas first, and later, according to the model assumed, we state the hypotheses to be tested and calculate the appropriate EMS values.

The sample identity corresponding to Eq. (12.5.15) is

$$\begin{aligned}(x_{ijkm} - \bar{x}_{\cdots\cdot}) = {} & (\bar{x}_{i\cdots} - \bar{x}_{\cdots\cdot}) + (\bar{x}_{\cdot j\cdot\cdot} - \bar{x}_{\cdots\cdot}) + (\bar{x}_{\cdot\cdot k\cdot} - \bar{x}_{\cdots\cdot}) \\ & + (\bar{x}_{ij\cdot\cdot} - \bar{x}_{i\cdots} - \bar{x}_{\cdot j\cdot\cdot} + \bar{x}_{\cdots\cdot}) \\ & + (\bar{x}_{i\cdot k\cdot} - \bar{x}_{i\cdots} - \bar{x}_{\cdot\cdot k\cdot} + \bar{x}_{\cdots\cdot}) \\ & + (\bar{x}_{\cdot jk\cdot} - \bar{x}_{\cdot j\cdot\cdot} - \bar{x}_{\cdot\cdot k\cdot} + \bar{x}_{\cdots\cdot}) \\ & + (\bar{x}_{ijk\cdot} - \bar{x}_{ij\cdot\cdot} - \bar{x}_{i\cdot k\cdot} - \bar{x}_{\cdot jk\cdot} + \bar{x}_{i\cdots} \\ & + \bar{x}_{\cdot j\cdot\cdot} + \bar{x}_{\cdot\cdot k\cdot} - \bar{x}_{\cdots\cdot}) \\ & + (x_{ijkm} - \bar{x}_{ijk\cdot})\end{aligned} \tag{12.5.16}$$

Squaring both sides of Eq. (12.5.16) and summing over i, j, k, and m we obtain

$$\sum_{i=1}^{I}\sum_{j=1}^{J}\sum_{k=1}^{K}\sum_{m=1}^{M}(x_{ijkm} - \bar{x}_{\cdots\cdot})^2 = \sum_{i=1}^{I}\sum_{j=1}^{J}\sum_{k=1}^{K}\sum_{m=1}^{M}(\bar{x}_{i\cdots} - \bar{x}_{\cdots\cdot})^2 + \sum_{i=1}^{I}\sum_{j=1}^{J}\sum_{k=1}^{K}\sum_{m=1}^{M}(\bar{x}_{\cdot j\cdot\cdot} - \bar{x}_{\cdots\cdot})^2$$

$$+\sum_{i=1}^{I}\sum_{j=1}^{J}\sum_{k=1}^{K}\sum_{m=1}^{M}(\bar{x}_{..k.}-\bar{x}_{....})^2$$

$$+\sum_{i=1}^{I}\sum_{j=1}^{J}\sum_{k=1}^{K}\sum_{m=1}^{M}(\bar{x}_{ij..}-\bar{x}_{i...}-\bar{x}_{.j..}+\bar{x}_{....})^2$$

$$+\sum_{i=1}^{I}\sum_{j=1}^{J}\sum_{k=1}^{K}\sum_{m=1}^{M}(\bar{x}_{i.k.}-\bar{x}_{i...}-\bar{x}_{..k.}+\bar{x}_{....})^2$$

$$+\sum_{i=1}^{I}\sum_{j=1}^{J}\sum_{k=1}^{K}\sum_{m=1}^{M}(\bar{x}_{.jk.}-\bar{x}_{.j..}-\bar{x}_{..k.}+\bar{x}_{....})^2$$

$$+\sum_{i=1}^{I}\sum_{j=1}^{J}\sum_{k=1}^{K}\sum_{m=1}^{M}(\bar{x}_{ijk.}-\bar{x}_{ij..}-\bar{x}_{i.k.}-\bar{x}_{.jk.}+\bar{x}_{i...}+\bar{x}_{.j..}+\bar{x}_{..k.}-\bar{x}_{....})^2$$

$$+\sum_{i=1}^{I}\sum_{j=1}^{J}\sum_{k=1}^{K}\sum_{m=1}^{M}(x_{ijkm}-\bar{x}_{ijk.})^2$$

$$+\sum_{i=1}^{I}\sum_{j=1}^{J}\sum_{k=1}^{K}\sum_{m=1}^{M}\left\{\binom{8}{2}\text{ cross-product terms}\right\} \qquad (12.5.17)$$

The 28 cross-product terms of Eq. (12.5.17) when summed over i, j, k, and m will yield each a value of zero.

Now letting the left-hand side of Eq. (12.5.17) to be equal to SS and the terms of the right-hand side equal to SS_i, SS_j, SS_k, SS_{ij}, SS_{ik}, SS_{ijk}, and $SS_{(ijk)m}$ respectively, we obtain

$$SS = SS_i + SS_j + SS_k + SS_{ij} + SS_{ik} + SS_{jk} + SS_{ijk} + SS_{(ijk)m} \qquad (12.5.18)$$

In order to calculate the mean squares for analysis of variance, the degrees of freedom associated with each sums of squares should be indicated. They are as follows:

Sums of Squares	D.F.
SS_i	$I-1$
SS_j	$J-1$
SS_k	$K-1$
SS_{ij}	$(I-1)(J-1) = IJ - I - J + 1$
SS_{ik}	$(I-1)(K-1) = IK - I - K + 1$
SS_{jk}	$(J-1)(K-1) = JK - J - K + 1$
SS_{ijk}	$(I-1)(J-1)(K-1) = IJK - IJ - IK - JK + I + J + K - 1$
$SS_{(ijk)m}$	$IJK(M-1) = IJKM - IJK$
SS	$IJKM - 1$

It is apparent that the d.f. values and the subscript letters of the sums of squares are related. Also there is an isomorphism, a one-to-one correspondence, between the degrees of freedoms and the sums of squares formulas. The number of terms, numerical signs of the terms, and the subscript letters of the sums of squares have a one-to-one correspondence with the degrees-of-freedom terms in the expanded form. Let us illustrate this notion with a few cases.

$$SS_i = \sum_{i=1}^{I}\sum_{j=1}^{J}\sum_{k=1}^{K}\sum_{m=1}^{M} (\overline{x}_{i\ldots} - \overline{x}_{\ldots\ldots})^2$$

$$\text{D.F.} = I - 1$$

For another,

$$SS_{ij} = \sum_{i=1}^{I}\sum_{j=1}^{J}\sum_{k=1}^{K}\sum_{m=1}^{M} (\overline{x}_{ij\cdot\cdot} - \overline{x}_{i\ldots} - \overline{x}_{\cdot j\cdot\cdot} + \overline{x}_{\ldots\ldots})^2$$

$$\text{D.F.} = IJ - I - J + 1$$

Notice that the degrees of freedom associated with $\overline{x}_{\ldots\ldots}$ is 1.

There is also an isomorphism between the degrees of freedoms and the *computational formulas* for the sums of squares, which is analogous to the above. The only extension is that each term of the formula is divided by the letter corresponding to the dot that appears in the T terms. For example,

$$SS_i = \frac{\sum_{i=1}^{I} T^2_{i\ldots}}{JKM} - \frac{T^2_{\ldots\ldots}}{IJKM}$$

$$\text{D.F.} = I - 1$$

As before, the subscript i corresponds to I, – corresponds to –, and 1 is associated with $T^2_{\ldots\ldots}$, which is divided by $IJKM$ because the subscripts corresponding to these letters are dotted. For SS_{ij} the formula becomes

$$SS_{ij} = \frac{\sum_{i=1}^{I}\sum_{j=1}^{J} T^2_{ij\cdot\cdot}}{KM} - \frac{\sum_{i=1}^{I} T^2_{i\ldots}}{JKM} - \frac{\sum_{j=1}^{J} T^2_{\cdot j\cdot\cdot}}{IKM} + \frac{T^2_{\ldots\ldots}}{IJKM}$$

$$\text{D.F.} = IJ - I - J + 1$$

The following steps may be useful in determining the entries for the Anova table.

1. Write the subscript letters for the terms of the mathematical model. For example, for the three-way classification we have:

i
j
k
ij
ik
jk
ijk
$(ijk)m$

2. The degrees of freedom corresponding to each of the above is 1 less than the total levels associated with the letters, *except* for nested terms, for which 1 is not subtracted. Then, for the above list, the degrees of freedom will be the following respectively:

$$
\begin{aligned}
&I - 1 &&= I - 1\\
&J - 1 &&= J - 1\\
&K - 1 &&= K - 1\\
&(I - 1)(J - 1) &&= IJ - I - J + 1\\
&(I - 1)(K - 1) &&= IK - I - K + 1\\
&(J - 1)(K - 1) &&= JK - J - K + 1\\
&(I - 1)(J - 1)(K - 1) &&= IJK - IJ - IK - JK + I + J + K - 1\\
&IJK(M - 1) &&= IJKM - IJK
\end{aligned}
$$

The total degrees of freedom are as expected $IJKM - 1$.

3. Using the above described methods, write the formulas for the sums of squares terms.

4. Obtain mean squares by dividing sums of squares by their corresponding degrees of freedom.

5. Use the Hicks or Schultz method to obtain expected mean squares.

6. Determine the proper F-ratios.

Next let us consider testing of hypotheses. As in the two-way classification, here too there are three mathematical models to discuss—the fixed-, random-, and mixed-effects models.

(a) Fixed-Effects Model. The underlying assumptions for this model are stated by Eq. (12.5.15) and it may be shown that

$$\sum_{i=1}^{I} \alpha_i = \sum_{j=1}^{J} \beta_j = \sum_{k=1}^{K} \gamma_k = 0 \tag{12.5.19}$$

$$
\begin{aligned}
\sum_{i=1}^{I} (\alpha\beta)_{ij} = \sum_{j=1}^{J} (\alpha\beta)_{ij} = \sum_{i=1}^{I} (\alpha\gamma)_{ik} = \sum_{k=1}^{K} (\beta\gamma)_{jk}\\
= \sum_{j=1}^{J} (\beta\gamma)_{jk} = \sum_{k=1}^{K} (\beta\gamma)_{jk} = 0
\end{aligned} \tag{12.5.20}
$$

$$\sum_{i=1}^{I} (\alpha\beta\gamma)_{ijk} = \sum_{j=1}^{J} (\alpha\beta\gamma)_{ijk} = \sum_{k=1}^{K} (\alpha\beta\gamma)_{ijk} = 0 \tag{12.5.21}$$

hold true. Then Eq. (12.5.19) is the basis for testing the treatment effects for the variables A, B, and C; Eq. (12.5.20) is the basis for testing 2nd order interaction effects for all possible combinations of factors A, B, and C; and Eq. (12.5.21) is the basis for testing the 3rd order interaction effects. The following are examples of possible hypotheses to consider.

1. H_0: $\alpha_i = 0$ $\quad i = 1, 2, \ldots, I$
   H_1: $\alpha_i \neq 0$ $\quad$ for at least one i

which is the hypothesis of equality of treatment effect for factor A.

2. $H_0: (\alpha\beta)_{ij} = 0 \qquad i = 1, 2, \ldots, I; j = 1, 2, \ldots, J$
   $H_1: (\alpha\beta)_{ij} \neq 0 \qquad$ for some i and j

which is the hypothesis of no interaction between factors A and B.

3. $H_0: (\alpha\beta\gamma)_{ijk} = 0 \qquad i = 1, 2, \ldots, I; j = 1, 2, \ldots, J; k = 1, 2, \ldots, K$
   $H_1: (\alpha\beta\gamma)_{ijk} \neq 0 \qquad$ for some i, j, k

which is the hypothesis of no interaction between factors A, B, and C.

Now the next item to obtain for the analysis of variance is the expected mean square for each sums of squares. This is accomplished readily by using the Hicks method described in the nested classification. For this model we obtain the following:

	F	F	F	R
	i	j	k	m
α_i	0	J	K	M
β_j	I	0	K	M
γ_k	I	J	0	M
$(\alpha\beta)_{ij}$	0	0	K	M
$(\alpha\gamma)_{ik}$	0	J	0	M
$(\beta\gamma)_{jk}$	I	0	0	M
$(\alpha\beta\gamma)_{ijk}$	0	0	0	M
$\epsilon_{(ijk)m}$	1	1	1	1

Therefore,

$$MS_A = \frac{E[SS_i]}{I - 1} = JKM\sigma_\alpha^2 + \sigma^2$$

$$MS_B = \frac{E[SS_j]}{J - 1} = IKM\sigma_\beta^2 + \sigma^2$$

$$MS_C = \frac{E[SS_k]}{K - 1} = IJM\sigma_k^2 + \sigma^2$$

$$MS_{AB} = \frac{E[SS_{ij}]}{(I - 1)(J - 1)} = KM\sigma_{\alpha\beta}^2 + \sigma^2$$

$$MS_{AC} = \frac{E[SS_{jk}]}{(I - 1)(K - 1)} = JM\sigma_{\alpha\gamma}^2 + \sigma^2$$

$$MS_{BC} = \frac{E[SS_{jk}]}{(J - 1)(K - 1)} = IM\sigma_{\beta\gamma}^2 + \sigma^2$$

$$MS_{ABC} = \frac{E[SS_{ijk}]}{(I - 1)(J - 1)(K - 1)} = M\sigma_{\alpha\beta\gamma}^2 + \sigma^2$$

$$MS_E = \frac{E[SS_{(ijk)m}]}{IJK(M - 1)} = \sigma^2$$

From these EMS values it is evident that all the F-ratios have MS_E for denominator. The completion of the Anova table is a matter of following the above-mentioned steps.

Let us indicate here the special case where there are one observation in each of the IJK cells, that is, $M = 1$. Then MS_E is undefined and the denominator of the F-ratios will be MS_{ABC}.

(b) Random-Effects Model. The underlying assumptions for this model is identical with that of fixed-effects case given by Eq. (12.5.15). However, it is further assumed that each factor effect and each interaction effect is independently normal distributed with means zero and variances σ^2_α, σ^2_β, σ^2_γ, $\sigma^2_{\alpha\beta}$, $\sigma^2_{\alpha\gamma}$, $\sigma^2_{\beta\gamma}$, and $\sigma^2_{\alpha\beta\gamma}$, respectively and that the errors are independently normal distributed with mean zero and variance σ^2.

With these assumptions various hypotheses can be tested among which are:

1. H_0: $\sigma^2_\beta = 0$
   H_1: $\sigma^2_\beta \neq 0$

which is the hypothesis of equality of treatment effects for factor B.

2. H_0: $\sigma^2_{\beta\gamma} = 0$
   H_1: $\sigma^2_{\beta\gamma} \neq 0$

which is the hypothesis of no interaction between factor B and C.

As mentioned earlier, the degrees of freedom, the sums of squares, and the mean-square entries of the Anova table are identical with the fixed-effects model. We need the expected mean square values which are calculated by the Hicks method.

	R	R	R	R
	i	j	k	m
α_i	1	J	K	M
β_j	I	1	K	M
γ_k	I	J	1	M
$(\alpha\beta)_{ij}$	1	1	K	M
$(\alpha\gamma)_{ik}$	1	J	1	M
$(\beta\gamma)_{jk}$	I	1	1	M
$(\alpha\beta\gamma)_{ijk}$	1	1	1	M
$\epsilon_{(ijk)m}$	1	1	1	1

Therefore

$$MS_A = \frac{E[SS_i]}{I-1} = JKM\sigma^2_\alpha + KM\sigma^2_{\alpha\beta} + JM\sigma^2_{\alpha\gamma} + M\sigma^2_{\alpha\beta\gamma} + \sigma^2$$

$$MS_B = \frac{E[SS_j]}{J-1} = IKM\sigma^2_\beta + KM\sigma^2_{\alpha\beta} + IM\sigma^2_{\beta\gamma} + M\sigma^2_{\alpha\beta\gamma} + \sigma^2$$

$$MS_C = \frac{E[SS_k]}{K-1} = IJM\sigma^2_\gamma + JM\sigma^2_{\alpha\gamma} + IM\sigma^2_{\beta\gamma} + M\sigma^2_{\alpha\beta\gamma} + \sigma^2$$

$$MS_{AB} = \frac{E[SS_{ij}]}{(I-1)(J-1)} = KM\sigma^2_{\alpha\beta} + M\sigma^2_{\alpha\beta\gamma} + \sigma^2$$

$$MS_{AC} = \frac{E[SS_{ik}]}{(I-1)(K-1)} = JM\sigma^2_{\alpha\gamma} + M\sigma^2_{\alpha\beta\gamma} + \sigma^2$$

$$MS_{BC} = \frac{E[SS_{jk}]}{(J-1)(K-1)} = IM\sigma^2_{\beta\gamma} + M\sigma^2_{\alpha\beta\gamma} + \sigma^2$$

$$MS_{ABC} = \frac{E[SS_{ijk}]}{(I-1)(J-1)(K-1)} = M\sigma^2_{\alpha\beta\gamma} + \sigma^2$$

$$MS_E = \frac{E[SS_{(ijk)m}]}{IJK(M-1)} = \sigma^2$$

Here too, it is evident that the third-order interaction is tested against error term, and the second-order interactions are tested against the third-order interaction. That is, the proper F-ratios are:

$$\frac{MS_{AB}}{MS_{ABC}}, \quad \frac{MS_{AC}}{MS_{ABC}}, \quad \frac{MS_{BC}}{MS_{ABC}} \quad \text{and} \quad \frac{MS_{ABC}}{MS_E}$$

However, it is also evident that no proper F-ratio is readily available in the list for the testing of hypotheses related to the main effects. For example, to test factor A effect, the numerator of the F-ratio would be MS_A, but no comparable MS is available for the denominator. If the null hypothesis is true, that is, if $\sigma^2_\alpha = 0$, then MS_A and $[MS_{AB} + MS_{AC} - MS_{ABC}]$ will have comparable values. Therefore the proper F-ratio would be

$$\frac{MS_A}{[MS_{AB} + MS_{AC} - MS_{ABC}]}$$

If factor B effect were to be tested, and the null hypothesis is true—namely, $\sigma^2_\beta = 0$, then MS_B and $MS_{AB} + MS_{BC} - MS_{ABC}$ will have comparable values. Hence the proper F-ratio would be

$$\frac{MS_B}{[MS_{AB} + MS_{BC} - MS_{ABC}]}$$

Similarly for factor C effect the proper F-ratio would be

$$\frac{MS_C}{[MS_{AC} + MS_{BC} - MS_{ABC}]}$$

There is a pattern that facilitates the choosing of terms for the denominators. The sum of subscript letters of the denominator MS terms with the proper numerical sign is equal to the numerator MS subscript. For example, $(A + B) + (A + C) - (A + B + C) = A$, holds true for factor A. Similarly, $(A + B) + (B + C) - (A + B + C) = B$ and $(A + C) + (B + C) - (A + B + C) = C$.

The only difficulty with this method is that the F-ratios obtained are not exact but approximate. Satterthwaite has shown that

$$F[n_1, n_2] = \frac{MS_A}{MS_{AB} + MS_{AC} - MS_{ABC}}$$

is approximately F-distributed, where $n_1 = I - 1$ and

$$n_2 = \frac{[MS_{AB} + MS_{AC} - MS_{ABC}]^2}{\dfrac{MS^2_{AB}}{(I-1)(J-1)} + \dfrac{MS^2_{AC}}{(I-1)(J-1)} + \dfrac{MS^2_{ABC}}{(I-1)(J-1)(K-1)}}$$

This may be used to test factor A effect.

To test factor B effect, interchange A with B in the above. Similarly, to test the factor C effect, interchange A with C. Notice that n_2 is not necessarily an integer and it is common practice to round n_2 to the nearest integer.

(c) Mixed-Effects Model. Since the basic principles of the mixed-effects model was discussed in detail in Sec. 12.3, we present here a numerical example. Just for review, however, let us mention that the source of variations, the sums of squares, and the mean squares in this model are identical with the fixed or random models. The differences come in the expected mean squares and consequently the F-ratios change. Also, we have indicated two methods, Hicks' and Schultz's, for deriving EMS values. Hence, the writing out of the entries for Anova table becomes rather elementary. We conclude our discussion of this section with an example.

Example 12.5.1 In an industrial experiment there are three factors to consider. Factor A is human operators, factor B is types of machines and factor C is time interval at which machines are stopped. There are $I = 6$ subjects, $J = 2$ machines and $K = 3$ time intervals. The experiment is replicated $M = 5$ times.

(a) Find the expected mean squares for this experiment.

(b) Write the Anova table showing the correct F-ratios for each hypothesis.

Solution: Using the Hicks method we obtain the following configuration.

	R	F	F	R
	i	j	k	m
α_i	1	J	K	M
β_j	I	0	K	M
γ_k	I	J	0	M
$(\alpha\beta)_{ij}$	1	0	K	M
$(\alpha\gamma)_{ik}$	1	J	0	M
$(\beta\gamma)_{jk}$	I	0	0	M
$(\alpha\beta\gamma)_{ijk}$	1	0	0	M
$\epsilon_{(ijk)m}$	1	1	1	1

From this table we compute the EMS values which along with the F-ratios are tabulated below.

Source	Degrees of Freedom	Expected Mean Squares	F-Ratio
(1) A	$I - 1 = 6 - 1 = 5$	$JKM\sigma_\alpha^2 + \sigma^2$	(1) ÷ (8)
(2) B	$J - 1 = 2 - 1 = 1$	$IKM\sigma_\beta^2 + KM\sigma_{\alpha\beta}^2 + \sigma^2$	(2) ÷ (4)
(3) C	$K - 1 = 3 - 1 = 2$	$IJM\sigma_\gamma^2 + JM\sigma_{\alpha\gamma}^2 + \sigma^2$	(3) ÷ (5)
(4) AB	$5 \times 1 = 5$	$KM\sigma_{\alpha\beta}^2 + \sigma^2$	(4) ÷ (8)
(5) AC	$5 \times 2 = 10$	$JM\sigma_{\alpha\gamma}^2 + \sigma^2$	(5) ÷ (8)
(6) BC	$1 \times 2 = 2$	$IM\sigma_{\beta\gamma}^2 + M\sigma_{\alpha\beta\gamma}^2 + \sigma^2$	(6) ÷ (7)
(7) ABC	$5 \times 1 \times 2 = 10$	$M\sigma_{\alpha\beta\gamma}^2 + \sigma^2$	(7) ÷ (8)
(8) Error	$6 \times 2 \times 3(5 - 1) = 144$	σ^2	
Total	$180 - 1 = 179$		

PROBLEMS

12.5.1 Derive the expected mean squares in Example 12.5.1 using the Schultz method.

12.5.2 Suppose in Example 12.5.1 the various sums of squares are as follows:

$$
\begin{aligned}
SS_i &= 17{,}943.40 \\
SS_j &= 665.08 \\
SS_k &= 1{,}299.60 \\
SS_{ij} &= 5{,}975.80 \\
SS_{ik} &= 1{,}874.80 \\
SS_{jk} &= 4{,}254.10 \\
SS_{ijk} &= 7{,}674.80 \\
\text{Error} &= 52{,}067.20
\end{aligned}
$$

Analyze the data.

12.5.3 Repeat Prob. 12.5.2 assuming factors A, B, and C are (a) all fixed, (b) all random.

12.5.4 Using the method described in this section, write out for a four-way classification analysis of variance (a) the degrees of freedom, (b) the corresponding definitional sums of squares, (c) the corresponding computational sums of squares, and (d) the EMS for the fixed-effects model, using the Schultz method.

Bibliography

PROBABILITY THEORY

Drake, A. W., *Fundamentals of Applied Probability Theory* (New York: McGraw-Hill, 1967).

Dwass, M., *First Steps in Probability* (New York: McGraw-Hill, 1967).

Feller, W., *An Introduction to Probability Theory and Its Applications*, Vol. I, 3d ed. (New York: Wiley, 1968).

Goldberg, S., *Probability: an Introduction* (Englewood Cliffs, N. J.: Prentice-Hall, 1960).

Harris, B., *Theory of Probability* (Reading, Mass.: Addison-Wesley, 1966).

Papoulis, A., *Probability, Random Variables, and Stochastic Processes* (New York: McGraw-Hill, 1965).

Parzen, E., *Modern Probability Theory and Its Applications* (New York: Wiley, 1960).

Thorp, E. O., *Elementary Probability* (New York: Wiley, 1966).

THEORETICAL STATISTICS

Brunk, H. D., *An Introduction to Mathematical Statistics* (Waltham, Mass.: Blaisdell, 1965).

Fisz, M., *Probability Theory and Mathematical Statistics* (New York: Wiley, 1963).

Graybill, F. A., and A. M. Mood, *Introduction to the Theory of Statistics* (New York: McGraw-Hill, 1963).

Hoel, P. G., *Introduction to Mathematical Statistics*, 3d ed. (New York: Wiley, 1962).

Lindgren, B. W., *Statistical Theory*, 2d ed. (New York: Macmillan, 1968).

Whitney, D. R., *Elements of Mathematical Statistics* (New York: Holt, 1959).

Wilks, S. S., *Mathematical Statistics* (New York: Wiley, 1962).

APPLIED STATISTICS

Bennett, C. A., and N. L. Franklin, *Statistical Analysis in Chemistry and the Chemical Industry* (New York: Wiley, 1954).

Bowker, A. A., and G. J. Lieberman, *Engineering Statistics* (Englewood Cliffs, N. J.: Prentice-Hall, 1959).

Brownlee, K. A., *Statistical Theory and Methodology in Science and Engineering*, 2d ed. (New York: Wiley, 1965).

Chew, V., *Experimental Designs in Industry* (New York: Wiley, 1958).

Cochran, W. G., and G. M. Cox, *Experimental Designs* (New York: Wiley, 1950).

Guenther, W. C., *Analysis of Variance* (Englewood Cliffs, N. J.: Prentice-Hall, 1964).

Hald, A., *Statistical Theory with Engineering Applications* (New York: Wiley, 1952).

Hicks, C. R., "Fundamentals of Analysis of Variance," Part II, *Industrial Quality Control* (September 1956), pp. 5–8.

Hicks, C. R., *Fundamental Concepts in the Design of Experiments* (New York: Holt, 1964).

Kempthrone, O., *The Design and Analysis of Experiments* (New York: Wiley, 1960).

Satterthwaite, F. E., "An Approximate Distribution of Estimates of Variance Components," *Biometrics*, Vol. 2 (1946), pp. 110–114.

Scheffé, H., *The Analysis of Variance* (New York: Wiley, 1959).

Schultz, E. F., Jr., "Rules of Thumb for Determining Expectations of Mean Squares in Analysis of Variance," *Biometrics*, Vol. 11 (1955), pp. 123–148.

APPENDIX

Statistical Tables

Tables II and III were adapted with the permission of the publisher from William H. Beyer (Ed.), *Handbook of Tables for Probability and Statistics*, Chemical Rubber Company, Cleveland, Ohio, Tables III.2 and III.4.

Tables V, VI, VII, and VIII were reproduced with the permission of the Trustees of *Biometrika*, from E. S. Pearson and H. O. Hartley (Eds.), *Biometrika Tables for Statisticians*, Cambridge University Press, Cambridge, 1962, Tables 8, 12, 18, and 31.

TABLE I
Random Digits

75847	41210	71729	73632	07132	81250	70465	08789	69529	90722
53750	24414	30909	91210	18262	93945	36272	58300	80554	19921
27627	07519	44570	19042	69707	27539	18889	03808	79078	12500
09143	06640	09697	26562	74824	46289	84556	76269	87044	55566
55611	08398	46239	99023	53005	73730	42691	65039	38310	42480
56552	73437	40995	72753	36127	07519	04335	81542	27434	32617
91676	87988	02537	59765	76533	81347	50652	70996	19213	37890
38316	65039	32548	70605	23071	89941	39397	09472	21566	40625
61795	04394	34032	47070	35433	34960	36005	85937	24650	51269
09578	73535	27921	75292	41710	44921	60756	95800	62084	96093
73058	22753	24066	52832	80863	28125	00601	19628	38821	77734
98649	16992	13500	00000	94921	87500	14294	43359	92597	53417
40993	28613	06324	82910	07386	47460	66926	26953	71063	59863
75569	21386	75755	12695	45202	02636	82548	33984	70164	06250
94903	56445	90776	97753	17401	48925	20523	07128	25772	46093
05235	96191	15550	65917	27382	44628	58377	44140	15251	46484
75107	91015	44606	32324	10781	73828	13016	11328	88101	56250
58526	61132	36173	33984	69089	84375	81881	71386	32639	89257
36188	59863	55354	37011	12525	75683	02305	05371	37862	67089
90806	76269	80989	86816	45851	56250	11456	29882	47397	70507
85266	96777	56540	16113	87513	79394	83617	43164	20601	07421
77956	66503	19446	16699	79968	13964	73579	58984	88352	53906
22205	32226	61062	74414	94825	92773	43063	11035	72733	76464
63337	89062	67610	10742	18694	09179	99362	67089	23228	63769
52706	17675	85946	65527	53506	71386	58128	17382	72434	32617
08083	12988	71019	04296	31676	75781	79172	48535	61002	80761
63178	83300	25988	76953	45721	80175	27462	76855	38873	90136
34928	71093	75865	84472	96737	30468	75301	26953	04950	31738
28678	95507	84900	75683	86191	77246	45659	91210	71973	87695
87365	11230	68656	00585	86008	91113	13465	94238	79179	44335
45939	33105	82849	73144	49260	74218	27419	18945	06902	46582
58615	96679	26938	35449	37335	08300	50524	90234	59061	76757
68842	40722	61416	25976	10202	14843	37944	70214	92164	79492
58531	73828	98758	30078	45663	81835	19657	59277	60849	24316
88612	79296	99132	81250	17340	08789	70994	75097	35143	67675
99960	32714	18837	15820	45778	80859	23347	04589	98118	77441
38945	43457	08136	84082	26670	16601	63549	92675	55941	77246
82964	59960	51459	96093	73351	19628	00345	21484	14048	33984
88523	43750	08367	91992	47459	96093	45226	19628	77591	30859
59528	68652	68536	62109	28675	41503	41687	37792	79125	00000
81347	65625	13301	95000	53079	49218	39347	90039	50737	18261
50373	65722	12807	86132	45963	37890	76402	58789	48777	95410
34010	00976	61251	95312	04505	85937	03166	13769	49141	72363
74556	03027	07791	38183	09641	47949	93841	43066	25276	61132
52384	27734	56510	49804	25415	52734	48136	47460	03449	70703
10681	51855	89630	73730	25211	18164	53682	12890	99425	04882
84677	97851	66729	85839	73466	67480	09995	11718	10707	76367
10005	73730	40347	90039	28080	93261	84821	89941	23576	78222
02517	33398	29174	56054	34772	21679	65536	98730	12052	00195
19164	42871	40780	15136	04582	03125	32998	65722	15639	89257

TABLE I (*continued*)

55673	58398	09179	44335	53751	83105	50281	37207	86280	15136
24503	90625	19949	34082	22226	80664	23323	24218	07546	14257
15998	16894	90148	19335	41813	35449	16924	92675	03109	74121
60708	12988	66038	57421	34939	08691	02525	75683	31992	55371
34102	90527	95230	34667	79817	74902	37756	59179	95895	87402
03930	90820	84719	23828	70389	16015	42677	61230	66947	02148
24383	30078	47714	59960	53608	39843	99394	89746	16619	38476
73349	12109	75013	30566	89766	35742	80730	22460	76374	51171
06051	14746	66546	14257	30841	91894	88268	31054	94025	14648
67901	36718	74111	08398	76318	11523	17617	55371	58028	68652
57989	74609	82642	70019	22023	80371	45260	13183	91843	62792
38036	13281	08261	84082	52549	07226	68167	23632	19583	98437
62309	20410	10401	61132	72794	43359	03925	65917	20644	16503
03967	65136	33244	14062	12263	42773	00045	53222	55813	23242
13872	31445	39776	00097	46886	84082	49131	10351	44915	89355
89716	18652	68292	48046	48443	11523	46621	45996	09618	65234
15189	69726	21109	13085	79819	94628	64578	61328	13150	39062
27229	24804	88281	73828	57937	98828	50833	49609	26075	31738
02214	11132	27726	07421	53054	32128	35757	81250	96734	61914
42518	79882	28305	90820	31105	95703	11389	52636	32304	19921
38369	26269	74788	69628	47952	63671	59334	83886	02232	17773
48263	30566	51680	41992	64500	97656	65436	52343	85686	40136
76579	10156	03485	83984	51755	85937	85392	70019	91359	74121
31215	94238	53984	13085	85972	29003	66431	03027	25662	47558
62641	35742	65007	56835	49418	33496	51159	17968	01705	07812
13941	89453	89142	21191	61766	47949	85345	33691	13194	70214
68141	35742	03679	44335	15079	95605	81494	75097	08971	80175
19064	33105	18884	88769	28414	18457	52838	98925	07192	99316
05092	28515	61683	95996	78026	85546	76262	93945	44085	08300
47985	83984	64646	48437	41654	90722	82754	15039	82499	87792
78587	89062	24836	66992	82005	85937	48088	01269	11873	65722
42104	73632	73832	03125	69912	71972	26754	39453	91730	10253
52228	14941	50651	97753	10272	58300	97741	69921	35976	68457
68512	81738	38102	78320	21865	23437	09599	24316	78261	23046
37285	88867	50008	17871	51400	26855	44684	44824	64456	05468
17073	73046	19561	27929	85147	58300	99206	54296	17370	11718
37563	23242	35552	00195	84398	80371	58834	35058	92756	10351
76654	17480	19907	22656	08136	71875	25180	05371	73702	51464
88899	41406	97925	53710	80091	55273	80087	03613	24952	75878
99106	20117	92494	75097	86315	55175	56637	57324	76626	46484
81650	87890	14830	32226	34207	27539	69279	66308	99012	20703
45105	34667	02376	34277	08090	45410	60426	02539	22380	24902
96399	16992	47679	68750	27435	30273	03597	77832	18192	38281
74985	47363	50019	89746	94451	41601	71386	84082	21396	72851
90533	56933	46109	98535	66032	10449	55963	01269	42244	75097
82995	23925	25479	12597	24486	93847	12823	12011	32227	90527
07046	75292	19933	10546	24041	25976	72409	17968	01119	14062
61384	52148	22771	97265	78181	76269	67220	33691	60753	29589
17381	46972	76144	04296	92711	91406	37232	17773	94357	05566
19527	09960	67915	03906	41004	15039	38945	19042	05156	49414

TABLE I (*continued*)

60637	32910	01771	24023	21584	83886	86802	49023	00710	81542
76946	04492	82774	90234	35819	58007	50733	27636	02689	94140
36198	73046	79033	93554	69720	94726	85781	98242	43340	08789
53807	25097	26794	06738	76017	82226	51931	88476	34140	13671
49715	69824	81082	03125	70889	28222	47683	34960	72138	67187
99021	85058	62824	46289	00181	76269	18782	83691	82677	12402
42236	57226	83157	47070	05843	50585	31858	52050	97955	32226
43679	93164	02290	52734	60538	81835	99247	43652	16559	08203
37231	81152	89886	71875	49984	74121	65297	97363	94404	66308
00672	36328	07559	57031	79911	01074	76205	32226	40750	24414
39503	66210	22437	74414	98243	89648	66314	45312	02601	56250
57354	73632	31058	59375	33224	48730	72354	73632	36527	34375
90426	63574	40768	31054	60040	77148	19573	73046	37139	40429
61868	77441	34062	62207	03492	06542	27751	70898	65978	75976
04782	22656	76789	06250	66485	59570	91744	38476	26571	77734
62516	35742	39128	66210	44801	02539	87516	96777	22360	47363
55000	36621	91189	08691	48033	56933	47281	86035	71146	85058
91828	36914	51771	72851	79107	78808	71241	21093	43688	11035
02128	29589	80174	43847	91875	85449	31426	87988	28904	78515
41615	60058	02936	76757	49213	50097	50744	26269	36800	41503
23816	28417	26125	12207	10181	51855	86115	11230	09866	94335
46085	81542	70989	01367	65108	27636	78764	16015	76564	33105
23181	76269	80501	58691	85387	08496	22814	45312	96742	18750
34906	00585	98704	22363	85542	35839	18242	06542	81462	64648
17071	28906	89759	03320	91323	24218	85671	14257	90314	57519
72841	67480	80600	58593	93871	21582	88864	99023	77712	76855
42194	21386	66087	15820	28005	37109	62440	06347	07806	03027
88455	44433	78373	16894	03722	41210	39600	83007	08570	19042
16582	27539	20353	88183	60471	55761	78193	48144	10271	48437
84330	68847	27349	48730	56046	14257	57013	79394	69164	30664
90852	53906	39783	44726	37783	93554	29681	88476	27694	82421
71584	59472	35384	76562	42939	69726	66226	31835	26737	67089
87584	10644	41924	31640	71440	42968	75557	61718	34194	09179
08347	04589	92650	02441	81743	28613	42848	26660	50129	39453
31085	57128	62540	03906	28211	18164	74775	87890	91490	47851
27130	24902	79797	60742	91887	45117	72987	91503	65759	64355
30023	80371	01510	13183	34226	44042	03227	78320	01650	39062
46369	87304	38489	25781	39572	87597	67333	61816	43581	05468
95296	38671	85970	58105	45569	45800	67797	85156	10492	67578
84421	14257	31525	51269	32918	57910	38123	77929	78165	16113
64564	57519	41786	98730	95059	81445	98125	61035	22391	96777
39450	31738	71257	08007	37403	19824	82009	76562	95771	72851
88482	78808	12159	17968	27486	32812	26466	30859	75056	03027
11307	00683	24985	71777	01437	62207	49097	53417	35133	91113
80751	09863	31184	44824	69534	17968	05904	29687	73936	40136
43961	91406	44458	74023	09231	32324	87051	26953	37567	50488
87706	05468	30550	29296	28380	85937	46037	23144	77922	85156
06684	08203	92798	21777	90744	26269	18050	41503	41980	34667
55403	68652	14532	71484	01304	19921	20400	51269	29695	92285
99058	22753	06879	02832	72513	67187	76658	56933	73551	39160

TABLE II
Cumulative Terms, Binomial Distribution

n	m	p = .05	.10	.15	.20	.25	.30	.35	.40	.45	.50
2	1	.0975	.1900	.2775	.3600	.4375	.5100	.5775	.6400	.6975	.7500
	2	.0025	.0100	.0225	.0400	.0625	.0900	.1225	.1600	.2025	.2500
3	1	.1426	.2710	.3859	.4880	.5781	.6570	.7254	.7840	.8336	.8750
	2	.0072	.0280	.0608	.1040	.1562	.2160	.2818	.3520	.4252	.5000
	3	.0001	.0010	.0034	.0080	.0156	.0270	.0429	.0640	.0911	.1250
4	1	.1855	.3439	.4780	.5904	.6836	.7599	.8215	.8704	.9085	.9375
	2	.0140	.0523	.1095	.1808	.2617	.3483	.4370	.5248	.6090	.6875
	3	.0005	.0037	.0120	.0272	.0508	.0837	.1265	.1792	.2415	.3125
	4	.0000	.0001	.0005	.0016	.0039	.0081	.0150	.0256	.0410	.0625
5	1	.2262	.4095	.5563	.6723	.7627	.8319	.8840	.9222	.9497	.9688
	2	.0226	.0815	.1648	.2627	.3672	.4718	.5716	.6630	.7438	.8125
	3	.0012	.0086	.0266	.0579	.1035	.1631	.2352	.3174	.4069	.5000
	4	.0000	.0005	.0022	.0067	.0156	.0308	.0540	.0870	.1312	.1875
	5	.0000	.0000	.0001	.0003	.0010	.0024	.0053	.0102	.0185	.0312
6	1	.2649	.4686	.6229	.7379	.8220	.8824	.9246	.9533	.9723	.9844
	2	.0328	.1143	.2235	.3447	.4661	.5798	.6809	.7667	.8364	.8906
	3	.0022	.0158	.0473	.0989	.1694	.2557	.3529	.4557	.5585	.6562
	4	.0001	.0013	.0059	.0170	.0376	.0705	.1174	.1792	.2553	.3438
	5	.0000	.0001	.0004	.0016	.0046	.0109	.0223	.0410	.0692	.1094
	6	.0000	.0000	.0000	.0001	.0002	.0007	.0018	.0041	.0083	.0156
7	1	.3017	.5217	.6794	.7903	.8665	.9176	.9510	.9720	.9848	.9922
	2	.0444	.1497	.2834	.4233	.5551	.6706	.7662	.8414	.8976	.9375
	3	.0038	.0257	.0738	.1480	.2436	.3529	.4677	.5801	.6836	.7734
	4	.0002	.0027	.0121	.0333	.0706	.1260	.1998	.2898	.3917	.5000
	5	.0000	.0002	.0012	.0047	.0129	.0288	.0556	.0963	.1529	.2266
	6	.0000	.0000	.0001	.0004	.0013	.0038	.0090	.0188	.0357	.0625
	7	.0000	.0000	.0000	.0000	.0001	.0002	.0006	.0016	.0037	.0078
8	1	.3366	.5695	.7275	.8322	.8999	.9424	.9681	.9832	.9916	.9961
	2	.0572	.1869	.3428	.4967	.6329	.7447	.8309	.8936	.9368	.9648
	3	.0058	.0381	.1052	.2031	.3215	.4482	.5722	.6846	.7799	.8555
	4	.0004	.0050	.0214	.0563	.1138	.1941	.2936	.4059	.5230	.6367
	5	.0000	.0004	.0029	.0104	.0273	.0580	.1061	.1737	.2604	.3633
	6	.0000	.0000	.0002	.0012	.0042	.0113	.0253	.0498	.0885	.1445
	7	.0000	.0000	.0000	.0001	.0004	.0013	.0036	.0085	.0181	.0352
	8	.0000	.0000	.0000	.0000	.0000	.0001	.0002	.0007	.0017	.0039
9	1	.3698	.6126	.7684	.8658	.9249	.9596	.9793	.9899	.9954	.9980
	2	.0712	.2252	.4005	.5638	.6997	.8040	.8789	.9295	.9615	.9805
	3	.0084	.0530	.1409	.2618	.3993	.5372	.6627	.7682	.8505	.9102
	4	.0006	.0083	.0339	.0856	.1657	.2703	.3911	.5174	.6386	.7461
	5	.0000	.0009	.0056	.0196	.0489	.0988	.1717	.2666	.3786	.5000
	6	.0000	.0001	.0006	.0031	.0100	.0253	.0536	.0994	.1658	.2539
	7	.0000	.0000	.0000	.0003	.0013	.0043	.0112	.0250	.0498	.0898
	8	.0000	.0000	.0000	.0000	.0001	.0004	.0014	.0038	.0091	.0195
	9	.0000	.0000	.0000	.0000	.0000	.0000	.0001	.0003	.0008	.0020

Linear interpolation will be accurate at most to two decimal places.

TABLE II (*continued*)

n	m	.05	.10	.15	.20	.25	.30	.35	.40	.45	.50
						p					
10	1	.4013	.6513	.8031	.8926	.9437	.9718	.9865	.9940	.9975	.9990
	2	.0861	.2639	.4557	.6242	.7560	.8507	.9140	.9536	.9767	.9893
	3	.0115	.0702	.1798	.3222	.4744	.6172	.7384	.8327	.9004	.9453
	4	.0010	.0128	.0500	.1209	.2241	.3504	.4862	.6177	.7340	.8281
	5	.0001	.0016	.0099	.0328	.0781	.1503	.2485	.3669	.4956	.6230
	6	.0000	.0001	.0014	.0064	.0197	.0473	.0949	.1662	.2616	.3770
	7	.0000	.0000	.0001	.0009	.0035	.0106	.0260	.0548	.1020	.1719
	8	.0000	.0000	.0000	.0001	.0004	.0016	.0048	.0123	.0274	.0547
	9	.0000	.0000	.0000	.0000	.0000	.0001	.0005	.0017	.0045	.0107
	10	.0000	.0000	.0000	.0000	.0000	.0000	.0000	.0001	.0003	.0010
11	1	.4312	.6862	.8327	.9141	.9578	.9802	.9912	.9964	.9986	.9995
	2	.1019	.3026	.5078	.6779	.8029	.8870	.9394	.9698	.9861	.9941
	3	.0152	.0896	.2212	.3826	.5448	.6873	.7999	.8811	.9348	.9673
	4	.0016	.0185	.0694	.1611	.2867	.4304	.5744	.7037	.8089	.8867
	5	.0001	.0028	.0159	.0504	.1146	.2103	.3317	.4672	.6029	.7256
	6	.0000	.0003	.0027	.0117	.0343	.0782	.1487	.2465	.3669	.5000
	7	.0000	.0000	.0003	.0020	.0076	.0216	.0501	.0994	.1738	.2744
	8	.0000	.0000	.0000	.0002	.0012	.0043	.0122	.0293	.0610	.1133
	9	.0000	.0000	.0000	.0000	.0001	.0006	.0020	.0059	.0148	.0327
	10	.0000	.0000	.0000	.0000	.0000	.0000	.0002	.0007	.0022	.0059
	11	.0000	.0000	.0000	.0000	.0000	.0000	.0000	.0000	.0002	.0005
12	1	.4596	.7176	.8578	.9313	.9683	.9862	.9943	.9978	.9992	.9998
	2	.1184	.3410	.5565	.7251	.8416	.9150	.9576	.9804	.9917	.9968
	3	.0196	.1109	.2642	.4417	.6093	.7472	.8487	.9166	.9579	.9807
	4	.0022	.0256	.0922	.2054	.3512	.5075	.6533	.7747	.8655	.9270
	5	.0002	.0043	.0239	.0726	.1576	.2763	.4167	.5618	.6956	.8062
	6	.0000	.0005	.0046	.0194	.0544	.1178	.2127	.3348	.4731	.6128
	7	.0000	.0001	.0007	.0039	.0143	.0386	.0846	.1582	.2607	.3872
	8	.0000	.0000	.0001	.0006	.0028	.0095	.0255	.0573	.1117	.1938
	9	.0000	.0000	.0000	.0001	.0004	.0017	.0056	.0153	.0356	.0730
	10	.0000	.0000	.0000	.0000	.0000	.0002	.0008	.0028	.0079	.0193
	11	.0000	.0000	.0000	.0000	.0000	.0000	.0001	.0003	.0011	.0032
	12	.0000	.0000	.0000	.0000	.0000	.0000	.0000	.0000	.0001	.0002
13	1	.4867	.7458	.8791	.9450	.9762	.9903	.9963	.9987	.9996	.9999
	2	.1354	.3787	.6017	.7664	.8733	.9363	.9704	.9874	.9951	.9983
	3	.0245	.1339	.2704	.4983	.6674	.7975	.8868	.9421	.9731	.9888
	4	.0031	.0342	.0967	.2527	.4157	.5794	.7217	.8314	.9071	.9539
	5	.0003	.0065	.0260	.0991	.2060	.3457	.4995	.6470	.7721	.8666
	6	.0000	.0009	.0053	.0300	.0802	.1654	.2841	.4256	.5732	.7095
	7	.0000	.0001	.0013	.0070	.0243	.0624	.1295	.2288	.3563	.5000
	8	.0000	.0000	.0002	.0012	.0056	.0182	.0462	.0977	.1788	.2905
	9	.0000	.0000	.0000	.0002	.0010	.0040	.0126	.0321	.0698	.1334
	10	.0000	.0000	.0000	.0000	.0001	.0007	.0025	.0078	.0203	.0461
	11	.0000	.0000	.0000	.0000	.0000	.0001	.0003	.0013	.0041	.0112
	12	.0000	.0000	.0000	.0000	.0000	.0000	.0000	.0001	.0005	.0017
	13	.0000	.0000	.0000	.0000	.0000	.0000	.0000	.0000	.0000	.0001

TABLE II (*continued*)

n	m	.05	.10	.15	.20	.25	.30	.35	.40	.45	.50
						p					
14	1	.5123	.7712	.8972	.9560	.9822	.9932	.9976	.9992	.9998	.9999
	2	.1530	.4154	.6433	.8021	.8990	.9525	.9795	.9919	.9971	.9991
	3	.0301	.1584	.3521	.5519	.7189	.8392	.9161	.9602	.9830	.9935
	4	.0042	.0441	.1465	.3018	.4787	.6448	.7795	.8757	.9368	.9713
	5	.0004	.0092	.0467	.1298	.2585	.4158	.5773	.7207	.8328	.9102
	6	.0000	.0015	.0115	.0439	.1117	.2195	.3595	.5141	.6627	.7880
	7	.0000	.0002	.0022	.0116	.0383	.0933	.1836	.3075	.4539	.6047
	8	.0000	.0000	.0003	.0024	.0103	.0315	.0753	.1501	.2586	.3953
	9	.0000	.0000	.0000	.0004	.0022	.0083	.0243	.0583	.1189	.2120
	10	.0000	.0000	.0000	.0000	.0003	.0017	.0060	.0175	.0426	.0898
	11	.0000	.0000	.0000	.0000	.0000	.0002	.0011	.0039	.0114	.0287
	12	.0000	.0000	.0000	.0000	.0000	.0000	.0001	.0006	.0022	.0065
	13	.0000	.0000	.0000	.0000	.0000	.0000	.0000	.0001	.0003	.0009
	14	.0000	.0000	.0000	.0000	.0000	.0000	.0000	.0000	.0000	.0001
15	1	.5367	.7941	.9126	.9648	.9866	.9953	.9984	.9995	.9999	1.0000
	2	.1710	.4510	.6814	.8329	.9198	.9647	.9858	.9948	.9983	.9995
	3	.0362	.1841	.3958	.6020	.7639	.8732	.9383	.9729	.9893	.9963
	4	.0055	.0556	.1773	.3518	.5387	.7031	.8273	.9095	.9576	.9824
	5	.0006	.0127	.0617	.1642	.3135	.4845	.6481	.7827	.8796	.9408
	6	.0001	.0022	.0168	.0611	.1484	.2784	.4357	.5968	.7392	.8491
	7	.0000	.0003	.0036	.0181	.0566	.1311	.2452	.3902	.5478	.6964
	8	.0000	.0000	.0006	.0042	.0173	.0500	.1132	.2131	.3465	.5000
	9	.0000	.0000	.0001	.0008	.0042	.0152	.0422	.0950	.1818	.3036
	10	.0000	.0000	.0000	.0001	.0008	.0037	.0124	.0338	.0769	.1509
	11	.0000	.0000	.0000	.0000	.0001	.0007	.0028	.0093	.0255	.0592
	12	.0000	.0000	.0000	.0000	.0000	.0001	.0005	.0019	.0063	.0176
	13	.0000	.0000	.0000	.0000	.0000	.0000	.0001	.0003	.0011	.0037
	14	.0000	.0000	.0000	.0000	.0000	.0000	.0000	.0000	.0001	.0005
	15	.0000	.0000	.0000	.0000	.0000	.0000	.0000	.0000	.0000	.0000
16	1	.5599	.8147	.9257	.9719	.9900	.9967	.9990	.9997	.9999	1.0000
	2	.1892	.4853	.7161	.8593	.9365	.9739	.9902	.9967	.9990	.9997
	3	.0429	.2108	.4386	.6482	.8029	.9006	.9549	.9817	.9934	.9979
	4	.0070	.0684	.2101	.4019	.5950	.7541	.8661	.9349	.9719	.9894
	5	.0009	.0170	.0791	.2018	.3698	.5501	.7108	.8334	.9147	.9616
	6	.0001	.0033	.0235	.0817	.1897	.3402	.5100	.6712	.8024	.8949
	7	.0000	.0005	.0056	.0267	.0796	.1753	.3119	.4728	.6340	.7228
	8	.0000	.0001	.0011	.0070	.0271	.0744	.1594	.2839	.4371	.5982
	9	.0000	.0000	.0002	.0015	.0075	.0257	.0671	.1423	.2559	.4018
	10	.0000	.0000	.0000	.0002	.0016	.0071	.0229	.0583	.1241	.2272
	11	.0000	.0000	.0000	.0000	.0003	.0016	.0062	.0191	.0486	.1051
	12	.0000	.0000	.0000	.0000	.0000	.0003	.0013	.0049	.0149	.0384
	13	.0000	.0000	.0000	.0000	.0000	.0000	.0002	.0009	.0035	.0106
	14	.0000	.0000	.0000	.0000	.0000	.0000	.0000	.0001	.0006	.0021
	15	.0000	.0000	.0000	.0000	.0000	.0000	.0000	.0000	.0001	.0003
	16	.0000	.0000	.0000	.0000	.0000	.0000	.0000	.0000	.0000	.0000

TABLE II (*continued*)

n	m					p					
		.05	.10	.15	.20	.25	.30	.35	.40	.45	.50
17	1	.5819	.8332	.9369	.9775	.9925	.9977	.9993	.9998	1.0000	1.0000
	2	.2078	.5182	.7475	.8818	.9499	.9807	.9933	.9979	.9994	.9999
	3	.0503	.2382	.4802	.6904	.8363	.9226	.9673	.9877	.9959	.9988
	4	.0088	.0826	.2444	.4511	.6470	.7981	.8972	.9536	.9816	.9936
	5	.0012	.0221	.0987	.2418	.4261	.6113	.7652	.8740	.9404	.9755
	6	.0001	.0047	.0319	.1057	.2347	.4032	.5803	.7361	.8529	.9283
	7	.0000	.0008	.0083	.0377	.1071	.2248	.3812	.5522	.7098	.8338
	8	.0000	.0001	.0017	.0109	.0402	.1046	.2128	.3595	.5257	.6855
	9	.0000	.0000	.0003	.0026	.0124	.0403	.0994	.1989	.3374	.5000
	10	.0000	.0000	.0000	.0005	.0031	.0127	.0383	.0919	.1834	.3145
	11	.0000	.0000	.0000	.0001	.0006	.0032	.0120	.0348	.0826	.1662
	12	.0000	.0000	.0000	.0000	.0001	.0007	.0030	.0106	.0301	.0717
	13	.0000	.0000	.0000	.0000	.0000	.0001	.0006	.0025	.0086	.0245
	14	.0000	.0000	.0000	.0000	.0000	.0000	.0000	.0005	.0019	.0064
	15	.0000	.0000	.0000	.0000	.0000	.0000	.0000	.0001	.0003	.0012
	16	.0000	.0000	.0000	.0000	.0000	.0000	.0000	.0000	.0000	.0001
	17	.0000	.0000	.0000	.0000	.0000	.0000	.0000	.0000	.0000	.0000
18	1	.6028	.8499	.9464	.9820	.9944	.9984	.9996	.9999	1.0000	1.0000
	2	.2265	.5497	.7759	.9009	.9605	.9858	.9954	.9987	.9997	.9999
	3	.0581	.2662	.5203	.7287	.8647	.9400	.9764	.9918	.9975	.9993
	4	.0109	.0982	.2798	.4990	.6943	.8354	.9217	.9672	.9880	.9962
	5	.0015	.0282	.1206	.2836	.4813	.6673	.8114	.9058	.9589	.9846
	6	.0002	.0064	.0419	.1329	.2825	.4656	.6450	.7912	.8923	.9519
	7	.0000	.0012	.0118	.0513	.1390	.2783	.4509	.6257	.7742	.8811
	8	.0000	.0002	.0027	.0163	.0569	.1407	.2717	.4366	.6085	.7597
	9	.0000	.0000	.0005	.0043	.0193	.0596	.1391	.2632	.4222	.5927
	10	.0000	.0000	.0001	.0009	.0054	.0210	.0597	.1347	.2527	.4073
	11	.0000	.0000	.0000	.0002	.0012	.0061	.0212	.0576	.1280	.2403
	12	.0000	.0000	.0000	.0000	.0002	.0014	.0062	.0203	.0537	.1189
	13	.0000	.0000	.0000	.0000	.0000	.0003	.0014	.0058	.0183	.0481
	14	.0000	.0000	.0000	.0000	.0000	.0000	.0003	.0013	.0049	.0154
	15	.0000	.0000	.0000	.0000	.0000	.0000	.0000	.0002	.0010	.0038
	16	.0000	.0000	.0000	.0000	.0000	.0000	.0000	.0000	.0001	.0007
	17	.0000	.0000	.0000	.0000	.0000	.0000	.0000	.0000	.0000	.0001
	18	.0000	.0000	.0000	.0000	.0000	.0000	.0000	.0000	.0000	.0000
19	1	.6226	.8649	.9544	.9856	.9958	.9989	.9997	.9999	1.0000	1.0000
	2	.2453	.5797	.8015	.9171	.9690	.9896	.9969	.9992	.9998	1.0000
	3	.0665	.2946	.5587	.7631	.8887	.9538	.9830	.9945	.9985	.9996
	4	.0132	.1150	.3159	.5449	.7369	.8668	.9409	.9770	.9923	.9978
	5	.0020	.0352	.1444	.3267	.5346	.7178	.8500	.9304	.9720	.9904
	6	.0002	.0086	.0537	.1631	.3322	.5261	.7032	.8371	.9223	.9682
	7	.0000	.0017	.0163	.0676	.1749	.3345	.5188	.6919	.8273	.9165
	8	.0000	.0003	.0041	.0233	.0775	.1820	.3344	.5122	.6831	.8204
	9	.0000	.0000	.0008	.0067	.0287	.0839	.1855	.3325	.5060	.6762
	10	.0000	.0000	.0001	.0016	.0089	.0326	.0875	.1861	.3290	.5000

TABLE II (*continued*)

n	m	p = .05	.10	.15	.20	.25	.30	.35	.40	.45	.50
19	11	.0000	.0000	.0000	.0003	.0023	.0105	.0347	.0885	.1841	.3238
	12	.0000	.0000	.0000	.0000	.0005	.0028	.0114	.0352	.0871	.1796
	13	.0000	.0000	.0000	.0000	.0001	.0006	.0031	.0116	.0342	.0835
	14	.0000	.0000	.0000	.0000	.0000	.0001	.0007	.0031	.0109	.0318
	15	.0000	.0000	.0000	.0000	.0000	.0000	.0001	.0006	.0028	.0096
	16	.0000	.0000	.0000	.0000	.0000	.0000	.0000	.0001	.0005	.0022
	17	.0000	.0000	.0000	.0000	.0000	.0000	.0000	.0000	.0001	.0004
	18	.0000	.0000	.0000	.0000	.0000	.0000	.0000	.0000	.0000	.0000
	19	.0000	.0000	.0000	.0000	.0000	.0000	.0000	.0000	.0000	.0000
20	1	.6415	.8784	.9612	.9885	.9968	.9992	.9998	1.0000	1.0000	1.0000
	2	.2642	.6083	.8244	.9308	.9757	.9924	.9979	.9995	.9999	1.0000
	3	.0755	.3231	.5951	.7939	.9087	.9645	.9879	.9964	.9991	.9998
	4	.0159	.1330	.3523	.5886	.7748	.8929	.9556	.9840	.9951	.9987
	5	.0026	.0432	.1702	.3704	.5852	.7625	.8818	.9490	.9811	.9941
	6	.0003	.0113	.0673	.1958	.3828	.5836	.7546	.8744	.9447	.9793
	7	.0000	.0024	.0219	.0867	.2142	.3920	.5834	.7500	.8701	.9423
	8	.0000	.0004	.0059	.0321	.1018	.2277	.3990	.5841	.7480	.8684
	9	.0000	.0001	.0013	.0100	.0409	.1133	.2376	.4044	.5857	.7483
	10	.0000	.0000	.0002	.0026	.0139	.0480	.1218	.2447	.4086	.5881
	11	.0000	.0000	.0000	.0006	.0039	.0171	.0532	.1275	.2493	.4119
	12	.0000	.0000	.0000	.0001	.0009	.0051	.0196	.0565	.1308	.2517
	13	.0000	.0000	.0000	.0000	.0002	.0013	.0060	.0210	.0580	.1316
	14	.0000	.0000	.0000	.0000	.0000	.0003	.0015	.0065	.0214	.0577
	15	.0000	.0000	.0000	.0000	.0000	.0000	.0003	.0016	.0064	.0207
	16	.0000	.0000	.0000	.0000	.0000	.0000	.0000	.0003	.0015	.0059
	17	.0000	.0000	.0000	.0000	.0000	.0000	.0000	.0000	.0003	.0013
	18	.0000	.0000	.0000	.0000	.0000	.0000	.0000	.0000	.0000	.0002
	19	.0000	.0000	.0000	.0000	.0000	.0000	.0000	.0000	.0000	.0000
	20	.0000	.0000	.0000	.0000	.0000	.0000	.0000	.0000	.0000	.0000

TABLE III
Cumulative Terms, Poisson Distribution

m	k: 0.1	0.2	0.3	0.4	0.5	0.6	0.7	0.8	0.9	1.0
0	1.0000	1.0000	1.0000	1.0000	1.0000	1.0000	1.0000	1.0000	1.0000	1.0000
1	.0952	.1813	.2592	.3297	.3935	.4512	.5034	.5507	.5934	.6321
2	.0047	.0175	.0369	.0616	.0902	.1219	.1558	.1912	.2275	.2642
3	.0002	.0011	.0036	.0079	.0144	.0231	.0341	.0474	.0629	.0803
4	.0000	.0001	.0003	.0008	.0018	.0034	.0058	.0091	.0135	.0190
5	.0000	.0000	.0000	.0001	.0002	.0004	.0008	.0014	.0023	.0037
6	.0000	.0000	.0000	.0000	.0000	.0000	.0001	.0002	.0003	.0006
7	.0000	.0000	.0000	.0000	.0000	.0000	.0000	.0000	.0000	.0001

m	k: 1.1	1.2	1.3	1.4	1.5	1.6	1.7	1.8	1.9	2.0
0	1.0000	1.0000	1.0000	1.0000	1.0000	1.0000	1.0000	1.0000	1.0000	1.0000
1	.6671	.6988	.7275	.7534	.7769	.7981	.8173	.8347	.8504	.8647
2	.3010	.3374	.3732	.4082	.4422	.4751	.5068	.5372	.5663	.5940
3	.0996	.1205	.1429	.1665	.1912	.2166	.2428	.2694	.2963	.3233
4	.0257	.0338	.0431	.0537	.0656	.0788	.0932	.1087	.1253	.1429
5	.0054	.0077	.0107	.0143	.0186	.0237	.0296	.0364	.0441	.0527
6	.0010	.0015	.0022	.0032	.0045	.0060	.0080	.0104	.0132	.0166
7	.0001	.0003	.0004	.0006	.0009	.0013	.0019	.0026	.0034	.0045
8	.0000	.0000	.0001	.0001	.0002	.0003	.0004	.0006	.0008	.0011
9	.0000	.0000	.0000	.0000	.0000	.0000	.0001	.0001	.0002	.0002

m	k: 2.1	2.2	2.3	2.4	2.5	2.6	2.7	2.8	2.9	3.0
0	1.0000	1.0000	1.0000	1.0000	1.0000	1.0000	1.0000	1.0000	1.0000	1.0000
1	.8775	.8892	.8997	.9093	.9179	.9257	.9328	.9392	.9450	.9502
2	.6204	.6454	.6691	.6916	.7127	.7326	.7513	.7689	.7854	.8009
3	.3504	.3773	.4040	.4303	.4562	.4816	.5064	.5305	.5540	.5768
4	.1614	.1806	.2007	.2213	.2424	.2640	.2859	.3081	.3304	.3528
5	.0621	.0725	.0838	.0959	.1088	.1226	.1371	.1523	.1682	.1847
6	.0204	.0249	.0300	.0357	.0420	.0490	.0567	.0651	.0742	.0839
7	.0059	.0075	.0094	.0116	.0142	.0172	.0206	.0244	.0287	.0335
8	.0015	.0020	.0026	.0033	.0042	.0053	.0066	.0081	.0099	.0119
9	.0003	.0005	.0006	.0009	.0011	.0015	.0019	.0024	.0031	.0038
10	.0001	.0001	.0001	.0002	.0003	.0004	.0005	.0007	.0009	.0011
11	.0000	.0000	.0000	.0000	.0001	.0001	.0001	.0002	.0002	.0003
12	.0000	.0000	.0000	.0000	.0000	.0000	.0000	.0000	.0001	.0001

m	k: 3.1	3.2	3.3	3.4	3.5	3.6	3.7	3.8	3.9	4.0
0	1.0000	1.0000	1.0000	1.0000	1.0000	1.0000	1.0000	1.0000	1.0000	1.0000
1	.9550	.9592	.9631	.9666	.9698	.9727	.9753	.9776	.9798	.9817
2	.8153	.8288	.8414	.8532	.8641	.8743	.8838	.8926	.9008	.9084
3	.5988	.6201	.6406	.6603	.6792	.6973	.7146	.7311	.7469	.7619
4	.3752	.3975	.4197	.4416	.4634	.4848	.5058	.5265	.5468	.5665

TABLE III (*continued*)

m	3.1	3.2	3.3	3.4	3.5	3.6	3.7	3.8	3.9	5.0
					k					
5	.2018	.2194	.2374	.2558	.2746	.2936	.3128	.3322	.3516	.3712
6	.0943	.1054	.1171	.1295	.1424	.1559	.1699	.1844	.1994	.2149
7	.0388	.0446	.0510	.0579	.0653	.0733	.0818	.0909	.1005	.1107
8	.0142	.0168	.0198	.0231	.0267	.0308	.0352	.0401	.0454	.0511
9	.0047	.0057	.0069	.0083	.0099	.0117	.0137	.0160	.0185	.0214
10	.0014	.0018	.0022	.0027	.0033	.0040	.0048	.0058	.0069	.0081
11	.0004	.0005	.0006	.0008	.0010	.0013	.0016	.0019	.0023	.0028
12	.0001	.0001	.0002	.0002	.0003	.0004	.0005	.0006	.0007	.0009
13	.0000	.0000	.0000	.0001	.0001	.0001	.0001	.0002	.0002	.0003
14	.0000	.0000	.0000	.0000	.0000	.0000	.0000	.0000	.0001	.0001

m	4.1	4.2	4.3	4.4	4.5	4.6	4.7	4.8	4.9	5.0
					k					
0	1.0000	1.0000	1.0000	1.0000	1.0000	1.0000	1.0000	1.0000	1.0000	1.0000
1	.9834	.9850	.9864	.9877	.9889	.9899	.9909	.9918	.9926	.9933
2	.9155	.9220	.9281	.9337	.9389	.9437	.9482	.9523	.9561	.9596
3	.7762	.7898	.8026	.8149	.8264	.8374	.8477	.8575	.8667	.8753
4	.5858	.6046	.6228	.6406	.6577	.6743	.6903	.7058	.7207	.7350
5	.3907	.4102	.4296	.4488	.4679	.4868	.5054	.5237	.5418	.5595
6	.2307	.2469	.2633	.2801	.2971	.3142	.3316	.3490	.3665	.3840
7	.1214	.1325	.1442	.1564	.1689	.1820	.1954	.2092	.2233	.2378
8	.0573	.0639	.0710	.0786	.0866	.0951	.1040	.1133	.1231	.1334
9	.0245	.0279	.0317	.0358	.0403	.0451	.0503	.0558	.0618	.0681
10	.0095	.0111	.0129	.0149	.0171	.0195	.0222	.0251	.0283	.0318
11	.0034	.0041	.0048	.0057	.0067	.0078	.0090	.0104	.0120	.0137
12	.0011	.0014	.0017	.0020	.0024	.0029	.0034	.0040	.0047	.0055
13	.0003	.0004	.0005	.0007	.0008	.0010	.0012	.0014	.0017	.0020
14	.0001	.0001	.0002	.0002	.0003	.0003	.0004	.0005	.0006	.0007
15	.0000	.0000	.0000	.0001	.0001	.0001	.0001	.0001	.0002	.0002
16	.0000	.0000	.0000	.0000	.0000	.0000	.0000	.0000	.0001	.0001

m	5.1	5.2	5.3	5.4	5.5	5.6	5.7	5.8	5.9	6.0
					k					
0	1.0000	1.0000	1.0000	1.0000	1.0000	1.0000	1.0000	1.0000	1.0000	1.0000
1	.9939	.9945	.9950	.9955	.9959	.9963	.9967	.9970	.9973	.9975
2	.9628	.9658	.9686	.9711	.9734	.9756	.9776	.9794	.9811	.9826
3	.8835	.8912	.8984	.9052	.9116	.9176	.9232	.9285	.9334	.9380
4	.7487	.7619	.7746	.7867	.7983	.8094	.8200	.8300	.8396	.8488
5	.5769	.5939	.6105	.6267	.6425	.6579	.6728	.6873	.7013	.7149
6	.4016	.4191	.4365	.4539	.4711	.4881	.5050	.5217	.5381	.5543
7	.2526	.2676	.2829	.2983	.3140	.3297	.3456	.3616	.3776	.3937
8	.1440	.1551	.1665	.1783	.1905	.2030	.2159	.2290	.2424	.2560
9	.0748	.0819	.0894	.0974	.1056	.1143	.1234	.1328	.1426	.1528

TABLE III (*continued*)

m	k: 5.1	5.2	5.3	5.4	5.5	5.6	5.7	5.8	5.9	6.0
10	.0356	.0397	.0441	.0488	.0538	.0591	.0648	.0708	.0772	.0839
11	.0156	.0177	.0200	.0225	.0253	.0282	.0314	.0349	.0386	.0426
12	.0063	.0073	.0084	.0096	.0110	.0125	.0141	.0160	.0179	.0201
13	.0024	.0028	.0033	.0038	.0045	.0051	.0059	.0068	.0078	.0088
14	.0008	.0010	.0012	.0014	.0017	.0020	.0023	.0027	.0031	.0036
15	.0003	.0003	.0004	.0005	.0006	.0007	.0009	.0010	.0012	.0014
16	.0001	.0001	.0001	.0002	.0002	.0002	.0003	.0004	.0004	.0005
17	.0000	.0000	.0000	.0001	.0001	.0001	.0001	.0001	.0001	.0002
18	.0000	.0000	.0000	.0000	.0000	.0000	.0000	.0000	.0000	.0001

m	k: 6.1	6.2	6.3	6.4	6.5	6.6	6.7	6.8	6.9	7.0
0	1.0000	1.0000	1.0000	1.0000	1.0000	1.0000	1.0000	1.0000	1.0000	1.0000
1	.9978	.9980	.9982	.9983	.9985	.9986	.9988	.9989	.9990	.9991
2	.9841	.9854	.9866	.9877	.9887	.9897	.9905	.9913	.9920	.9927
3	.9423	.9464	.9502	.9537	.9570	.9600	.9629	.9656	.9680	.9704
4	.8575	.8658	.8736	.8811	.8882	.8948	.9012	.9072	.9129	.9182
5	.7281	.7408	.7531	.7649	.7763	.7873	.7978	.8080	.8177	.8270
6	.5702	.5859	.6012	.6163	.6310	.6453	.6594	.6730	.6863	.6993
7	.4098	.4258	.4418	.4577	.4735	.4892	.5047	.5201	.5353	.5503
8	.2699	.2840	.2983	.3127	.3272	.3419	.3567	.3715	.3864	.4013
9	.1633	.1741	.1852	.1967	.2084	.2204	.2327	.2452	.2580	.2709
10	.0910	.0984	.1061	.1142	.1226	.1314	.1404	.1498	.1505	.1695
11	.0469	.0514	.0563	.0614	.0668	.0726	.0786	.0849	.0916	.0985
12	.0224	.0250	.0277	.0307	.0339	.0373	.0409	.0448	.0490	.0534
13	.0100	.0113	.0127	.0143	.0160	.0179	.0199	.0221	.0245	.0270
14	.0042	.0048	.0055	.0063	.0071	.0080	.0091	.0102	.0115	.0128
15	.0016	.0019	.0022	.0026	.0030	.0034	.0039	.0044	.0050	.0057
16	.0006	.0007	.0008	.0010	.0012	.0014	.0016	.0018	.0021	.0024
17	.0002	.0003	.0003	.0004	.0004	.0005	.0006	.0007	.0008	.0010
18	.0001	.0001	.0001	.0001	.0002	.0002	.0002	.0003	.0003	.0004
19	.0000	.0000	.0000	.0000	.0001	.0001	.0001	.0001	.0001	.0001

m	k: 7.1	7.2	7.3	7.4	7.5	7.6	7.7	7.8	7.9	8.0
0	1.0000	1.0000	1.0000	1.0000	1.0000	1.0000	1.0000	1.0000	1.0000	1.0000
1	.9992	.9993	.9993	.9994	.9994	.9995	.9995	.9996	.9996	.9997
2	.9933	.9939	.9944	.9949	.9953	.9957	.9961	.9964	.9967	.9970
3	.9725	.9745	.9764	.0781	.9797	.9812	.9826	.9839	.9851	.9862
4	.9233	.9281	.9326	.9368	.9409	.9446	.9482	.9515	.9547	.9576
5	.8359	.8445	.8527	.8605	.8679	.8751	.8819	.8883	.8945	.9004
6	.7119	.7241	.7360	.7474	.7586	.7693	.7797	.7897	.7994	.8088
7	.5651	.5796	.5940	.6080	.6218	.6354	.6486	.6616	.6743	.6866
8	.4162	.4311	.4459	.4607	.4754	.4900	.5044	.5188	.5330	.5470
9	.2840	.2973	.3108	.3243	.3380	.3518	.3657	.3796	.3935	.4075
10	.1798	.1904	.2012	.2123	.2236	.2351	.2469	.2589	.2710	.2834
11	.1058	.1133	.1212	.1293	.1378	.1465	.1555	.1648	.1743	.1841
12	.0580	.0629	.0681	.0735	.0792	.0852	.0915	.0980	.1048	.1119
13	.0297	.0327	.0358	.0391	.0427	.0464	.0504	.0546	.0591	.0638
14	.0143	.0159	.0176	.0195	.0216	.0238	.0261	.0286	.0313	.0342

TABLE III (*continued*)

m	k: 7.1	7.2	7.3	7.4	7.5	7.6	7.7	7.8	7.9	8.0
15	.0065	.0073	.0082	.0092	.0103	.0114	.0127	.0141	.0156	.0173
16	.0028	.0031	.0036	.0041	.0046	.0052	.0059	.0066	.0074	.0082
17	.0011	.0013	.0015	.0017	.0020	.0022	.0026	.0029	.0033	.0037
18	.0004	.0005	.0006	.0007	.0008	.0009	.0011	.0012	.0014	.0016
19	.0002	.0002	.0002	.0003	.0003	.0004	.0004	.0005	.0006	.0006
20	.0001	.0001	.0001	.0001	.0001	.0001	.0002	.0002	.0002	.0003
21	.0000	.0000	.0000	.0000	.0000	.0000	.0001	.0001	.0001	.0001

m	k: 8.1	8.2	8.3	8.4	8.5	8.6	8.7	8.8	8.9	9.0
0	1.0000	1.0000	1.0000	1.0000	1.0000	1.0000	1.0000	1.0000	1.0000	1.0000
1	.9997	.9997	.9998	.9998	.9998	.9998	.9998	.9998	.9999	.9999
2	.9972	.9975	.9977	.9979	.9981	.9982	.9984	.9985	.9987	.9988
3	.9873	.9882	.9891	.9900	.9907	.9914	.9921	.9927	.9932	.9938
4	.9604	.9630	.9654	.9677	.9699	.9719	.9738	.9756	.9772	.9788
5	.9060	.9113	.9163	.9211	.9256	.9299	.9340	.9379	.9416	.9450
6	.8178	.8264	.8347	.8427	.8504	.8578	.8648	.8716	.8781	.8843
7	.6987	.7104	.7219	.7330	.7438	.7543	.7645	.7744	.7840	.7932
8	.5609	.5746	.5881	.6013	.6144	.6272	.6398	.6522	.6643	.6761
9	.4214	.4353	.4493	.4631	.4769	.4906	.5042	.5177	.5311	.5443
10	.2959	.3085	.3212	.3341	.3470	.3600	.3731	.3863	.3994	.4126
11	.1942	.2045	.2150	.2257	.2366	.2478	.2591	.2706	.2822	.2940
12	.1193	.1269	.1348	.1429	.1513	.1600	.1689	.1780	.1874	.1970
13	.0687	.0739	.0793	.0850	.0909	.0971	.1035	.1102	.1171	.1242
14	.0372	.0405	.0439	.0476	.0514	.0555	.0597	.0642	.0689	.0739
15	.0190	.0209	.0229	.0251	.0274	.0299	.0325	.0353	.0383	.0415
16	.0092	.0102	.0113	.0125	.0138	.0152	.0168	.0184	.0202	.0220
17	.0042	.0047	.0053	.0059	.0066	.0074	.0082	.0091	.0101	.0111
18	.0018	.0021	.0023	.0027	.0030	.0034	.0038	.0043	.0048	.0053
19	.0008	.0009	.0010	.0011	.0013	.0015	.0017	.0019	.0022	.0024
20	.0003	.0003	.0004	.0005	.0005	.0006	.0007	.0008	.0009	.0011
21	.0001	.0001	.0002	.0002	.0002	.0002	.0003	.0003	.0004	.0004
22	.0000	.0000	.0001	.0001	.0001	.0001	.0001	.0001	.0002	.0002
23	.0000	.0000	.0000	.0000	.0000	.0000	0000	.0000	.0001	.0001

m	k: 9.1	9.2	9.3	9.4	9.5	9.6	9.7	9.8	9.9	10
0	1.0000	1.0000	1.0000	1.0000	1.0000	1.0000	1.0000	1.0000	1.0000	1.0000
1	.9999	.9999	:9999	.9999	.9999	.9999	.9999	.9999	1.0000	1.0000
2	.9989	.9990	.9991	.9991	.9992	.9993	.9993	.9994	.9995	.9995
3	.9942	.9947	.9951	.9955	.9958	.9962	.9965	.9967	.9970	.9972
4	.9802	.9816	.9828	.9840	.9851	.9862	.9871	.9880	.9889	.9897
5	.9483	.9514	.9544	.9571	.9597	.9622	.9645	.9667	.9688	.9707
6	.8902	.8959	.9014	.9065	.9115	.9162	.9207	.9250	.9290	.9329
7	.8022	.8108	.8192	.8273	.8351	.8426	.8498	.8567	.8634	.8699
8	.6877	.6990	.7101	.7208	.7313	.7416	.7515	.7612	.7706	.7798
9	.5574	.5704	.5832	.5958	.6082	.6204	.6324	.6442	.6558	.6672

TABLE III (*continued*)

m	k = 9.1	9.2	9.3	9.4	9.5	9.6	9.7	9.8	9.9	10
10	.4258	.4389	.4521	.4651	.4782	.4911	.5040	.5168	.5295	.5421
11	.3059	.3180	.3301	.3424	.3547	.3671	.3795	.3920	.4045	.4170
12	.2068	.2168	.2270	.2374	.2480	.2588	.2697	.2807	.2919	.3032
13	.1316	.1393	.1471	.1552	.1636	.1721	.1809	.1899	.1991	.2084
14	.0790	.0844	.0900	.0958	.1019	.1081	.1147	.1214	.1284	.1355
15	.0448	.0483	.0520	.0559	.0600	.0643	.0688	.0735	.0784	.0835
16	.0240	.0262	.0285	.0309	.0335	.0362	.0391	.0421	.0454	.0487
17	.0122	.0135	.0148	.0162	.0177	.0194	.0211	.0230	.0249	.0270
18	.0059	.0066	.0073	.0081	.0089	.0098	.0108	.0119	.0130	.0143
19	.0027	.0031	.0034	.0038	.0043	.0048	.0053	.0059	.0065	.0072
20	.0012	.0014	.0015	.0017	.0020	.0022	.0025	.0028	.0031	.0035
21	.0005	.0006	.0007	.0008	.0009	.0010	.0011	.0013	.0014	.0016
22	.0002	.0002	.0003	.0003	.0004	.0004	.0005	.0005	.0006	.0007
23	.0001	.0001	.0001	.0001	.0001	.0002	.0002	.0002	.0003	.0003
24	.0000	.0000	.0000	.0000	.0001	.0001	.0001	.0001	.0001	.0001

m	k = 11	12	13	14	15	16	17	18	19	20
0	1.0000	1.0000	1.0000	1.0000	1.0000	1.0000	1.0000	1.0000	1.0000	1.0000
1	1.0000	1.0000	1.0000	1.0000	1.0000	1.0000	1.0000	1.0000	1.0000	1.0000
2	.9998	.9999	1.0000	1.0000	1.0000	1.0000	1.0000	1.0000	1.0000	1.0000
3	.9988	.9995	.9998	.9999	1.0000	1.0000	1.0000	1.0000	1.0000	1.0000
4	.9951	.9977	.9990	.9995	.9998	.9999	1.0000	1.0000	1.0000	1.0000
5	.9849	.9924	.9963	.9982	.9991	.9996	.9998	.9999	1.0000	1.0000
6	.9625	.9797	.9893	.9945	.9972	.9986	.9993	.9997	.9998	.9999
7	.9214	.9542	.9741	.9858	.9924	.9960	.9979	.9990	.9995	.9997
8	.8568	.9105	.9460	.9684	.9820	.9900	.9946	.9971	.9985	.9992
9	.7680	.8450	.9002	.9379	.9626	.9780	.9874	.9929	.9961	.9979
10	.6595	.7576	.8342	.8906	.9301	.9567	.9739	.9846	.9911	.9950
11	.5401	.6528	.7483	.8243	.8815	.9226	.9509	.9696	.9817	.9892
12	.4207	.5384	.6468	.7400	.8152	.8730	.9153	.9451	.9653	.9786
13	.3113	.4240	.5369	.6415	.7324	.8069	.8650	.9083	.9394	.9610
14	.2187	.3185	.4270	.5356	.6368	.7255	.7991	.8574	.9016	.9339
15	.1460	.2280	.3249	.4296	.5343	.6325	.7192	.7919	.8503	.8951
16	.0926	.1556	.2364	.3306	.4319	.5333	.6285	.7133	.7852	.8435
17	.0559	.1013	.1645	.2441	.3359	.4340	.5323	.6250	.7080	.7789
18	.0322	.0630	.1095	.1728	.2511	.3407	.4360	.5314	.6216	.7030
19	.0177	.0374	.0698	.1174	.1805	.2577	.3450	.4378	.5305	.6186
20	.0093	.0213	.0427	.0765	.1248	.1878	.2637	.3491	.4394	.5297
21	.0047	.0116	.0250	.0479	.0830	.1318	.1945	.2693	.3528	.4409
22	.0023	.0061	.0141	.0288	.0531	.0892	.1385	.2009	.2745	.3563
23	.0010	.0030	.0076	.0167	.0327	.0582	.0953	.1449	.2069	.2794
24	.0005	.0015	.0040	.0093	.0195	.0367	.0633	.1011	.1510	.2125
25	.0002	.0007	.0020	.0050	.0112	.0223	.0406	.0683	.1067	.1568
26	.0001	.0003	.0010	.0026	.0062	.0131	.0252	.0446	.0731	.1122
27	.0000	.0001	.0005	.0013	.0033	.0075	.0152	.0282	.0486	.0779
28	.0000	.0001	.0002	.0006	.0017	.0041	.0088	.0173	.0313	.0525
29	.0000	.0000	.0001	.0003	.0009	.0022	.0050	.0103	.0195	.0343

TABLE III (*continued*)

m	k 11	12	13	14	15	16	17	18	19	20
30	.0000	.0000	.0000	.0001	.0004	.0011	.0027	.0059	.0118	.0218
31	.0000	.0000	.0000	.0001	.0002	.0006	.0014	.0033	.0070	.0135
32	.0000	.0000	.0000	.0000	.0001	.0003	.0007	.0018	.0040	.0081
33	.0000	.0000	.0000	.0000	.0000	.0001	.0004	.0010	.0022	.0047
34	.0000	.0000	.0000	.0000	.0000	.0001	.0002	.0005	.0012	.0027
35	.0000	.0000	.0000	.0000	.0000	.0000	.0001	.0002	.0006	.0015
36	.0000	.0000	.0000	.0000	.0000	.0000	.0000	.0001	.0003	.0008
37	.0000	.0000	.0000	.0000	.0000	.0000	.0000	.0001	.0002	.0004
38	.0000	.0000	.0000	.0000	.0000	.0000	.0000	.0000	.0001	.0002
39	.0000	.0000	.0000	.0000	.0000	.0000	.0000	.0000	.0000	.0001
40	.0000	.0000	.0000	.0000	.0000	.0000	.0000	.0000	.0000	.0001

TABLE IV*a*

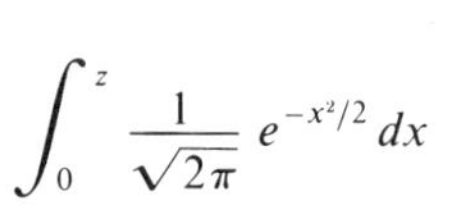

$$\int_0^z \frac{1}{\sqrt{2\pi}} e^{-x^2/2}\, dx$$

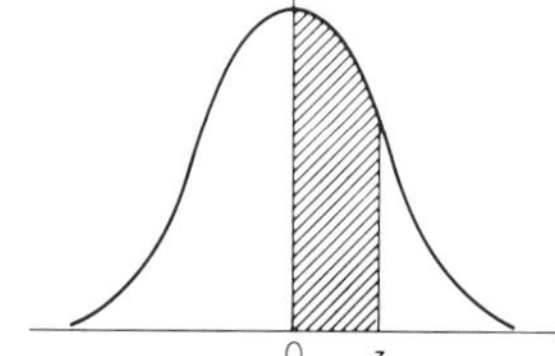

z	.00	.01	.02	.03	.04	.05	.06	.07	.08	.09
0.0	.00000	.00399	.00798	.01197	.01595	.01994	.02392	.02790	.03188	.03586
0.1	.03983	.04379	.04776	.05172	.05567	.05962	.06356	.06749	.07142	.07535
0.2	.07926	.08317	.08706	.09095	.09483	.09871	.10257	.10642	.11026	.11409
0.3	.11791	.12172	.12552	.12930	.13307	.13683	.14058	.14431	.14803	.15173
0.4	.15542	.15910	.16276	.16640	.17003	.17364	.17724	.18082	.18439	.18793
0.5	.19146	.19497	.19847	.20194	.20540	.20884	.21226	.21566	.21904	.22240
0.6	.22575	.22907	.23237	.23565	.23891	.24215	.24537	.24857	.25175	.25490
0.7	.25804	.26115	.26424	.26730	.27035	.27337	.27637	.27935	.28230	.28524
0.8	.28814	.29103	.29389	.29673	.29955	.30234	.30511	.30785	.31057	.31327
0.9	.31594	.31859	.32121	.32381	.32639	.32894	.33147	.33398	.33646	.33891
1.0	.34134	.34375	.34614	.34849	.35083	.35314	.35543	.35769	.35993	.36214
1.1	.36433	.36650	.36864	.37076	.37286	.37493	.37698	.37900	.38100	.38298
1.2	.38493	.38686	.38877	.39065	.39251	.39435	.39616	.39796	.39973	.40147
1.3	.40320	.40490	.40658	.40824	.40988	.41149	.41308	.41466	.41621	.41773
1.4	.41924	.42073	.42220	.42364	.42507	.42647	.42785	.42922	.43056	.43189
1.5	.43319	.43448	.43574	.43699	.43822	.43943	.44062	.44179	.44295	.44408
1.6	.44520	.44630	.44738	.44845	.44950	.45053	.45154	.45254	.45352	.45449
1.7	.45543	.45637	.45728	.45818	.45907	.45994	.46080	.46164	.46246	.46327
1.8	.46407	.46485	.46562	.46637	.46712	.46784	.46856	.46926	.46995	.47062
1.9	.47128	.47193	.47257	.47320	.47381	.47441	.47500	.47558	.47615	.47670
2.0	.47725	.47778	.47830	.47882	.47932	.47982	.48030	.48077	.48124	.48169
2.1	.48213	.48257	.48300	.48341	.48382	.48422	.48461	.48500	.48537	.48574
2.2	.48610	.48645	.48679	.48713	.48745	.48777	.48809	.48840	.48870	.48899
2.3	.48927	.48955	.48983	.49010	.49036	.49061	.49086	.49110	.49134	.49157
2.4	.49180	.49202	.49224	.49245	.49266	.49286	.49305	.49324	.49342	.49361
2.5	.49379	.49396	.49413	.49430	.49446	.49461	.49477	.49491	.49506	.49520
2.6	.49534	.49547	.49560	.49573	.49585	.49597	.49609	.49621	.49632	.49643
2.7	.49653	.49663	.49673	.49683	.49693	.49702	.49711	.49720	.49728	.49736
2.8	.49744	.49752	.49760	.49767	.49774	.49781	.49788	.49795	.49801	.49807
2.9	.49813	.49819	.49825	.49830	.49836	.49841	.49846	.49851	.49856	.49860
3.0	.49865	.49869	.49873	.49878	.49882	.49885	.49889	.49893	.49896	.49900
3.1	.49903	.49906	.49909	.49912	.49915	.49918	.49921	.49924	.49926	.49929
3.2	.49931	.49933	.49936	.49938	.49940	.49942	.49944	.49946	.49948	.49950
3.3	.49952	.49953	.49955	.49956	.49958	.49959	.49961	.49962	.49964	.49965
3.4	.49966	.49967	.49969	.49970	.49971	.49972	.49973	.49974	.49975	.49976
3.5	.49977	.49977	.49978	.49979	.49980	.49981	.49981	.49982	.49983	.49983
3.6	.49984	.49985	.49985	.49986	.49986	.49987	.49987	.49988	.49988	.49989
3.7	.49989	.49990	.49990	.49990	.49991	.49991	.49991	.49992	.49992	.49992
3.8	.49993	.49993	.49993	.49993	.49994	.49994	.49994	.49994	.49995	.49995
3.9	.49995	.49995	.49995	.49996	.49996	.49996	.49996	.49996	.49996	.49997

TABLE IV*b*

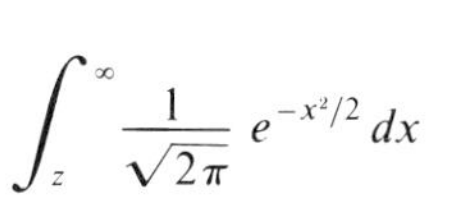

$$\int_z^\infty \frac{1}{\sqrt{2\pi}} e^{-x^2/2}\, dx$$

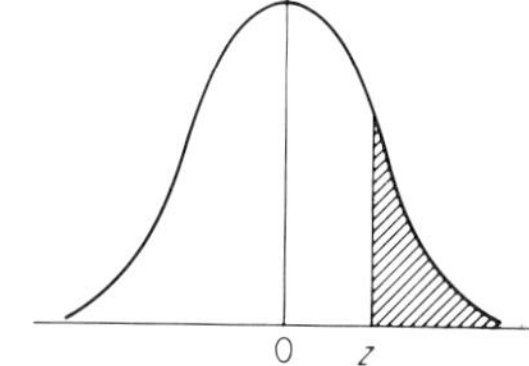

z	.00	.01	.02	.03	.04	.05	.06	.07	.08	.09
0.0	.50000	.49601	.49202	.48803	.48405	.48006	.47608	.47210	.46812	.46414
0.1	.46017	.45620	.45224	.44828	.44433	.44038	.43644	.43251	.42858	.42465
0.2	.42074	.41683	.41294	.40905	.40517	.40129	.39743	.39358	.38974	.38591
0.3	.38209	.37828	.37448	.37070	.36693	.36317	.35942	.35569	.35197	.34827
0.4	.34458	.34090	.33724	.33360	.32997	.32636	.32276	.31918	.31561	.31207
0.5	.30854	.30503	.30153	.29806	.29460	.29116	.28774	.28434	.28096	.27760
0.6	.27425	.27093	.26763	.26435	.26109	.25785	.25463	.25143	.24825	.24510
0.7	.24196	.23885	.23576	.23270	.22965	.22663	.22363	.22065	.21770	.21476
0.8	.21186	.20897	.20611	.20327	.20045	.19766	.19489	.19215	.18943	.18673
0.9	.18406	.18141	.17879	.17619	.17361	.17106	.16853	.16602	.16354	.16109
1.0	.15866	.15625	.15386	.15151	.14917	.14686	.14457	.14231	.14007	.13786
1.1	.13567	.13350	.13136	.12924	.12714	.12507	.12302	.12100	.11900	.11702
1.2	.11507	.11314	.11123	.10935	.10749	.10566	.10384	.10204	.10027	.09853
1.3	.09680	.09510	.09342	.09176	.09012	.08851	.08692	.08534	.08379	.08227
1.4	.08076	.07927	.07780	.07636	.07493	.07353	.07215	.07078	.06944	.06811
1.5	.06681	.06552	.06426	.06301	.06178	.06057	.05938	.05821	.05705	.05592
1.6	.05480	.05370	.05262	.05155	.05050	.04947	.04846	.04746	.04648	.04551
1.7	.04457	.04363	.04272	.04182	.04093	.04006	.03920	.03836	.03754	.03673
1.8	.03593	.03149	.03438	.03363	.03288	.03216	.03144	.03074	.03005	.02938
1.9	.02872	.02807	.02743	.02680	.02619	.02559	.02500	.02442	.02385	.02330
2.0	.02275	.02222	.02169	.02118	.02068	.02018	.01970	.01923	.01876	.01831
2.1	.01787	.01743	.01700	.01659	.01618	.01578	.01539	.01500	.01463	.01426
2.2	.01390	.01355	.01321	.01287	.01255	.01223	.01191	.01160	.01130	.01101
2.3	.01073	.01045	.01017	.00990	.00964	.00939	.00914	.00890	.00866	.00843
2.4	.00820	.00798	.00776	.00755	.00734	.00714	.00695	.00676	.00657	.00639
2.5	.00621	.00604	.00587	.00570	.00554	.00539	.00523	.00509	.00494	.00480
2.6	.00466	.00453	.00440	.00427	.00415	.00403	.00391	.00379	.00368	.00357
2.7	.00347	.00337	.00327	.00317	.00307	.00298	.00289	.00280	.00272	.00264
2.8	.00256	.00248	.00240	.00233	.00226	.00219	.00212	.00205	.00199	.00193
2.9	.00187	.00181	.00175	.00170	.00164	.00159	.00154	.00149	.00144	.00140
3.0	.00135	.00131	.00127	.00122	.00118	.00115	.00111	.00107	.00104	.00100
3.1	.00097	.00094	.00091	.00088	.00085	.00082	.00079	.00076	.00074	.00071
3.2	.00069	.00067	.00064	.00062	.00060	.00058	.00056	.00054	.00052	.00050
3.3	.00048	.00047	.00045	.00044	.00042	.00041	.00039	.00038	.00036	.00035
3.4	.00034	.00033	.00031	.00030	.00029	.00028	.00027	.00026	.00025	.00024
3.5	.00023	.00023	.00022	.00021	.00020	.00019	.00019	.00018	.00017	.00017
3.6	.00016	.00015	.00015	.00014	.00014	.00013	.00013	.00012	.00012	.00011
3.7	.00011	.00010	.00010	.00010	.00009	.00009	.00009	.00008	.00008	.00008
3.8	.00007	.00007	.00007	.00007	.00006	.00006	.00006	.00006	.00005	.00005
3.9	.00005	.00005	.00005	.00004	.00004	.00004	.00004	.00004	.00004	.00003

TABLE V
Percentage Points of the t-Distribution

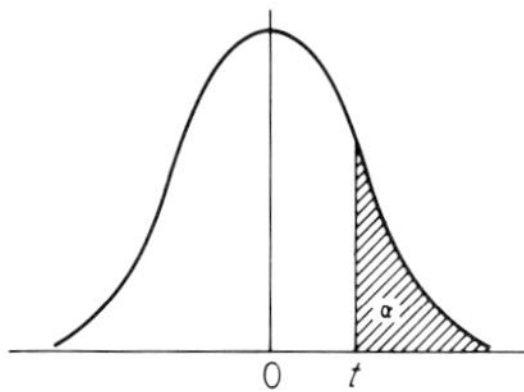

d.f.	α = 0·4	0·25	0·1	0·05	0·025	0·01	0·005	0·0025	0·001	0·0005
1	0·325	1·000	3·078	6·314	12·706	31·821	63·657	127·32	318·31	636·62
2	·289	0·816	1·886	2·920	4·303	6·965	9·925	14·089	22·327	31·598
3	·277	·765	1·638	2·353	3·182	4·541	5·841	7·453	10·214	12·924
4	·271	·741	1·533	2·132	2·776	3·747	4·604	5·598	7·173	8·610
5	0·267	0·727	1·476	2·015	2·571	3·365	4·032	4·773	5·893	6·869
6	·265	·718	1·440	1·943	2·447	3·143	3·707	4·317	5·208	5·959
7	·263	·711	1·415	1·895	2·365	2·998	3·499	4·029	4·785	5·408
8	·262	·706	1·397	1·860	2·306	2·896	3·355	3·833	4·501	5·041
9	·261	·703	1·383	1·833	2·262	2·821	3·250	3·690	4·297	4·781
10	0·260	0·700	1·372	1·812	2·228	2·764	3·169	3·581	4·144	4·587
11	·260	·697	1·363	1·796	2·201	2·718	3·106	3·497	4·025	4·437
12	·259	·695	1·356	1·782	2·179	2·681	3·055	3·428	3·930	4·318
13	·259	·694	1·350	1·771	2·160	2·650	3·012	3·372	3·852	4·221
14	·258	·692	1·345	1·761	2·145	2·624	2·977	3·326	3·787	4·140
15	0·258	0·691	1·341	1·753	2·131	2·602	2·947	3·286	3·733	4·073
16	·258	·690	1·337	1·746	2·120	2·583	2·921	3·252	3·686	4·015
17	·257	·689	1·333	1·740	2·110	2·567	2·898	3·222	3·646	3·965
18	·257	·688	1·330	1·734	2·101	2·552	2·878	3·197	3·610	3·922
19	·257	·688	1·328	1·729	2·093	2·539	2·861	3·174	3·579	3·883
20	0·257	0·687	1·325	1·725	2·086	2·528	2·845	3·153	3·552	3·850
21	·257	·686	1·323	1·721	2·080	2·518	2·831	3·135	3·527	3·819
22	·256	·686	1·321	1·717	2·074	2·508	2·819	3·119	3·505	3·792
23	·256	·685	1·319	1·714	2·069	2·500	2·807	3·104	3·485	3·767
24	·256	·685	1·318	1·711	2·064	2·492	2·797	3·091	3·467	3·745
25	0·256	0·684	1·316	1·708	2·060	2·485	2·787	3·078	3·450	3·725
26	·256	·684	1·315	1·706	2·056	2·479	2·779	3·067	3·435	3·707
27	·256	·684	1·314	1·703	2·052	2·473	2·771	3·057	3·421	3·690
28	·256	·683	1·313	1·701	2·048	2·467	2·763	3·047	3·408	3·674
29	·256	·683	1·311	1·699	2·045	2·462	2·756	3·038	3·396	3·659
30	0·256	0·683	1·310	1·697	2·042	2·457	2·750	3·030	3·385	3·646
40	·255	·681	1·303	1·684	2·021	2·423	2·704	2·971	3·307	3·551
60	·254	·679	1·296	1·671	2·000	2·390	2·660	2·915	3·232	3·460
120	·254	·677	1·289	1·658	1·980	2·358	2·617	2·860	3·160	3·373
∞	·253	·674	1·282	1·645	1·960	2·326	2·576	2·807	3·090	3·291

TABLE VI
Percentage Points of the χ^2-Distribution

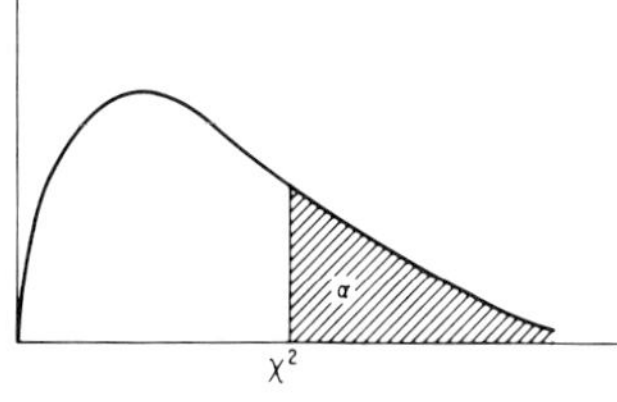

d.f. \ α	0·995	0·990	0·975	0·950	0·900	0·750	0·500
1	392704.10^{-10}	157088.10^{-9}	982069.10^{-9}	393214.10^{-8}	0·0157908	0·1015308	0·454936
2	0·0100251	0·0201007	0·0506356	0·102587	0·210721	0·575364	1·38629
3	0·0717218	0·114832	0·215795	0·351846	0·584374	1·212534	2·36597
4	0·206989	0·297109	0·484419	0·710723	1·063623	1·92256	3·35669
5	0·411742	0·554298	0·831212	1·145476	1·61031	2·67460	4·35146
6	0·675727	0·872090	1·23734	1·63538	2·20413	3·45460	5·34812
7	0·989256	1·239043	1·68987	2·16735	2·83311	4·25485	6·34581
8	1·34441	1·64650	2·17973	2·73264	3·48954	5·07064	7·34412
9	1·73493	2·08790	2·70039	3·32511	4·16816	5·89883	8·34283
10	2·15586	2·55821	3·24697	3·94030	4·86518	6·73720	9·34182
11	2·60322	3·05348	3·81575	4·57481	5·57778	7·58414	10·3410
12	3·07382	3·57057	4·40379	5·22603	6·30380	8·43842	11·3403
13	3·56503	4·10692	5·00875	5·89186	7·04150	9·29907	12·3398
14	4·07467	4·66043	5·62873	6·57063	7·78953	10·1653	13·3393
15	4·60092	5·22935	6·26214	7·26094	8·54676	11·0365	14·3389
16	5·14221	5·81221	6·90766	7·96165	9·31224	11·9122	15·3385
17	5·69722	6·40776	7·56419	8·67176	10·0852	12·7919	16·3382
18	6·26480	7·01491	8·23075	9·39046	10·8649	13·6753	17·3379
19	6·84397	7·63273	8·90652	10·1170	11·6509	14·5620	18·3377
20	7·43384	8·26040	9·59078	10·8508	12·4426	15·4518	19·3374
21	8·03365	8·89720	10·28293	11·5913	13·2396	16·3444	20·3372
22	8·64272	9·54249	10·9823	12·3380	14·0415	17·2396	21·3370
23	9·26043	10·19567	11·6886	13·0905	14·8480	18·1373	22·3369
24	9·88623	10·8564	12·4012	13·8484	15·6587	19·0373	23·3367
25	10·5197	11·5240	13·1197	14·6114	16·4734	19·9393	24·3366
26	11·1602	12·1981	13·8439	15·3792	17·2919	20·8434	25·3365
27	11·8076	12·8785	14·5734	16·1514	18·1139	21·7494	26·3363
28	12·4613	13·5647	15·3079	16·9279	18·9392	22·6572	27·3362
29	13·1211	14·2565	16·0471	17·7084	19·7677	23·5666	28·3361
30	13·7867	14·9535	16·7908	18·4927	20·5992	24·4776	29·3360
40	20·7065	22·1643	24·4330	26·5093	29·0505	33·6603	39·3353
50	27·9907	29·7067	32·3574	34·7643	37·6886	42·9421	49·3349
60	35·5345	37·4849	40·4817	43·1880	46·4589	52·2938	59·3347
70	43·2752	45·4417	48·7576	51·7393	55·3289	61·6983	69·3345
80	51·1719	53·5401	57·1532	60·3915	64·2778	71·1445	79·3343
90	59·1963	61·7541	65·6466	69·1260	73·2911	80·6247	89·3342
100	67·3276	70·0649	74·2219	77·9295	82·3581	90·1332	99·3341

TABLE VI (*continued*)

d.f. \ α	0·250	0·100	0·050	0·025	0·010	0·005	0·001
1	1·32330	2·70554	3·84146	5·02389	6·63490	7·87944	10·828
2	2·77259	4·60517	5·99146	7·37776	9·21034	10·5966	13·816
3	4·10834	6·25139	7·81473	9·34840	11·3449	12·8382	16·266
4	5·38527	7·77944	9·48773	11·1433	13·2767	14·8603	18·467
5	6·62568	9·23636	11·0705	12·8325	15·0863	16·7496	20·515
6	7·84080	10·6446	12·5916	14·4494	16·8119	18·5476	22·458
7	9·03715	12·0170	14·0671	16·0128	18·4753	20·2777	24·322
8	10·2189	13·3616	15·5073	17·5345	20·0902	21·9550	26·125
9	11·3888	14·6837	16·9190	19·0228	21·6660	23·5894	27·877
10	12·5489	15·9872	18·3070	20·4832	23·2093	25·1882	29·588
11	13·7007	17·2750	19·6751	21·9200	24·7250	26·7568	31·264
12	14·8454	18·5493	21·0261	23·3367	26·2170	28·2995	32·909
13	15·9839	19·8119	22·3620	24·7356	27·6882	29·8195	34·528
14	17·1169	21·0641	23·6848	26·1189	29·1412	31·3194	36·123
15	18·2451	22·3071	24·9958	27·4884	30·5779	32·8013	37·697
16	19·3689	23·5418	26·2962	28·8454	31·9999	34·2672	39·252
17	20·4887	24·7690	27·5871	30·1910	33·4087	35·7185	40·790
18	21·6049	25·9894	28·8693	31·5264	34·8053	37·1565	42·312
19	22·7178	27·2036	30·1435	32·8523	36·1909	38·5823	43·820
20	23·8277	28·4120	31·4104	34·1696	37·5662	39·9968	45·315
21	24·9348	29·6151	32·6706	35·4789	38·9322	41·4011	46·797
22	26·0393	30·8133	33·9244	36·7807	40·2894	42·7957	48·268
23	27·1413	32·0069	35·1725	38·0756	41·6384	44·1813	49·728
24	28·2412	33·1962	36·4150	39·3641	42·9798	45·5585	51·179
25	29·3389	34·3816	37·6525	40·6465	44·3141	46·9279	52·618
26	30·4346	35·5632	38·8851	41·9232	45·6417	48·2899	54·052
27	31·5284	36·7412	40·1133	43·1945	46·9629	49·6449	55·476
28	32·6205	37·9159	41·3371	44·4608	48·2782	50·9934	56·892
29	33·7109	39·0875	42·5570	45·7223	49·5879	52·3356	58·301
30	34·7997	40·2560	43·7730	46·9792	50·8922	53·6720	59·703
40	45·6160	51·8051	55·7585	59·3417	63·6907	66·7660	73·402
50	56·3336	63·1671	67·5048	71·4202	76·1539	79·4900	86·661
60	66·9815	74·3970	79·0819	83·2977	88·3794	91·9517	99·607
70	77·5767	85·5270	90·5312	95·0232	100·425	104·215	112·317
80	88·1303	96·5782	101·879	106·629	112·329	116·321	124·839
90	98·6499	107·565	113·145	118·136	124·116	128·299	137·208
100	109·141	118·498	124·342	129·561	135·807	140·169	149·449

TABLE VII
Percentage Points of the F-Distribution (Variance Ratio)
Upper 25 Percentage Points

n \ m	1	2	3	4	5	6	7	8	9	10	12	15	20	24	30	40	60	120	∞
1	5·83	7·50	8·20	8·58	8·82	8·98	9·10	9·19	9·26	9·32	9·41	9·49	9·58	9·63	9·67	9·71	9·76	9·80	9·85
2	2·57	3·00	3·15	3·23	3·28	3·31	3·34	3·35	3·37	3·38	3·39	3·41	3·43	3·43	3·44	3·45	3·46	3·47	3·48
3	2·02	2·28	2·36	2·39	2·41	2·42	2·43	2·44	2·44	2·44	2·45	2·46	2·46	2·46	2·47	2·47	2·47	2·47	2·47
4	1·81	2·00	2·05	2·06	2·07	2·08	2·08	2·08	2·08	2·08	2·08	2·08	2·08	2·08	2·08	2·08	2·08	2·08	2·08
5	1·69	1·85	1·88	1·89	1·89	1·89	1·89	1·89	1·89	1·89	1·89	1·89	1·88	1·88	1·88	1·88	1·87	1·87	1·87
6	1·62	1·76	1·78	1·79	1·79	1·78	1·78	1·78	1·77	1·77	1·77	1·76	1·76	1·75	1·75	1·75	1·74	1·74	1·74
7	1·57	1·70	1·72	1·72	1·71	1·71	1·70	1·70	1·69	1·69	1·68	1·68	1·67	1·67	1·66	1·66	1·65	1·65	1·65
8	1·54	1·66	1·67	1·66	1·66	1·65	1·64	1·64	1·63	1·63	1·62	1·62	1·61	1·60	1·60	1·59	1·59	1·58	1·58
9	1·51	1·62	1·63	1·63	1·62	1·61	1·60	1·60	1·59	1·59	1·58	1·57	1·56	1·56	1·55	1·54	1·54	1·53	1·53
10	1·49	1·60	1·60	1·59	1·59	1·58	1·57	1·56	1·56	1·55	1·54	1·53	1·52	1·52	1·51	1·51	1·50	1·49	1·48
11	1·47	1·58	1·58	1·57	1·56	1·55	1·54	1·53	1·53	1·52	1·51	1·50	1·49	1·49	1·48	1·47	1·47	1·46	1·45
12	1·46	1·56	1·56	1·55	1·54	1·53	1·52	1·51	1·51	1·50	1·49	1·48	1·47	1·46	1·45	1·45	1·44	1·43	1·42
13	1·45	1·55	1·55	1·53	1·52	1·51	1·50	1·49	1·49	1·48	1·47	1·46	1·45	1·44	1·43	1·42	1·42	1·41	1·40
14	1·44	1·53	1·53	1·52	1·51	1·50	1·49	1·48	1·47	1·46	1·45	1·44	1·43	1·42	1·41	1·41	1·40	1·39	1·38
15	1·43	1·52	1·52	1·51	1·49	1·48	1·47	1·46	1·46	1·45	1·44	1·43	1·41	1·41	1·40	1·39	1·38	1·37	1·36
16	1·42	1·51	1·51	1·50	1·48	1·47	1·46	1·45	1·44	1·44	1·43	1·41	1·40	1·39	1·38	1·37	1·36	1·35	1·34
17	1·42	1·51	1·50	1·49	1·47	1·46	1·45	1·44	1·43	1·43	1·41	1·40	1·39	1·38	1·37	1·36	1·35	1·34	1·33
18	1·41	1·50	1·49	1·48	1·46	1·45	1·44	1·43	1·42	1·42	1·40	1·39	1·38	1·37	1·36	1·35	1·34	1·33	1·32
19	1·41	1·49	1·49	1·47	1·46	1·44	1·43	1·42	1·41	1·41	1·40	1·38	1·37	1·36	1·35	1·34	1·33	1·32	1·30
20	1·40	1·49	1·48	1·47	1·45	1·44	1·43	1·42	1·41	1·40	1·39	1·37	1·36	1·35	1·34	1·33	1·32	1·31	1·29
21	1·40	1·48	1·48	1·46	1·44	1·43	1·42	1·41	1·40	1·39	1·38	1·37	1·35	1·34	1·33	1·32	1·31	1·30	1·28
22	1·40	1·48	1·47	1·45	1·44	1·42	1·41	1·40	1·39	1·39	1·37	1·36	1·34	1·33	1·32	1·31	1·30	1·29	1·28
23	1·39	1·47	1·47	1·45	1·43	1·42	1·41	1·40	1·39	1·38	1·37	1·35	1·34	1·33	1·32	1·31	1·30	1·28	1·27
24	1·39	1·47	1·46	1·44	1·43	1·41	1·40	1·39	1·38	1·38	1·36	1·35	1·33	1·32	1·31	1·30	1·29	1·28	1·26
25	1·39	1·47	1·46	1·44	1·42	1·41	1·40	1·39	1·38	1·37	1·36	1·34	1·33	1·32	1·31	1·29	1·28	1·27	1·25
26	1·38	1·46	1·45	1·44	1·42	1·41	1·39	1·38	1·37	1·37	1·35	1·34	1·32	1·31	1·30	1·29	1·28	1·26	1·25
27	1·38	1·46	1·45	1·43	1·42	1·40	1·39	1·38	1·37	1·36	1·35	1·33	1·32	1·31	1·30	1·28	1·27	1·26	1·24
28	1·38	1·46	1·45	1·43	1·41	1·40	1·39	1·38	1·37	1·36	1·34	1·33	1·31	1·30	1·29	1·28	1·27	1·25	1·24
29	1·38	1·45	1·45	1·43	1·41	1·40	1·38	1·37	1·36	1·35	1·34	1·32	1·31	1·30	1·29	1·27	1·26	1·25	1·23
30	1·38	1·45	1·44	1·42	1·41	1·39	1·38	1·37	1·36	1·35	1·34	1·32	1·30	1·29	1·28	1·27	1·26	1·24	1·23
40	1·36	1·44	1·42	1·40	1·39	1·37	1·36	1·35	1·34	1·33	1·31	1·30	1·28	1·26	1·25	1·24	1·22	1·21	1·19
60	1·35	1·42	1·41	1·38	1·37	1·35	1·33	1·32	1·31	1·30	1·29	1·27	1·25	1·24	1·22	1·21	1·19	1·17	1·15
120	1·34	1·40	1·39	1·37	1·35	1·33	1·31	1·30	1·29	1·28	1·26	1·24	1·22	1·21	1·19	1·18	1·16	1·13	1·10
∞	1·32	1·39	1·37	1·35	1·33	1·31	1·29	1·28	1·27	1·25	1·24	1·22	1·19	1·18	1·16	1·14	1·12	1·08	1·00

Note: m represents the numerator degrees of freedom; n represents the denominator degrees of freedom.

TABLE VII (*continued*)
Upper 10 Percentage Points

n \ m	1	2	3	4	5	6	7	8	9	10	12	15	20	24	30	40	60	120	∞
1	39·86	49·50	53·59	55·83	57·24	58·20	58·91	59·44	59·86	60·19	60·71	61·22	61·74	62·00	62·26	62·53	62·79	63·06	63·33
2	8·53	9·00	9·16	9·24	9·29	9·33	9·35	9·37	9·38	9·39	9·41	9·42	9·44	9·45	9·46	9·47	9·47	9·48	9·49
3	5·54	5·46	5·39	5·34	5·31	5·28	5·27	5·25	5·24	5·23	5·22	5·20	5·18	5·18	5·17	5·16	5·15	5·14	5·13
4	4·54	4·32	4·19	4·11	4·05	4·01	3·98	3·95	3·94	3·92	3·90	3·87	3·84	3·83	3·82	3·80	3·79	3·78	3·76
5	4·06	3·78	3·62	3·52	3·45	3·40	3·37	3·34	3·32	3·30	3·27	3·24	3·21	3·19	3·17	3·16	3·14	3·12	3·10
6	3·78	3·46	3·29	3·18	3·11	3·05	3·01	2·98	2·96	2·94	2·90	2·87	2·84	2·82	2·80	2·78	2·76	2·74	2·72
7	3·59	3·26	3·07	2·96	2·88	2·83	2·78	2·75	2·72	2·70	2·67	2·63	2·59	2·58	2·56	2·54	2·51	2·49	2·47
8	3·46	3·11	2·92	2·81	2·73	2·67	2·62	2·59	2·56	2·54	2·50	2·46	2·42	2·40	2·38	2·36	2·34	2·32	2·29
9	3·36	3·01	2·81	2·69	2·61	2·55	2·51	2·47	2·44	2·42	2·38	2·34	2·30	2·28	2·25	2·23	2·21	2·18	2·16
10	3·29	2·92	2·73	2·61	2·52	2·46	2·41	2·38	2·35	2·32	2·28	2·24	2·20	2·18	2·16	2·13	2·11	2·08	2·06
11	3·23	2·86	2·66	2·54	2·45	2·39	2·34	2·30	2·27	2·25	2·21	2·17	2·12	2·10	2·08	2·05	2·03	2·00	1·97
12	3·18	2·81	2·61	2·48	2·39	2·33	2·28	2·24	2·21	2·19	2·15	2·10	2·06	2·04	2·01	1·99	1·96	1·93	1·90
13	3·14	2·76	2·56	2·43	2·35	2·28	2·23	2·20	2·16	2·14	2·10	2·05	2·01	1·98	1·96	1·93	1·90	1·88	1·85
14	3·10	2·73	2·52	2·39	2·31	2·24	2·19	2·15	2·12	2·10	2·05	2·01	1·96	1·94	1·91	1·89	1·86	1·83	1·80
15	3·07	2·70	2·49	2·36	2·27	2·21	2·16	2·12	2·09	2·06	2·02	1·97	1·92	1·90	1·87	1·85	1·82	1·79	1·76
16	3·05	2·67	2·46	2·33	2·24	2·18	2·13	2·09	2·06	2·03	1·99	1·94	1·89	1·87	1·84	1·81	1·78	1·75	1·72
17	3·03	2·64	2·44	2·31	2·22	2·15	2·10	2·06	2·03	2·00	1·96	1·91	1·86	1·84	1·81	1·78	1·75	1·72	1·69
18	3·01	2·62	2·42	2·29	2·20	2·13	2·08	2·04	2·00	1·98	1·93	1·89	1·84	1·81	1·78	1·75	1·72	1·69	1·66
19	2·99	2·61	2·40	2·27	2·18	2·11	2·06	2·02	1·98	1·96	1·91	1·86	1·81	1·79	1·76	1·73	1·70	1·67	1·63
20	2·97	2·59	2·38	2·25	2·16	2·09	2·04	2·00	1·96	1·94	1·89	1·84	1·79	1·77	1·74	1·71	1·68	1·64	1·61
21	2·96	2·57	2·36	2·23	2·14	2·08	2·02	1·98	1·95	1·92	1·87	1·83	1·78	1·75	1·72	1·69	1·66	1·62	1·59
22	2·95	2·56	2·35	2·22	2·13	2·06	2·01	1·97	1·93	1·90	1·86	1·81	1·76	1·73	1·70	1·67	1·64	1·60	1·57
23	2·94	2·55	2·34	2·21	2·11	2·05	1·99	1·95	1·92	1·89	1·84	1·80	1·74	1·72	1·69	1·66	1·62	1·59	1·55
24	2·93	2·54	2·33	2·19	2·10	2·04	1·98	1·94	1·91	1·88	1·83	1·78	1·73	1·70	1·67	1·64	1·61	1·57	1·53
25	2·92	2·53	2·32	2·18	2·09	2·02	1·97	1·93	1·89	1·87	1·82	1·77	1·72	1·69	1·66	1·63	1·59	1·56	1·52
26	2·91	2·52	2·31	2·17	2·08	2·01	1·96	1·92	1·88	1·86	1·81	1·76	1·71	1·68	1·65	1·61	1·58	1·54	1·50
27	2·90	2·51	2·30	2·17	2·07	2·00	1·95	1·91	1·87	1·85	1·80	1·75	1·70	1·67	1·64	1·60	1·57	1·53	1·49
28	2·89	2·50	2·29	2·16	2·06	2·00	1·94	1·90	1·87	1·84	1·79	1·74	1·69	1·66	1·63	1·59	1·56	1·52	1·48
29	2·89	2·50	2·28	2·15	2·06	1·99	1·93	1·89	1·86	1·83	1·78	1·73	1·68	1·65	1·62	1·58	1·55	1·51	1·47
30	2·88	2·49	2·28	2·14	2·05	1·98	1·93	1·88	1·85	1·82	1·77	1·72	1·67	1·64	1·61	1·57	1·54	1·50	1·46
40	2·84	2·44	2·23	2·09	2·00	1·93	1·87	1·83	1·79	1·76	1·71	1·66	1·61	1·57	1·54	1·51	1·47	1·42	1·38
60	2·79	2·39	2·18	2·04	1·95	1·87	1·82	1·77	1·74	1·71	1·66	1·60	1·54	1·51	1·48	1·44	1·40	1·35	1·29
120	2·75	2·35	2·13	1·99	1·90	1·82	1·77	1·72	1·68	1·65	1·60	1·55	1·48	1·45	1·41	1·37	1·32	1·26	1·19
∞	2·71	2·30	2·08	1·94	1·85	1·77	1·72	1·67	1·63	1·60	1·55	1·49	1·42	1·38	1·34	1·30	1·24	1·17	1·00

TABLE VII (*continued*)
Upper 5 Percentage Points

n \ m	1	2	3	4	5	6	7	8	9	10	12	15	20	24	30	40	60	120	∞
1	161·4	199·5	215·7	224·6	230·2	234·0	236·8	238·9	240·5	241·9	243·9	245·9	248·0	249·1	250·1	251·1	252·2	253·3	254·3
2	18·51	19·00	19·16	19·25	19·30	19·33	19·35	19·37	19·38	19·40	19·41	19·43	19·45	19·45	19·46	19·47	19·48	19·49	19·50
3	10·13	9·55	9·28	9·12	9·01	8·94	8·89	8·85	8·81	8·79	8·74	8·70	8·66	8·64	8·62	8·59	8·57	8·55	8·53
4	7·71	6·94	6·59	6·39	6·26	6·16	6·09	6·04	6·00	5·96	5·91	5·86	5·80	5·77	5·75	5·72	5·69	5·66	5·63
5	6·61	5·79	5·41	5·19	5·05	4·95	4·88	4·82	4·77	4·74	4·68	4·62	4·56	4·53	4·50	4·46	4·43	4·40	4·36
6	5·99	5·14	4·76	4·53	4·39	4·28	4·21	4·15	4·10	4·06	4·00	3·94	3·87	3·84	3·81	3·77	3·74	3·70	3·67
7	5·59	4·74	4·35	4·12	3·97	3·87	3·79	3·73	3·68	3·64	3·57	3·51	3·44	3·41	3·38	3·34	3·30	3·27	3·23
8	5·32	4·46	4·07	3·84	3·69	3·58	3·50	3·44	3·39	3·35	3·28	3·22	3·15	3·12	3·08	3·04	3·01	2·97	2·93
9	5·12	4·26	3·86	3·63	3·48	3·37	3·29	3·23	3·18	3·14	3·07	3·01	2·94	2·90	2·86	2·83	2·79	2·75	2·71
10	4·96	4·10	3·71	3·48	3·33	3·22	3·14	3·07	3·02	2·98	2·91	2·85	2·77	2·74	2·70	2·66	2·62	2·58	2·54
11	4·84	3·98	3·59	3·36	3·20	3·09	3·01	2·95	2·90	2·85	2·79	2·72	2·65	2·61	2·57	2·53	2·49	2·45	2·40
12	4·75	3·89	3·49	3·26	3·11	3·00	2·91	2·85	2·80	2·75	2·69	2·62	2·54	2·51	2·47	2·43	2·38	2·34	2·30
13	4·67	3·81	3·41	3·18	3·03	2·92	2·83	2·77	2·71	2·67	2·60	2·53	2·46	2·42	2·38	2·34	2·30	2·25	2·21
14	4·60	3·74	3·34	3·11	2·96	2·85	2·76	2·70	2·65	2·60	2·53	2·46	2·39	2·35	2·31	2·27	2·22	2·18	2·13
15	4·54	3·68	3·29	3·06	2·90	2·79	2·71	2·64	2·59	2·54	2·48	2·40	2·33	2·29	2·25	2·20	2·16	2·11	2·07
16	4·49	3·63	3·24	3·01	2·85	2·74	2·66	2·59	2·54	2·49	2·42	2·35	2·28	2·24	2·19	2·15	2·11	2·06	2·01
17	4·45	3·59	3·20	2·96	2·81	2·70	2·61	2·55	2·49	2·45	2·38	2·31	2·23	2·19	2·15	2·10	2·06	2·01	1·96
18	4·41	3·55	3·16	2·93	2·77	2·66	2·58	2·51	2·46	2·41	2·34	2·27	2·19	2·15	2·11	2·06	2·02	1·97	1·92
19	4·38	3·52	3·13	2·90	2·74	2·63	2·54	2·48	2·42	2·38	2·31	2·23	2·16	2·11	2·07	2·03	1·98	1·93	1·88
20	4·35	3·49	3·10	2·87	2·71	2·60	2·51	2·45	2·39	2·35	2·28	2·20	2·12	2·08	2·04	1·99	1·95	1·90	1·84
21	4·32	3·47	3·07	2·84	2·68	2·57	2·49	2·42	2·37	2·32	2·25	2·18	2·10	2·05	2·01	1·96	1·92	1·87	1·81
22	4·30	3·44	3·05	2·82	2·66	2·55	2·46	2·40	2·34	2·30	2·23	2·15	2·07	2·03	1·98	1·94	1·89	1·84	1·78
23	4·28	3·42	3·03	2·80	2·64	2·53	2·44	2·37	2·32	2·27	2·20	2·13	2·05	2·01	1·96	1·91	1·86	1·81	1·76
24	4·26	3·40	3·01	2·78	2·62	2·51	2·42	2·36	2·30	2·25	2·18	2·11	2·03	1·98	1·94	1·89	1·84	1·79	1·73
25	4·24	3·39	2·99	2·76	2·60	2·49	2·40	2·34	2·28	2·24	2·16	2·09	2·01	1·96	1·92	1·87	1·82	1·77	1·71
26	4·23	3·37	2·98	2·74	2·59	2·47	2·39	2·32	2·27	2·22	2·15	2·07	1·99	1·95	1·90	1·85	1·80	1·75	1·69
27	4·21	3·35	2·96	2·73	2·57	2·46	2·37	2·31	2·25	2·20	2·13	2·06	1·97	1·93	1·88	1·84	1·79	1·73	1·67
28	4·20	3·34	2·95	2·71	2·56	2·45	2·36	2·29	2·24	2·19	2·12	2·04	1·96	1·91	1·87	1·82	1·77	1·71	1·65
29	4·18	3·33	2·93	2·70	2·55	2·43	2·35	2·28	2·22	2·18	2·10	2·03	1·94	1·90	1·85	1·81	1·75	1·70	1·64
30	4·17	3·32	2·92	2·69	2·53	2·42	2·33	2·27	2·21	2·16	2·09	2·01	1·93	1·89	1·84	1·79	1·74	1·68	1·62
40	4·08	3·23	2·84	2·61	2·45	2·34	2·25	2·18	2·12	2·08	2·00	1·92	1·84	1·79	1·74	1·69	1·64	1·58	1·51
60	4·00	3·15	2·76	2·53	2·37	2·25	2·17	2·10	2·04	1·99	1·92	1·84	1·75	1·70	1·65	1·59	1·53	1·47	1·39
120	3·92	3·07	2·68	2·45	2·29	2·17	2·09	2·02	1·96	1·91	1·83	1·75	1·66	1·61	1·55	1·50	1·43	1·35	1·25
∞	3·84	3·00	2·60	2·37	2·21	2·10	2·01	1·94	1·88	1·83	1·75	1·67	1·57	1·52	1·46	1·39	1·32	1·22	1·00

TABLE VII (*continued*)
Upper 2.5 Percentage Points

n \ m	1	2	3	4	5	6	7	8	9	10	12	15	20	24	30	40	60	120	∞
1	647·8	799·5	864·2	899·6	921·8	937·1	948·2	956·7	963·3	968·6	976·7	984·9	993·1	997·2	1001	1006	1010	1014	1018
2	38·51	39·00	39·17	39·25	39·30	39·33	39·36	39·37	39·39	39·40	39·41	39·43	39·45	39·46	39·46	39·47	39·48	39·49	39·50
3	17·44	16·04	15·44	15·10	14·88	14·73	14·62	14·54	14·47	14·42	14·34	14·25	14·17	14·12	14·08	14·04	13·99	13·95	13·90
4	12·22	10·65	9·98	9·60	9·36	9·20	9·07	8·98	8·90	8·84	8·75	8·66	8·56	8·51	8·46	8·41	8·36	8·31	8·26
5	10·01	8·43	7·76	7·39	7·15	6·98	6·85	6·76	6·68	6·62	6·52	6·43	6·33	6·28	6·23	6·18	6·12	6·07	6·02
6	8·81	7·26	6·60	6·23	5·99	5·82	5·70	5·60	5·52	5·46	5·37	5·27	5·17	5·12	5·07	5·01	4·96	4·90	4·85
7	8·07	6·54	5·89	5·52	5·29	5·12	4·99	4·90	4·82	4·76	4·67	4·57	4·47	4·42	4·36	4·31	4·25	4·20	4·14
8	7·57	6·06	5·42	5·05	4·82	4·65	4·53	4·43	4·36	4·30	4·20	4·10	4·00	3·95	3·89	3·84	3·78	3·73	3·67
9	7·21	5·71	5·08	4·72	4·48	4·32	4·20	4·10	4·03	3·96	3·87	3·77	3·67	3·61	3·56	3·51	3·45	3·39	3·33
10	6·94	5·46	4·83	4·47	4·24	4·07	3·95	3·85	3·78	3·72	3·62	3·52	3·42	3·37	3·31	3·26	3·20	3·14	3·08
11	6·72	5·26	4·63	4·28	4·04	3·88	3·76	3·66	3·59	3·53	3·43	3·33	3·23	3·17	3·12	3·06	3·00	2·94	2·88
12	6·55	5·10	4·47	4·12	3·89	3·73	3·61	3·51	3·44	3·37	3·28	3·18	3·07	3·02	2·96	2·91	2·85	2·79	2·72
13	6·41	4·97	4·35	4·00	3·77	3·60	3·48	3·39	3·31	3·25	3·15	3·05	2·95	2·89	2·84	2·78	2·72	2·66	2·60
14	6·30	4·86	4·24	3·89	3·66	3·50	3·38	3·29	3·21	3·15	3·05	2·95	2·84	2·79	2·73	2·67	2·61	2·55	2·49
15	6·20	4·77	4·15	3·80	3·58	3·41	3·29	3·20	3·12	3·06	2·96	2·86	2·76	2·70	2·64	2·59	2·52	2·46	2·40
16	6·12	4·69	4·08	3·73	3·50	3·34	3·22	3·12	3·05	2·99	2·89	2·79	2·68	2·63	2·57	2·51	2·45	2·38	2·32
17	6·04	4·62	4·01	3·66	3·44	3·28	3·16	3·06	2·98	2·92	2·82	2·72	2·62	2·56	2·50	2·44	2·38	2·32	2·25
18	5·98	4·56	3·95	3·61	3·38	3·22	3·10	3·01	2·93	2·87	2·77	2·67	2·56	2·50	2·44	2·38	2·32	2·26	2·19
19	5·92	4·51	3·90	3·56	3·33	3·17	3·05	2·96	2·88	2·82	2·72	2·62	2·51	2·45	2·39	2·33	2·27	2·20	2·13
20	5·87	4·46	3·86	3·51	3·29	3·13	3·01	2·91	2·84	2·77	2·68	2·57	2·46	2·41	2·35	2·29	2·22	2·16	2·09
21	5·83	4·42	3·82	3·48	3·25	3·09	2·97	2·87	2·80	2·73	2·64	2·53	2·42	2·37	2·31	2·25	2·18	2·11	2·04
22	5·79	4·38	3·78	3·44	3·22	3·05	2·93	2·84	2·76	2·70	2·60	2·50	2·39	2·33	2·27	2·21	2·14	2·08	2·00
23	5·75	4·35	3·75	3·41	3·18	3·02	2·90	2·81	2·73	2·67	2·57	2·47	2·36	2·30	2·24	2·18	2·11	2·04	1·97
24	5·72	4·32	3·72	3·38	3·15	2·99	2·87	2·78	2·70	2·64	2·54	2·44	2·33	2·27	2·21	2·15	2·08	2·01	1·94
25	5·69	4·29	3·69	3·35	3·13	2·97	2·85	2·75	2·68	2·61	2·51	2·41	2·30	2·24	2·18	2·12	2·05	1·98	1·91
26	5·66	4·27	3·67	3·33	3·10	2·94	2·82	2·73	2·65	2·59	2·49	2·39	2·28	2·22	2·16	2·09	2·03	1·95	1·88
27	5·63	4·24	3·65	3·31	3·08	2·92	2·80	2·71	2·63	2·57	2·47	2·36	2·25	2·19	2·13	2·07	2·00	1·93	1·85
28	5·61	4·22	3·63	3·29	3·06	2·90	2·78	2·69	2·61	2·55	2·45	2·34	2·23	2·17	2·11	2·05	1·98	1·91	1·83
29	5·59	4·20	3·61	3·27	3·04	2·88	2·76	2·67	2·59	2·53	2·43	2·32	2·21	2·15	2·09	2·03	1·96	1·89	1·81
30	5·57	4·18	3·59	3·25	3·03	2·87	2·75	2·65	2·57	2·51	2·41	2·31	2·20	2·14	2·07	2·01	1·94	1·87	1·79
40	5·42	4·05	3·46	3·13	2·90	2·74	2·62	2·53	2·45	2·39	2·29	2·18	2·07	2·01	1·94	1·88	1·80	1·72	1·64
60	5·29	3·93	3·34	3·01	2·79	2·63	2·51	2·41	2·33	2·27	2·17	2·06	1·94	1·88	1·82	1·74	1·67	1·58	1·48
120	5·15	3·80	3·23	2·89	2·67	2·52	2·39	2·30	2·22	2·16	2·05	1·94	1·82	1·76	1·69	1·61	1·53	1·43	1·31
∞	5·02	3·69	3·12	2·79	2·57	2·41	2·29	2·19	2·11	2·05	1·94	1·83	1·71	1·64	1·57	1·48	1·39	1·27	1·00

TABLE VII (*continued*)
Upper 1 Percentage Points

n \ m	1	2	3	4	5	6	7	8	9	10	12	15	20	24	30	40	60	120	∞
1	4052	4999·5	5403	5625	5764	5859	5928	5981	6022	6056	6106	6157	6209	6235	6261	6287	6313	6339	6366
2	98·50	99·00	99·17	99·25	99·30	99·33	99·36	99·37	99·39	99·40	99·42	99·43	99·45	99·46	99·47	99·47	99·48	99·49	99·50
3	34·12	30·82	29·46	28·71	28·24	27·91	27·67	27·49	27·35	27·23	27·05	26·87	26·69	26·60	26·50	26·41	26·32	26·22	26·13
4	21·20	18·00	16·69	15·98	15·52	15·21	14·98	14·80	14·66	14·55	14·37	14·20	14·02	13·93	13·84	13·75	13·65	13·56	13·46
5	16·26	13·27	12·06	11·39	10·97	10·67	10·46	10·29	10·16	10·05	9·89	9·72	9·55	9·47	9·38	9·29	9·20	9·11	9·02
6	13·75	10·92	9·78	9·15	8·75	8·47	8·26	8·10	7·98	7·87	7·72	7·56	7·40	7·31	7·23	7·14	7·06	6·97	6·88
7	12·25	9·55	8·45	7·85	7·46	7·19	6·99	6·84	6·72	6·62	6·47	6·31	6·16	6·07	5·99	5·91	5·82	5·74	5·65
8	11·26	8·65	7·59	7·01	6·63	6·37	6·18	6·03	5·91	5·81	5·67	5·52	5·36	5·28	5·20	5·12	5·03	4·95	4·86
9	10·56	8·02	6·99	6·42	6·06	5·80	5·61	5·47	5·35	5·26	5·11	4·96	4·81	4·73	4·65	4·57	4·48	4·40	4·31
10	10·04	7·56	6·55	5·99	5·64	5·39	5·20	5·06	4·94	4·85	4·71	4·56	4·41	4·33	4·25	4·17	4·08	4·00	3·91
11	9·65	7·21	6·22	5·67	5·32	5·07	4·89	4·74	4·63	4·54	4·40	4·25	4·10	4·02	3·94	3·86	3·78	3·69	3·60
12	9·33	6·93	5·95	5·41	5·06	4·82	4·64	4·50	4·39	4·30	4·16	4·01	3·86	3·78	3·70	3·62	3·54	3·45	3·36
13	9·07	6·70	5·74	5·21	4·86	4·62	4·44	4·30	4·19	4·10	3·96	3·82	3·66	3·59	3·51	3·43	3·34	3·25	3·17
14	8·86	6·51	5·56	5·04	4·69	4·46	4·28	4·14	4·03	3·94	3·80	3·66	3·51	3·43	3·35	3·27	3·18	3·09	3·00
15	8·68	6·36	5·42	4·89	4·56	4·32	4·14	4·00	3·89	3·80	3·67	3·52	3·37	3·29	3·21	3·13	3·05	2·96	2·87
16	8·53	6·23	5·29	4·77	4·44	4·20	4·03	3·89	3·78	3·69	3·55	3·41	3·26	3·18	3·10	3·02	2·93	2·84	2·75
17	8·40	6·11	5·18	4·67	4·34	4·10	3·93	3·79	3·68	3·59	3·46	3·31	3·16	3·08	3·00	2·92	2·83	2·75	2·65
18	8·29	6·01	5·09	4·58	4·25	4·01	3·84	3·71	3·60	3·51	3·37	3·23	3·08	3·00	2·92	2·84	2·75	2·66	2·57
19	8·18	5·93	5·01	4·50	4·17	3·94	3·77	3·63	3·52	3·43	3·30	3·15	3·00	2·92	2·84	2·76	2·67	2·58	2·49
20	8·10	5·85	4·94	4·43	4·10	3·87	3·70	3·56	3·46	3·37	3·23	3·09	2·94	2·86	2·78	2·69	2·61	2·52	2·42
21	8·02	5·78	4·87	4·37	4·04	3·81	3·64	3·51	3·40	3·31	3·17	3·03	2·88	2·80	2·72	2·64	2·55	2·46	2·36
22	7·95	5·72	4·82	4·31	3·99	3·76	3·59	3·45	3·35	3·26	3·12	2·98	2·83	2·75	2·67	2·58	2·50	2·40	2·31
23	7·88	5·66	4·76	4·26	3·94	3·71	3·54	3·41	3·30	3·21	3·07	2·93	2·78	2·70	2·62	2·54	2·45	2·35	2·26
24	7·82	5·61	4·72	4·22	3·90	3·67	3·50	3·36	3·26	3·17	3·03	2·89	2·74	2·66	2·58	2·49	2·40	2·31	2·21
25	7·77	5·57	4·68	4·18	3·85	3·63	3·46	3·32	3·22	3·13	2·99	2·85	2·70	2·62	2·54	2·45	2·36	2·27	2·17
26	7·72	5·53	4·64	4·14	3·82	3·59	3·42	3·29	3·18	3·09	2·96	2·81	2·66	2·58	2·50	2·42	2·33	2·23	2·13
27	7·68	5·49	4·60	4·11	3·78	3·56	3·39	3·26	3·15	3·06	2·93	2·78	2·63	2·55	2·47	2·38	2·29	2·20	2·10
28	7·64	5·45	4·57	4·07	3·75	3·53	3·36	3·23	3·12	3·03	2·90	2·75	2·60	2·52	2·44	2·35	2·26	2·17	2·06
29	7·60	5·42	4·54	4·04	3·73	3·50	3·33	3·20	3·09	3·00	2·87	2·73	2·57	2·49	2·41	2·33	2·23	2·14	2·03
30	7·56	5·39	4·51	4·02	3·70	3·47	3·30	3·17	3·07	2·98	2·84	2·70	2·55	2·47	2·39	2·30	2·21	2·11	2·01
40	7·31	5·18	4·31	3·83	3·51	3·29	3·12	2·99	2·89	2·80	2·66	2·52	2·37	2·29	2·20	2·11	2·02	1·92	1·80
60	7·08	4·98	4·13	3·65	3·34	3·12	2·95	2·82	2·72	2·63	2·50	2·35	2·20	2·12	2·03	1·94	1·84	1·73	1·60
120	6·85	4·79	3·95	3·48	3·17	2·96	2·79	2·66	2·56	2·47	2·34	2·19	2·03	1·95	1·86	1·76	1·66	1·53	1·38
∞	6·63	4·61	3·78	3·32	3·02	2·80	2·64	2·51	2·41	2·32	2·18	2·04	1·88	1·79	1·70	1·59	1·47	1·32	1·00

TABLE VII (*continued*)
Upper 0.5 Percentage Points

n \ m	1	2	3	4	5	6	7	8	9	10	12	15	20	24	30	40	60	120	∞
1	16211	20000	21615	22500	23056	23437	23715	23925	24091	24224	24426	24630	24836	24940	25044	25148	25253	25359	25465
2	198·5	199·0	199·2	199·2	199·3	199·3	199·4	199·4	199·4	199·4	199·4	199·4	199·4	199·5	199·5	199·5	199·5	199·5	199·5
3	55·55	49·80	47·47	46·19	45·39	44·84	44·43	44·13	43·88	43·69	43·39	43·08	42·78	42·62	42·47	42·31	42·15	41·99	41·83
4	31·33	26·28	24·26	23·15	22·46	21·97	21·62	21·35	21·14	20·97	20·70	20·44	20·17	20·03	19·89	19·75	19·61	19·47	19·32
5	22·78	18·31	16·53	15·56	14·94	14·51	14·20	13·96	13·77	13·62	13·38	13·15	12·90	12·78	12·66	12·53	12·40	12·27	12·14
6	18·63	14·54	12·92	12·03	11·46	11·07	10·79	10·57	10·39	10·25	10·03	9·81	9·59	9·47	9·36	9·24	9·12	9·00	8·88
7	16·24	12·40	10·88	10·05	9·52	9·16	8·89	8·68	8·51	8·38	8·18	7·97	7·75	7·65	7·53	7·42	7·31	7·19	7·08
8	14·69	11·04	9·60	8·81	8·30	7·95	7·69	7·50	7·34	7·21	7·01	6·81	6·61	6·50	6·40	6·29	6·18	6·06	5·95
9	13·61	10·11	8·72	7·96	7·47	7·13	6·88	6·69	6·54	6·42	6·23	6·03	5·83	5·73	5·62	5·52	5·41	5·30	5·19
10	12·83	9·43	8·08	7·34	6·87	6·54	6·30	6·12	5·97	5·85	5·66	5·47	5·27	5·17	5·07	4·97	4·86	4·75	4·64
11	12·23	8·91	7·60	6·88	6·42	6·10	5·86	5·68	5·54	5·42	5·24	5·05	4·86	4·76	4·65	4·55	4·44	4·34	4·23
12	11·75	8·51	7·23	6·52	6·07	5·76	5·52	5·35	5·20	5·09	4·91	4·72	4·53	4·43	4·33	4·23	4·12	4·01	3·90
13	11·37	8·19	6·93	6·23	5·79	5·48	5·25	5·08	4·94	4·82	4·64	4·46	4·27	4·17	4·07	3·97	3·87	3·76	3·65
14	11·06	7·92	6·68	6·00	5·56	5·26	5·03	4·86	4·72	4·60	4·43	4·25	4·06	3·96	3·86	3·76	3·66	3·55	3·44
15	10·80	7·70	6·48	5·80	5·37	5·07	4·85	4·67	4·54	4·42	4·25	4·07	3·88	3·79	3·69	3·58	3·48	3·37	3·26
16	10·58	7·51	6·30	5·64	5·21	4·91	4·69	4·52	4·38	4·27	4·10	3·92	3·73	3·64	3·54	3·44	3·33	3·22	3·11
17	10·38	7·35	6·16	5·50	5·07	4·78	4·56	4·39	4·25	4·14	3·97	3·79	3·61	3·51	3·41	3·31	3·21	3·10	2·98
18	10·22	7·21	6·03	5·37	4·96	4·66	4·44	4·28	4·14	4·03	3·86	3·68	3·50	3·40	3·30	3·20	3·10	2·99	2·87
19	10·07	7·09	5·92	5·27	4·85	4·56	4·34	4·18	4·04	3·93	3·76	3·59	3·40	3·31	3·21	3·11	3·00	2·89	2·78
20	9·94	6·99	5·82	5·17	4·76	4·47	4·26	4·09	3·96	3·85	3·68	3·50	3·32	3·22	3·12	3·02	2·92	2·81	2·69
21	9·83	6·89	5·73	5·09	4·68	4·39	4·18	4·01	3·88	3·77	3·60	3·43	3·24	3·15	3·05	2·95	2·84	2·73	2·61
22	9·73	6·81	5·65	5·02	4·61	4·32	4·11	3·94	3·81	3·70	3·54	3·36	3·18	3·08	2·98	2·88	2·77	2·66	2·55
23	9·63	6·73	5·58	4·95	4·54	4·26	4·05	3·88	3·75	3·64	3·47	3·30	3·12	3·02	2·92	2·82	2·71	2·60	2·48
24	9·55	6·66	5·52	4·89	4·49	4·20	3·99	3·83	3·69	3·59	3·42	3·25	3·06	2·97	2·87	2·77	2·66	2·55	2·43
25	9·48	6·60	5·46	4·84	4·43	4·15	3·94	3·78	3·64	3·54	3·37	3·20	3·01	2·92	2·82	2·72	2·61	2·50	2·38
26	9·41	6·54	5·41	4·79	4·38	4·10	3·89	3·73	3·60	3·49	3·33	3·15	2·97	2·87	2·77	2·67	2·56	2·45	2·33
27	9·34	6·49	5·36	4·74	4·34	4·06	3·85	3·69	3·56	3·45	3·28	3·11	2·93	2·83	2·73	2·63	2·52	2·41	2·29
28	9·28	6·44	5·32	4·70	4·30	4·02	3·81	3·65	3·52	3·41	3·25	3·07	2·89	2·79	2·69	2·59	2·48	2·37	2·25
29	9·23	6·40	5·28	4·66	4·26	3·98	3·77	3·61	3·48	3·38	3·21	3·04	2·86	2·76	2·66	2·56	2·45	2·33	2·21
30	9·18	6·35	5·24	4·62	4·23	3·95	3·74	3·58	3·45	3·34	3·18	3·01	2·82	2·73	2·63	2·52	2·42	2·30	2·18
40	8·83	6·07	4·98	4·37	3·99	3·71	3·51	3·35	3·22	3·12	2·95	2·78	2·60	2·50	2·40	2·30	2·18	2·06	1·93
60	8·49	5·79	4·73	4·14	3·76	3·49	3·29	3·13	3·01	2·90	2·74	2·57	2·39	2·29	2·19	2·08	1·96	1·83	1·69
120	8·18	5·54	4·50	3·92	3·55	3·28	3·09	2·93	2·81	2·71	2·54	2·37	2·19	2·09	1·98	1·87	1·75	1·61	1·43
∞	7·88	5·30	4·28	3·72	3·35	3·09	2·90	2·74	2·62	2·52	2·36	2·19	2·00	1·90	1·79	1·67	1·53	1·36	1·00

TABLE VIII
Percentage Points of the Ratio s^2_{max}/s^2_{min}
Upper 5 Percentage Points

n \ k	2	3	4	5	6	7	8	9	10	11	12
2	39·0	87·5	142	202	266	333	403	475	550	626	704
3	15·4	27·8	39·2	50·7	62·0	72·9	83·5	93·9	104	114	124
4	9·60	15·5	20·6	25·2	29·5	33·6	37·5	41·1	44·6	48·0	51·4
5	7·15	10·8	13·7	16·3	18·7	20·8	22·9	24·7	26·5	28·2	29·9
6	5·82	8·38	10·4	12·1	13·7	15·0	16·3	17·5	18·6	19·7	20·7
7	4·99	6·94	8·44	9·70	10·8	11·8	12·7	13·5	14·3	15·1	15·8
8	4·43	6·00	7·18	8·12	9·03	9·78	10·5	11·1	11·7	12·2	12·7
9	4·03	5·34	6·31	7·11	7·80	8·41	8·95	9·45	9·91	10·3	10·7
10	3·72	4·85	5·67	6·34	6·92	7·42	7·87	8·28	8·66	9·01	9·34
12	3·28	4·16	4·79	5·30	5·72	6·09	6·42	6·72	7·00	7·25	7·48
15	2·86	3·54	4·01	4·37	4·68	4·95	5·19	5·40	5·59	5·77	5·93
20	2·46	2·95	3·29	3·54	3·76	3·94	4·10	4·24	4·37	4·49	4·59
30	2·07	2·40	2·61	2·78	2·91	3·02	3·12	3·21	3·29	3·36	3·39
60	1·67	1·85	1·96	2·04	2·11	2·17	2·22	2·26	2·30	2·33	2·36
∞	1·00	1·00	1·00	1·00	1·00	1·00	1·00	1·00	1·00	1·00	1·00

Upper 1 Percentage Points

n \ k	2	3	4	5	6	7	8	9	10	11	12
2	199	448	729	1036	1362	1705	2063	2432	2813	3204	3605
3	47·5	85	120	151	184	21(6)	24(9)	28(1)	31(0)	33(7)	36(1)
4	23·2	37	49	59	69	79	89	97	106	113	120
5	14·9	22	28	33	38	42	46	50	54	57	60
6	11·1	15·5	19·1	22	25	27	30	32	34	36	37
7	8·89	12·1	14·5	16·5	18·4	20	22	23	24	26	27
8	7·50	9·9	11·7	13·2	14·5	15·8	16·9	17·9	18·9	19·8	21
9	6·54	8·5	9·9	11·1	12·1	13·1	13·9	14·7	15·3	16·0	16·6
10	5·85	7·4	8·6	9·6	10·4	11·1	11·8	12·4	12·9	13·4	13·9
12	4·91	6·1	6·9	7·6	8·2	8·7	9·1	9·5	9·9	10·2	10·6
15	4·07	4·9	5·5	6·0	6·4	6·7	7·1	7·3	7·5	7·8	8·0
20	3·32	3·8	4·3	4·6	4·9	5·1	5·3	5·5	5·6	5·8	5·9
30	2·63	3·0	3·3	3·4	3·6	3·7	3·8	3·9	4·0	4·1	4·2
60	1·96	2·2	2·3	2·4	2·4	2·5	2·5	2·6	2·6	2·7	2·7
∞	1·00	1·0	1·0	1·0	1·0	1·0	1·0	1·0	1·0	1·0	1·0

Note: s^2_{max} is the largest and s^2_{min} the smallest in a set of k independent mean squares, each based on n degrees of freedom.

Values in the column $k = 2$ and in the rows $n = 2$ and ∞ are exact. Elsewhere the third digit may be in error by a few units for the 5 percentage points and several units for the 1 percentage points. The third digit figures in brackets for $n = 3$ are the most uncertain.

TABLE IX

VALUES OF $\zeta = \frac{1}{2} \ln \frac{1 + r}{1 - r}$

r	.00	.01	.02	.03	.04	.05	.06	.07	.08	.09
.00	.0000	.0100	.0200	.0300	.0400	.0500	.0601	.0701	.0802	.0902
.10	.1003	.1104	.1206	.1307	.1409	.1511	.1614	.1717	.1819	.1923
.20	.2027	.2132	.2236	.2342	.2448	.2554	.2661	.2769	.2877	.2986
.30	.3095	.3205	.3316	.3428	.3541	.3654	.3769	.3884	.4001	.4118
.40	.4236	.4356	.4477	.4599	.4722	.4847	.4973	.5101	.5230	.5361
.50	.5493	.5627	.5763	.5901	.6041	.6184	.6328	.6475	.6625	.6777
.60	.6931	.7089	.7250	.7414	.7582	.7753	.7928	.8107	.8291	.8480
.70	.8673	.8872	.9076	.9287	.9505	.9730	.9962	1.0203	1.0454	1.0714
	.000	.001	.002	.003	.004	.005	.006	.007	.008	.009
.80	1.0986	1.1014	1.1042	1.1070	1.1098	1.1127	1.1155	1.1184	1.1212	1.1241
.81	1.1270	1.1299	1.1329	1.1358	1.1388	1.1417	1.1447	1.1477	1.1507	1.1538
.82	1.1568	1.1599	1.1629	1.1660	1.1692	1.1723	1.1754	1.1786	1.1817	1.1849
.83	1.1881	1.1914	1.1946	1.1979	1.2011	1.2044	1.2077	1.2111	1.2144	1.2178
.84	1.2212	1.2246	1.2280	1.2315	1.2349	1.2384	1.2419	1.2454	1.2490	1.2526
.85	1.2562	1.2598	1.2634	1.2671	1.2707	1.2745	1.2782	1.2819	1.2857	1.2895
.86	1.2933	1.2972	1.3011	1.3050	1.3089	1.3129	1.3169	1.3209	1.3249	1.3290
.87	1.3331	1.3372	1.3414	1.3456	1.3498	1.3540	1.3583	1.3626	1.3670	1.3714
.88	1.3757	1.3802	1.3847	1.3892	1.3938	1.3984	1.4030	1.4077	1.4124	1.4171
.89	1.4219	1.4268	1.4316	1.4365	1.4415	1.4465	1.4516	1.4566	1.4618	1.4670
.90	1.4722	1.4775	1.4828	1.4882	1.4937	1.4992	1.5047	1.5103	1.5160	1.5217
.91	1.5275	1.5334	1.5393	1.5453	1.5513	1.5574	1.5636	1.5698	1.5762	1.5826
.92	1.5890	1.5956	1.6022	1.6089	1.6157	1.6226	1.6296	1.6366	1.6438	1.6510
.93	1.6584	1.6658	1.6734	1.6811	1.6888	1.6967	1.7047	1.7129	1.7211	1.7295
.94	1.7480	1.7467	1.7555	1.7645	1.7738	1.7828	1.7923	1.8019	1.8117	1.8216
.95	1.8318	1.8421	1.8527	1.8635	1.8745	1.8857	1.8972	1.9090	1.9210	1.9333
.96	1.9459	1.9588	1.9721	1.9857	1.9996	2.0139	2.0287	2.0439	2.0595	2.0756
.97	2.0923	2.1095	2.1273	2.1457	2.1649	2.1847	2.2054	2.2269	2.2494	2.2729
.98	2.2976	2.3236	2.3507	2.3796	2.4101	2.4427	2.4774	2.5147	2.5550	2.5987
.99	2.6467	2.6996	2.7587	2.8257	2.9031	2.9945	3.1063	3.2504	3.4534	3.8002

TABLE X

VALUES OF $r = \dfrac{e^{2\zeta} - 1}{e^{2\zeta} + 1}$

ζ	.00	.01	.02	.03	.04	.05	.06	.07	.08	.09
.0	.00000	.00999	.01999	.02999	.03997	.04995	.05992	.06988	.07982	.08975
.1	.09966	.10955	.11942	.12927	.13909	.14888	.15864	.16838	.17808	.18774
.2	.19737	.20696	.21651	.22602	.23549	.24491	.25429	.26362	.27290	.28213
.3	.29131	.30043	.30950	.31852	.32747	.33637	.34521	.35399	.36270	.37136
.4	.37994	.38847	.39693	.40532	.41364	.42189	.43008	.43819	.44624	.45421
.5	.46211	.46994	.47770	.48538	.49298	.50052	.50797	.51535	.52266	.52989
.6	.53704	.54412	.55112	.55805	.56489	.57166	.57836	.58497	.59151	.59798
.7	.60436	.61067	.61690	.62306	.62914	.63514	.64107	.64692	.65270	.65840
.8	.66403	.66959	.67506	.68047	.68580	.69106	.69625	.70137	.70641	.71139
.9	.71629	.72113	.72589	.73059	.73522	.73978	.74427	.74870	.75306	.75736
1.0	.76159	.76576	.76986	.77390	.77788	.78180	.78566	.78946	.79319	.79687
1.1	.80049	.80406	.80756	.81101	.81441	.81775	.82103	.82427	.82745	.83057
1.2	.83365	.83667	.83965	.84257	.84545	.84828	.85106	.85379	.85648	.85912
1.3	.86172	.86427	.86678	.86924	.87167	.87405	.87639	.87869	.88095	.88317
1.4	.88535	.88749	.88959	.89166	.89369	.89569	.89765	.89957	.90146	.90332
1.5	.90514	.90693	.90869	.91042	.91212	.91378	.91542	.91702	.91860	.92014
1.6	.92166	.92316	.92462	.92606	.92747	.92885	.93021	.93155	.93286	.93414
1.7	.93540	.93664	.93786	.93905	.94022	.94137	.94250	.94360	.94469	.94576
1.8	.94680	.94783	.94883	.94982	.95079	.95174	.95267	.95359	.95449	.95537
1.9	.95623	.95708	.95791	.95873	.95953	.96031	.96108	.96184	.96258	.96331
2.0	.96402	.96472	.96541	.96608	.96674	.96739	.96803	.96865	.96926	.96986
2.1	.97045	.97102	.97159	.97214	.97269	.97322	.97374	.97426	.97476	.97525
2.2	.97574	.97621	.97668	.97713	.97758	.97802	.97845	.97887	.97929	.97969
2.3	.98009	.98048	.98086	.98124	.98161	.98197	.98232	.98267	.98301	.98334
2.4	.98367	.98399	.98430	.98461	.98492	.98521	.98550	.98579	.98607	.98634
2.5	.98661	.98687	.98713	.98738	.98763	.98788	.98811	.98835	.98858	.98880
2.6	.98902	.98924	.98945	.98966	.98986	.99006	.99026	.99045	.99064	.99082
2.7	.99100	.99118	.99135	.99152	.99169	.99185	.99202	.99217	.99233	.99248
2.8	.99263	.99277	.99291	.99305	.99319	.99333	.99346	.99359	.99371	.99384
2.9	.99396	.99408	.99419	.99431	.99442	.99453	.99464	.99474	.99485	.99495

Principal Notations and Abbreviations

α	significance level of a test
Anova	analysis of variance
$C(n,k)$ or $\binom{n}{k}$	number of combinations of n things taken k at a time
ρ	correlation coefficient (product-moment), population
r	correlation coefficient (product-moment), sample
r_s	Spearman's correlation coefficient
$\text{cov}(x,y)$	covariance of x and y, population
s_{xy}	covariance of x and y, sample
d.f. or D.F.	degrees of freedom
$F(x:\alpha, \beta, \gamma, \ldots)$	distribution function of x with parameters $\alpha, \beta, \gamma, \ldots$
$E[g(x)]$	expectation of $g(x)$
EMS	expected mean square
$F_{i/j}$	fractile
$F_x(t)$	factorial moment-generating function
H_0	hypothesis (null)
$M_x(t)$ or m.g.f.	moment-generating function
μ	mean, population
$\bar{x}$	mean, sample
m_k	kth moment about the origin
μ_k	kth moment about the mean
$\tilde{x}$	median
MS	mean square
O.C.	operating characteristic
$P(n,k)$	number of permutations of n things taken k at a time
$P(E)$	probability of event E
$P(A \mid B)$	conditional probability of event A given event B
p.f.	probability function
$f(x:\alpha, \beta, \gamma, \ldots)$	probability function of x with parameters $\alpha, \beta, \gamma, \ldots$
r.v.	random variable
σ	standard deviation, population

s	standard deviation, sample
s_e	standard error of estimate
SS or ss	sums of squares
σ^2	variance, population
s^2	variance, sample

Index